CW00725632

Osiris
A RESEARCH JOURNAl
TO THE HISTORY OF SCIENCE
AND ITS CULTURAL INFLUENCES

EDITOR
MARGARET W. ROSSITER

ASSOCIATE EDITOR
KATHRYN M. OLESKO

MANAGING EDITOR
STEPHEN P. WELDON

MANUSCRIPT EDITORS
DIANA KENNEY AND
CHRISTIE A. LERCH

EDITORIAL OFFICE
DEPARTMENT OF SCIENCE AND TECHNOLOGY STUDIES
CORNELL UNIVERSITY
726 UNIVERSITY AVENUE
ITHACA, NEW YORK 14850 USA

SUGGESTIONS FOR CONTRIBUTORS TO OSIRIS

OSIRIS is devoted to thematic issues, often conceived and compiled by guest editors.

1. Manuscripts should be **typewritten** or processed on a **letter-quality** printer and **double-spaced** throughout, including quotations and notes, on paper of standard size or weight. Margins should be wider than usual to allow space for instructions to the typesetter. The right-hand margin should be left ragged (not justified) to maintain even spacing and readability.

2. Bibliographic information should be given in **footnotes** (not parenthetically in the text), typed separately from the main body of the manuscript, **double-** or even **triple-spaced,** numbered consecutively throughout the article, and keyed to reference numbers typed above the line in the text.

   a. References to **books** should include author's full name; complete title of the book, underlined (italics); place of publication and publisher's name for books published after 1900; date of publication, including the original date when a reprint is being cited; page numbers cited. *Example:*

   [1]Joseph Needham, *Science and Civilisation in China,* 5 vols., Vol. I: *Introductory Orientations* (Cambridge: Cambridge Univ. Press, 1954), p. 7.

   b. References to articles in **periodicals** should include author's name; title of article, in quotes; title of periodical, underlined; year; volume number, Arabic; number of issue if pagination requires it; page numbers of article; number of particular page cited. Journal titles are spelled out in full on first citation and abbreviated subsequently. *Example:*

   [2]John C. Greene, "Reflections of the Progress of Darwin Studies," *Journal of the History of Biology,* 1975, 8:243–272, on p. 270; and Dov Ospovat, "God and Natural Selection: The Darwinian Idea of Design," *J. Hist. Biol.,* 1980, *13*:169–174, on p. 171.

   c. When first citing a reference, please give the title in full. For succeeding citations, please use an abbreviated version of the title with the author's last name. *Example:*

   [3]Greene, "Reflections" (cit. n. 2), p. 250.

3. Please mark clearly for the typesetter all unusual alphabets, special characters, mathematics, and chemical formulae, and include all diacritical marks.

4. A small number of **figures** may be used to illustrate an article. Line drawings should be directly reproducible; glossy prints should be furnished for all halftone illustrations.

5. Manuscripts should be submitted to OSIRIS with the understanding that upon publication **copyright** will be transferred to the History of Science Society. That understanding precludes OSIRIS from considering material that has been submitted or accepted for publication elsewhere.

OSIRIS (SSN 0369-7827) is published once a year.

Subscriptions are $39 (hardcover) and $25 (paperback).

Address subscriptions, single issue orders, claims for missing issues, and advertising inquiries to *Osiris,* The University of Chicago Press, Journals Division, P.O. Box 37005, Chicago, Illinois 60637.

Postmaster: Send address changes to *Osiris,* The University of Chicago Press, Journals Division, P.O. Box 37005, Chicago, Illinois 60637.

Osiris is indexed in major scientific and historical indexing services, including *Biological Abstracts, Current Contexts, Historical Abstracts,* and *America: History and Life.*

Hardcover edition, ISBN 0-226-50078-0
Paperback edition, ISBN 0-226-50079-9

Nature and Empire: Science and the Colonial Enterprise

Edited by Roy MacLeod

Osiris

A RESEARCH JOURNAL
DEVOTED TO THE HISTORY OF SCIENCE
AND ITS CULTURAL INFLUENCES
SECOND SERIES   VOLUME 15   2000

Cover: Frontispiece to James Rennell, *Memoir of a Map of Hindoostan* (London: for the author, 1783) (British Library, India Office Collections, V51).

Introduction

By Roy MacLeod*

I. CHANGING PERSPECTIVES

DURING THE LAST THREE DECADES, HISTORIANS OF SCIENCE have come increasingly to appreciate the role of science and technology in the making of nations, and in the development of world systems in trade and commerce. For centuries, knowledge has been a companion of commerce, and both have followed the flag. From the fifteenth century, that flag was commonly European. Led by Iberia, through exploration and discovery, learning and commerce together began to comprehend the world. By the seventeenth century, the sea-trading European powers, including the Netherlands, France, and England, found strategic advantage in commerce and colonization. With both came reciprocal advantages for the improvement of natural knowledge. In England, the unfolding of empire coincided with the Scientific Revolution. Strategic advantage turned upon this new understanding of nature. During the early eighteenth century, and continuing through the Enlightenment, the needs of European navigation paid tribute to the solution of problems central to natural philosophy. Both pursuits drove international rivalry in voyages of discovery and exploration. Both also served a growing European interest in establishing colonies, whether by conquest, trade, or settlement.

Until fifty years ago, the history of science and the history of colonialism lived in separate spheres. Their relationship was uncontested, in part because it was taken for granted. Imperial history was, *in sensu strictu,* political history, the history of military conquest, administration, and trade. The interests of science were present, of course, but in a conceptually subordinate capacity; as extensions of metropolitan institutions, perhaps, in which scientists were independent explorers or *fonctionnaires.* However, from the 1960s, with the beginning of the "end of empire" and progressive decolonization, came an increasing interest in science and technology as instruments of postcolonial development. With this also came a growing interest in the methods by which the European colonial powers had—for centuries—successfully cultivated and employed science for economic advantage and political control. By the 1970s, in a context of growing political and intellectual dissatisfaction with the predominantly cognitive tendency in the history of science, this interest witnessed an historiographical turn that contributed, *inter alia,* to the emerging so-

*Department of History, University of Sydney, Sydney, NSW, Australia 2006. As guest editor for volume 15 of *Osiris,* Roy MacLeod would like to convey his thanks to all those who assisted in the final preparation of this volume, notably Margaret Rossiter, Stephen Weldon, and Jill Barnes.

1

cial history of science.[1] It is surely not coincidental that a period that saw renewed interest in the structure of scientific revolutions also saw a growing interest in the structure and practice of Western science in the world at large.

To see a relationship between the progress of science and the expansion of Europe was an idea whose time had come. It was not, of course, an especially novel idea, even in the New World. In 1893, before a large gathering of historians at, significantly, the World's Columbian Exposition in Chicago, Frederick Jackson Turner presented his influential interpretation of the significance of the frontier in American history. That empires, and characters, were shaped by a continually advancing line— "the outer edge of the wave—the meeting point between savagery and civilisation," as he put it,[2] was a common theme among both American and imperial historians. Science accompanied transoceanic colonization, and in shaping the economic destiny of the West, defined the relationship between the "old" and the "new."

Perhaps the first notable—certainly, the most quotable—manifestation of this interest among at least Anglo-American historians came with the epochal paper of 1967 by George Basalla on "The Spread of Western Science."[3] The Basalla model, and the critical response it evoked, are now well established in the literature. Famously, Basalla drew three curves, suggesting three stages in the passage of Western science into any non-European culture. A preliminary phase of exploration would lead to a period of colonial dependency, a period of adolescence, which in turn would be followed by a phase of maturity, scientific independence and national autonomy. Science is associated with nations, and an independent nation must have a national science.

Today, in an increasingly internationalized world, a global "knowledge economy" where information and its exchange feature so prominently, many historians are moving away from using the nation-state as the primary unit of analysis, and are looking for better ways of understanding relationships between peoples and cultures. In this, the history of imperialism and colonialism—conjuring images, scarcely a generation ago, that were reviled by revisionist scholars—has made a dramatic comeback. It is in the interdependent relationships of nations and peoples that we now look to find ways of approaching the complex problems of the globe. For these reasons, the history of imperial and colonial science has become a new "venue," reflecting a convergence of interests among scholars in world history, the history of medicine, the movement of global capital, and the history of environmental change.[4]

In the cool gaze of the new millennium, we can easily see how the Basalla model reflected (and why should it not!) certain intellectual fashions of its day, especially in its symmetry with the sociology of "center-periphery" relations and the political economy of modernization theory—then so beloved of major international founda-

[1] See, for example, Richard Drayton, "Science and the European Empires," *Journal of Imperial and Commonwealth History,* 1995, *23,* 3:503–510.

[2] [Frederick Jackson Turner], *Rereading Frederick Jackson Turner: "The Significance of the Frontier in American History" and Other Essays,* with commentary by John Mack Faragher (New Haven: Yale Univ. Press, 1998), p. 32.

[3] George Basalla, "The Spread of Western Science," *Science,* 1967, *156:*616–22; a vision regained, unaltered, in *idem,* "The Spread of Western Science Revisited," in *Mundialización de la ciencia y cultura nacional,* eds. Antonio Lafuente, Alberto Elena, and Mariá Luisa Ortega (Madrid: Ediciones Doce Calles, 1993), pp. 599–603.

[4] See, for example, Michael A. Osborne, "Introduction: The Social History of Science, Technoscience and Imperialism," *Science, Technology and Society,* 1999, *4,* 2:161–70.

tions, the United Nations, and the World Bank.[5] Its language fittingly described the advancing, inquisitive spirit of the American West and the Western world, a language fundamental to both scientific discovery and the Western image of itself. Thus emerged the Turneresque (or even Kiplingesque) "traits" of the conqueror of Nature, and of realms beyond the seas, by men whose lives revealed "that practical, inventive turn of mind, quick to find expedients; that masterful grasp of material things, lacking in the artistic but powerful to effect great ends; that restless, nervous energy; that dominant individualism, working for good and for evil, and withal that buoyancy and exuberance which comes with freedom."[6]

To many readers, the Basalla model reflected common sense. Its historical narrative fit—although perhaps too closely—the experience of the United States, with which other countries were meant to compare themselves and, if possible, to imitate. Its perspective was plainly diffusionist, at the time when much Western social science subscribed to diffusionism as a major explanatory factor in human history. It was a creation of its time.[7]

There is no question that the model served as a valuable heuristic device. However, it lacked explanatory power. Indeed, it seemed to embody a kind of genetic fallacy, in which the existence of an idea is to be taken as a sufficient explanation of its cause. The suggested trajectory was set out as linear and unidirectional; it made nothing of changing imperial strategies or their differences over time; it generalized widely over vastly dissimilar non-Western cultures; and perhaps most important, it dispensed with political agency "on the ground," except insofar as it represented cultural "resistance" to the apparent benefits of Western science. It failed to say much about Western science as a means of social control or cultural suppression. It failed, moreover, to say anything very much about "colonial science" itself. Nonetheless, like any good Popperian model, it prompted attempts to falsify it. Historians who had previously had little contact with the history of science—particularly its trajectory in former colonies—embarked on a trek to test the model's "predictions."

Within the last thirty years, there have been several attempts to replace these—in hindsight, rather shaky—foundations. Over the last decade, especially, momentum for change has increased. In 1990, a large international conference in Paris displayed a wide variety of attempts to "localize" the European experience of science, from the viewpoints of both "metropolitan" and "colonial" historians.[8] It was a moment, perhaps, when the study of European science first began to parallel revisionist stud-

---

[5] As enshrined for a generation by W. W. Rostow in *Stages of Economic Growth* (Cambridge: Cambridge Univ. Press, 1960). An alternative argument was proposed in Roy MacLeod, "On Visiting the 'Moving Metropolis': Reflections on the Architecture of Imperial Science," *Historical Records of Australian Science,* 1982, *5,* 3:1–16; and in *Scientific Colonialism: A Cross-Cultural Comparison,* eds. Nathan Reingold and Marc Rothenberg (Washington, D.C.: Smithsonian Institution Press, 1987). A decade on, the issue is taken up afresh in Richard Drayton, "Science, Medicine and the British Empire," in *The Oxford History of the British Empire,* vol. V: *Historiography,* ed. Robin W. Winks (Oxford: Oxford Univ. Press, 1999), pp. 269–70. For a recent contribution, see Dhruv Raina, "From West to Non-West? Basalla's Three-stage Model Revisited," *Science as Culture,* 1999, *8,* 4:497–516.

[6] [Turner], *Rereading Turner* (cit. n. 2), p. 59.

[7] For easy access to its higher simplifications, see J. M. Blaut, *The Colonizer's Model of the World: Geographical Diffusionism and Eurocentric History* (New York: The Guilford Press, 1993).

[8] Patrick Petitjean, Catherine Jami, and Anne-Marie Moulin, eds., *Science and Empires: Historical Studies about Scientific Development and European Expansion* (Dordrecht: Kluwer Academic Publishers, 1990).

ies of European nationalism.[9] What emerged was a growing awareness that the historical reality of European colonialism—the constitution of colonial empires, notably from the eighteenth century onwards—had to be more closely defined. Colonies once administratively categorized—as, for example, colonies of conquest, plantation, or settlement—were revisited as places within which the practice of science had a special meaning. Colonial science acquired a three-dimensional character and presence, whether the ultimate reference was Britain, France, Germany, or Spain, or for that matter, the United States.

The next few years brought several important additions to this more critical literature. In one, James McClellan produced an excellent account of French expansion in the eighteenth century, adopting, but qualifying the diffusionist interpretation with a rich discussion of contextual debate.[10] As he argued, the interests of France represented science as an instrument of empire—complicit, ironically, with both the teachings of the *philosophes* and the contradictions of slavery. Meanwhile, a more global perspective began to emerge, with Lewis Pyenson's highly influential insights into the history of European imperial science.[11] Pyenson looked systematically at the expansion of different European cultures,[12] inspiring among others the use of models of increasing—in some, almost ptolemaic—complexity.[13] By the mid-1990s, working largely from a *soi-disant* postcolonial environment in French Canada, he had executed a magnificently conceived trilogy, surveying the extension of the "exact" sciences into the colonial worlds occupied by Germany, the Netherlands, and France.[14]

It was a measure of the increasing sophistication of the field that one of Pyenson's leading arguments—that science was an agency of imperial culture, and of cultural imperialism, but that the exact sciences were not in turn shaped by the colonial experience—provoked strenuous debate.[15] Critics asserted that, on the contrary, colonial expansion, with its investment in geophysics, meteorology, and astronomy, was vital to the progress of the exact sciences in Europe; and that in any case, European imperialism underwrote the global exercise within which the exact sciences flourished (and without which they could not have benefited so quickly, or so well).[16]

[9] In the apt phrase of Drayton, "Science and the European Empires" (cit. n. 1), p. 508.

[10] James McClellan III, *Colonialism and Science: Saint Domingue in the Old Regime* (Baltimore and London: Johns Hopkins Univ. Press, 1992).

[11] Among these were: Lewis Pyenson, "Why Science May Serve Political Ends: Cultural Imperialism and the Mission to Civilize," *Berichte zur Wissenschaftgeschichte,* 1990, *13:*69–81; idem, "Habits of Mind: Geophysics at Shanghai and Algiers, 1920–1940," *Historical Studies in the Physical and Biological Sciences,* 1990, *21:*161–96; and idem, "Colonial Science and the Creation of a Postcolonial Scientific Tradition in Indonesia," *Akademika,* 1990, *37:*91–105.

[12] Lewis Pyenson, "Functionaries and Seekers in Latin America: Missionary Diffusion of the Exact Sciences, 1850–1930," *Quipu,* 1985, *2,* 3:387–420; idem, "Pure Learning and Political Economy: Science and European Expansion in the Age of Imperialism," in *New Trends in the History of Science: Proceedings of a Conference Held at the University of Utrecht,* eds. R. P. W. Visser et al. (Amsterdam: Rodopi, 1989), pp. 209–78; and idem, "Typologie des stratégies d'expansion en sciences exactes," in Petitjean, Jami, and Moulin, *Science and Empires* (cit. n. 8), pp. 211–18.

[13] See Xavier Polanco, ed., *Naissance et development de la science-monde: Production et reproduction des communautés scientifiques en Europe et en Amérique Latine* (Paris: La Decouverte, 1990).

[14] Lewis Pyenson, *Cultural Imperialism and Exact Sciences: German Expansion Overseas, 1900–1930* (New York: P. Lang, 1985); idem, *Empire of Reason: Exact Sciences in Indonesia, 1840–1940* (Leiden: E. J. Brill, 1989); and idem, *Civilizing Mission: Exact Sciences and French Overseas Expansion, 1830–1940* (Baltimore and London: Johns Hopkins Univ. Press, 1993).

[15] See especially Paolo Palladino and Michael Worboys, "Science and Imperialism," *Isis,* 1993, *84:*91–102; and Professor Pyenson's reply in "Cultural Imperialism and Exact Sciences Revisited," *Isis,* 1993, *84:*103–8. See also Roy MacLeod, "On Science and Colonialism," in *Science and Society*

More compellingly, scholars familiar with the ways of colonial life drew attention to the "other"—the "nonexact"—sciences (although the phrase is unappealing), including the social, natural history, and life sciences, which were everyday companions of colonial enterprise. In these, spectacularly in India and Africa, the level of intellectual interaction between ruler and ruled, expert and populace, was vital, everyday, and decisive. The ensuing relation was one of interdependence. Western science was, above all, a purveyor of solutions to the needs of imperial governments; at the same time, it could be, and was, assimilated and transformed by local and indigenous peoples into a body of knowledge for local empowerment. In India, colonial science became the basis of negotiated knowledge, which consolidated and ultimately transferred power to the colonized. In Latin America, colonial circumstances favored the creation of a criollo science, which found local ways of representing nature and generated its own discourse.[17] In Australasia, independent traditions were fostered *within* the ambit of colonial science, which sustained colonial Europeans on the march to nationhood.

Looking back, the diffusionist perspective, the center-periphery model, and even the "strong program" of imperial science have proved insufficiently accommodating to the sources and discoveries of recent research. Indeed, for the most part, these early strategies have, to borrow the words of Lakatos, failed to produce a progressive problem shift. What was needed, and what has now come about, is a revolution in approach, producing in its wake a new and deeper set of questions about the uses of knowledge and power. In the European empires from the eighteenth century, "metropolis" and "periphery" were, of course, geographically constituted—no one failed to grasp the differing narratives of science in different environments, particularly in subtropical and tropical climates. But relationships between metropolis and periphery were also socially constituted, and as such represented the combined effects of social, political, and economic relations among different cultures and peoples. It is these relationships that were to shape the nature of science "conveyed" and exchanged with Europe, and to empower new nations with Western science as an instrument of regional and international development. It is these relationships that were equally to constrain colonial development, rendering certain features of colonial dependency no less evident in the world today.

During the last five years, the history of imperial and colonial science has developed in maturity and sophistication. On the one hand, it is becoming clear how early and extensive imperial expansion in science and technology provided a basis for the "knowledge culture." The Iberian conquests of the fifteenth century onwards—and not merely the fatal "Columbian exchange"—are being revisited in fresh language, tailored to an understanding of nature's laws.[18] Thanks to a new interest in British imperial history, codified in part by the new *Oxford History of the British Empire,* we can trace how the early adventurers of Elizabethan England framed rules of colonial

---

in Ireland: The Social Context of Science and Technology in Ireland, 1800–1950, ed. Peter Bowler (Belfast: Queen's Univ. 1997), pp. 1–17; and *idem,* "Reading the Discourse of Colonial Science," in *Les Sciences coloniales: Les Sciences hors d'occident au XXème siècle,* vol. 2: *Figures et institutions,* ed. Patrick Petitjean (Paris: ORSTOM, 1996), pp. 87–98.

[16] Drayton, "Science and the European Empires" (cit. n. 1), p. 507.

[17] Juan-José Saldaña, "Cross Cultural Diffusion of Science: Latin America," *Cuadernos de Quipu,* 1987, 2:33–57.

[18] See, for example, Anthony Pagden, *European Encounters with the New World: From Renaissance to Romanticism* (New Haven: Yale Univ. Press, 1993).

engagement in which commerce held the key to the improvement of natural knowledge. Colonialism and science together combined the dictates of piety and patriotism. As science advanced, so did the "civilizing mission," in which "mother countries" became children of the same processes to which they gave birth.[19]

More recently, imperial perspectives have been taken up by cultural theorists, who have replaced the "model building" of the 1960s with multivalent perspectives colored by the complexities of contact. The current postmodern turn against canonical authority (and traditional historiography) has encouraged histories of multiple interactions, and a preference for nonlinear, nonprescriptive descriptions dealing with historically contingent practices of communication and exchange.[20] At the same time, the history of colonial science has proved relevant to those who see in Western colonialism the ultimate cause of contemporary dependency. To all these, the concept of "periphery," once apparently so obvious, is historiographically ambiguous and methodologically problematic. In purporting to explain, it has in fact evaded explanation. And as one reexamines non-Western cultures and their legacies of natural knowledge, even what have been thought of as simple conventions of periodization and historical narrative prove unreliable. Scholars as different as Roshdi Rashed and Joseph Needham have reminded us how far Western interpretations have failed to comprehend the character of non-Western science.[21] Until they do, our understanding of non-Western peoples, and their role in constituting science, remains incomplete.

The time has surely come to widen the research agenda and broaden theoretical approaches. First, we need scholarship from both imperial, European viewpoints, and from the perspective of colonial life. It is in the process of multiple engagements—between Europeans at home and abroad, between Europeans and indigenous peoples, and between Western and non-Western science—that the processes of colonial science developed. Second, we need studies that embrace the changing relationship between the history of technology and of colonial science; that show how fleets of trade followed flagships of science; that unravel the complex associations between imperial "promoters" and colonial knowledge. Third, we need to test similarities and differences between and among imperial systems, in their use of science and their privileging of contrasting scientific metaphors and mentalities. In the history of imperial and colonial science, France, Britain, Spain, Portugal, Germany, and the Netherlands present similarities and contrasts. We have yet to see a major comparison of their respective features, contributions, and legacies. Finally, while many authors have written vertically, so to speak, in relation to specific disciplines and cultural spheres, we lack a baseline for comparative studies, undertaken horizontally, across cultural frontiers. It is likely that much new insight awaits the study, say, of the transmission of science across different spaces, of "colonial" geol-

---

[19] To generalize Richard Drayton, "Knowledge and Empire," in *The Oxford History of the British Empire,* vol. II: *The Eighteenth Century,* ed. P. J. Marshall (Oxford: Oxford Univ. Press, 1999), p. 251.

[20] See Michel Paty, "Comparative History of Modern Science and the Context of Dependency," *Science, Technology and Society,* 1999, *4,* 2:171–203.

[21] See Roshdi Rashed, "Science classique et science moderne à l'epoque de l'expansion de la science européenne," in Petitjean, Jami, and Moulin, *Science and Empires* (cit. n. 8), pp. 19–32.

ogy across the Pacific and Indian Oceans, and of conceptual transfers on a "south-south" basis.

II. CIVILIZING PASSIONS

For most historians of science, European or American, colonial science is a foreign country—they do things differently there. Until recently, the same could be said of imperial and colonial historians, who showed little interest in science. Now, however, from the contours of the late twentieth century, comes a working partnership—a strategic alliance—for the twenty-first century, between imperial and colonial history and the history of science. Today, both find it necessary to understand how science and technology have driven political and economic change. Imperial historians have begun to revisit the history of European colonialism. They have revealed, for example, the proconsular roles of scientific expeditioners, administrators, and practitioners, and have traced the imperial mandates given the sciences—often with deliberate, far-reaching implications for science, as well as for commerce, at "home." For their part, historians of science are exploring sites—not excluding colonial museums—where the colonial nature of science was (and in some ways, is still being) "negotiated."

From these tendencies has emerged a new subdiscipline of "colonial science" (with its sister subjects, colonial medicine and colonial technology),[22] while interpretations of the experience are now routinely offered in courses on world history.[23] Impressive lists of empirical studies have been generated by scholars in former colonies,[24] while throughout the world, many have taken up the role of science in the construction of national identities.[25] Much of this work, however, has taken place within national linguistic and cultural traditions—as, for example, within the British, French, Dutch, or Iberian traditions. Much, also, has been biased towards the institutions of science, rather than towards the generation of scientific ideas, his-

---

[22] See, for example, David Arnold, *The Problem of Nature: Environment, Culture and European Expansion* (Oxford: Blackwell, 1996). See Michael Worboys, "Science and Imperialism Since 1870," in *The Cambridge History of Science,* vol. 8, ed. Ron Numbers (New York: Cambridge Univ. Press, forthcoming).

[23] Surely the intention of William K. Storey, ed., *Scientific Aspects of European Expansion,* vol. 6 of *An Expanding World: The European Impact on World History, 1450–1800* (Aldershot: Variorum, 1996), and James E. McClellan III and Harold Dorn, *Science and Technology in World History: An Introduction* (Baltimore: Johns Hopkins Univ. Press, 1999); and by now, the realization of Daniel Headrick's textbooks, *The Tools of Empire: Technology and European Imperialism in the Nineteenth Century* (New York: Oxford Univ. Press, 1981), and *The Tentacles of Progress: Technology Transfer in the Age of Imperialism, 1850–1940* (New York: Oxford Univ. Press, 1988).

[24] Notably in India and Latin America. For the largest democracy in Asia, see, for example, Satpal Sangwan, *Science, Technology and Colonisation: An Indian Experience, 1757–1857* (New Delhi: Anamika Prakashan, 1991) and Narender K. Sehgal *et al., Uncharted Terrains: Essays on Science Popularisation in Pre-Independence India* (New Delhi: Vigyan Prasar, 2000); and for the largest democracy in Latin America, see Maria Margaret Lopes, *O Brasil descobre a pesquisa científica* (São Paulo: Hucitec, 1997).

[25] See Ludmilla Jordanova, "Science and National Identity," in *Sciences et langues en Europe,* eds. Roger Chartier and Pietro Corsi (Paris: Ecole des Hautes Etudes en Sciences Sociales, 1996), pp. 221–31; Juan José Saldaña, ed., *Los Origenes de la ciencia nacional* (Mexico City: Cuadernos de Quipu, 4, 1992).

tory of technology or trade, or issues more central to geopolitical and diplomatic history.[26]

But despite this weight of courses and cases, there has still to come better understanding, among both colonial historians and historians of science, of what practitioners on the ground have known for generations: that ever since Europeans first engaged the world *outre-mer,* the traffic of ideas and institutions has always been reciprocal. That text has been taken to read relationships between Europeans at home and overseas. But it is now more significant to see the relationship among Europeans mediated by European encounters with indigenous peoples.[27] It requires little prescience to appreciate the value of Edward Said's reminder that the cultural consequences of colonialism were as profound and lasting for the colonizer as for the colonized.[28] Such received wisdom among imperial historians should occasion no surprise among historians of science. Indeed, the literature of scientific travel is replete with cross-references to reciprocality.[29] Precisely what the consequences of this traffic were, however, remain important objects of enquiry; as indeed, do the processes themselves—often informal and driven by motives that combined, or at least mixed, the pursuit of natural knowledge with the interests of statecraft and trade.

In certain contexts, the "binaries" of the diffusionist perspective have retained a degree of utility. For example, it can be useful to employ a bibliographical distinction between European institutions associated with the administration of science in colonial places (including the uses of science as an instrument of colonization); and the practice of colonial science as a category of ideology and self-reference, an activity conducted by colonials, whether formal or informal, with regard to either (or both) colonial aspirations and imperial goals.[30] But for the most part, such easy categories are now giving way to more complex readings of colonial science—no longer merely a phase, but rather a space, a complex of legacies, a combination of motives, and a role in the discourse of development.[31] How else, after all, to deal with the obvious differences in the relationship between Ottoman science and Euro-

[26] There are significant exceptions. See Ian Inkster, "Scientific Enterprise and the Colonial 'Model': Observations on Australian Experience in Historical Context," *Social Studies of Science,* 1985, 15:677–704; and Jan Todd, *Colonial Technology: Science and the Transfer of Innovation to Australia* (Sydney: Cambridge Univ. Press, 1995); and Ted Wheelwright and Greg Crough, "The Political Economy of Technology," in *The Commonwealth of Science: ANZAAS and the Scientific Enterprise in Australia, 1888–1988,* ed. Roy MacLeod (Melbourne: Oxford Univ. Press, 1988), pp. 326–42.

[27] A major Australasian landmark was Nicholas Thomas, *Entangled Objects: Exchange, Material Culture and Colonialism in the Pacific* (Cambridge, Mass.: Harvard Univ. Press, 1991), followed by his even more compelling *Possessions: Indigenous Art/Colonial Culture* (London: Thames and Hudson, 1999). For other vernaculars, see Ivan Karp and Steven D. Lavine, eds., *Exhibiting Cultures: The Poetics and Politics of Museum Display* (Washington, D.C.: Smithsonian Institution Press, 1991).

[28] Edward Said, *Culture and Imperialism* (New York: Knopf, 1993).

[29] See Barbara Stafford, *Voyage into Substance: Art, Science, Nature and the Illustrated Travel Account, 1760–1830* (Cambridge, Mass.: MIT Press, 1984), and for a closer view of the Pacific, Nicholas Thomas, *Colonialism's Culture: Anthropology, Travel and Government* (Melbourne: Melbourne Univ. Press, 1994); or for a more relaxed guide, see Peter Raby, *Bright Paradise: Victorian Scientific Travellers* (London: Pimlico, 1996).

[30] See, for example, R. A. Stafford, *Scientist of Empire: Sir Roderick Murchison, Scientific Exploration and Victorian Imperialism* (Cambridge: Cambridge Univ. Press, 1989).

[31] See Paty, *Comparative History* (cit. n. 19).

pean science, or between science in Europe and science in premodern China and Japan?[32]

In the last five years, several new lines of research have emerged. First, there is a growing tendency to see science in colonial contexts not as responding to or resisting Europe, but as part of a constructed colony's independent history. These may begin with empirical sketches drawn from Europe, but then focus upon the significance of "place," and the tangled experience of the colonial encounter. They remind us that colonial histories once written from Europe are in fact often histories about Europeans abroad, which must be rewritten in terms familiar to local narrators.[33]

The same applies to histories of scientific exploration and discovery, which have traditionally let the strategic and economic motives of European expansion speak for themselves.[34] In fact, the consequences of scientific voyaging were felt no less by those indigenous peoples who were recipients of scientific interest, who became "curiosities," or who traded in the commodities that were attractive to sailors or needed by settlers. A generation of Pacific studies has illuminated this mutual, often fatal attraction. Elsewhere, notably in India, a generation of subaltern scholars have labored to show how colonizers and colonized were each, enduringly, shaped by their experience of the "other," and where racial prejudice and exclusion were mandated by appeals to science.[35]

It is a commonplace that, to limit this history to a story of one-way traffic is to misrepresent historical complexity in linear terms. The colonial encounter was never "one to one," nor was it necessarily isolated, as it could involve behaviors transported from other places and other times; nor was it bound by mechanical rules of imperial administration, in which colonial linkages resembled "spokes of a wheel." On the contrary, in ways reminiscent of imperial Rome, there was a considerable circulation of scientists and administrators among widely dispersed colonial worlds. Rarely, it was long supposed, did these different elites communicate with each other. Now, however, libraries across the world are disclosing evidences of

---

[32] On which, see, *inter alia,* Alberto Elena, "Models of European Scientific Expansion: The Ottoman Empire as a Source of Evidence," in Petitjean, Jami, and Moulin, *Science and Empire* (cit. n. 8), pp. 259–68; and Morris Low and Nakamaya Shigeru, *Science, Technology and Society in Contemporary Japan* (Cambridge: Cambridge Univ. Press, 2000).

[33] See such case studies as Peter Hulme, *Colonial Encounters: Europe and the Native Caribbean, 1492–1797* (London: Routledge, 1986), and the essay of Wade Chambers that follows; see also David Wade Chambers, James E. McClellan, and H. Zogbaum, "Science/Nation/Culture in the Caribbean Basin," in *The Cambridge History of Science,* vol. 8, ed. Ron Numbers (Cambridge: Cambridge Univ. Press, forthcoming).

[34] See, for example, Franciso de Solano, "Viajes, comisiones y expediciones científicas españolas a ultramar durante el siglo XVIII," *Cuadernos Hispanoamericanos,* 1988, *2:*146–56; and Alan Frost, "Science for Political Purposes: European Explorations of the Pacific Ocean, 1764–1806," in *"Nature in its Greatest Extent": Western Science in the Pacific,* eds. Roy MacLeod and Philip F. Rehbock (Honolulu: Univ. of Hawaii Press, 1988), pp. 27–44.

[35] Cf. the influential work, beginning in the 1980s, surrounding Ranajit Guha, David Arnold, Partha Chatterjee, and Dipesh Chakrabarty. See Ranajit Guha, ed., *Subaltern Studies I: Writings on South Asian History and Society* (Delhi: Oxford Univ. Press, 1982) and Dipesh Chakrabarty, "Post Coloniality and the Artifice of History: Who Speaks for India's Pasts," *Representations,* Winter 1992, *37:*1–26. See also the work of Gyan Prakash, anticipated in *After Colonialism: Imperial Histories and Postcolonial Displacements* (Princeton: Princeton Univ. Press, 1995), and culminating in his recent *Another Reason: Science and the Imagination of Modern India* (Princeton: Princeton Univ. Press, 1999).

intercolonial exchange—of visits, as well as of periodicals, specimens, and commodities—demonstrating how ideas regularly moved between and among colonial
empires. Increasingly, significantly, an intellectual trade in ideas also developed
between the dependencies of Europe and the great powers of North America and
Asia.[36]

In the final analysis, the historiography of science, given to the study of rational
and universal enterprise, has tended to neglect, or even to suppress, the presence of
"other reasons." As such, it has tended also to deemphasize the uses of science in
relation to gender, race, and class. Today, locality and place are now being constituted as legitimate "centers" for historical reconstruction. Science occurs locally, it
is argued, before it is recognized universally. But between conception and recognition falls a long shadow. For most of us, the metropolis remains, after all, where the
action is. However, the modern injunction "to think globally and act locally" can
usefully be inverted. As we complete the turn of the century, we find in the study of
colonial science and its relationship to indigenous cultures new ways of seeing science in action—or rather in "inter-action"—as a highly textured activity, serving to
celebrate the diversity of knowledge among peoples, of places, and over time.

III. READING COLONIAL SCIENCE

Following the end of the Seven Years' War in 1763, and the subsequent readjustment
of colonial possessions throughout the world, the major powers of Europe entered
a period of intensive rivalry that, in turn, spurred the pursuit of natural knowledge.
By the end of the American Revolution, and the successful rebellion of Spain's
American colonies, the world was set for the expansion of science as a global
discourse. Within two generations, and certainly by the Treaty of Berlin—which
divided Africa among the European powers—science had become a metonym for
empire. Museums and learned societies throughout Europe represented the achievement of a rule of knowledge, coincident with the rule of law. In differing measure,
the colonies of Europe *outre-mer* were to embody both.

This period, which saw British ascendency established at sea, also saw important
shifts in the intellectual relations between the continental powers, the United States,
and the colonial worlds of Africa and Oceania, together with new accommodations
with the ancient civilizations of Asia. These, of course, reflected interests of strategy
and trade, prestige and profit. However, where colonies of conquest, plantation, or
of settlement ensued, European colonials incorporated the practices of European
science. Governmentality, in the language of Foucault, assigned to science a pastoral
influence in the regulation of colonial affairs. Whether favored by Iberian mercantilism or Manchester free trade, European science came to rely upon a more or less
continuing traffic with the overseas world. By the end of the nineteenth century, the
major powers of Europe had, in their different ways, become accustomed to a long,
and apparently indissoluble, marriage of convenience between strategic interests and
philosophical opportunity. The fruits of this marriage endured and evolved. Colonial
expansion enlarged universal knowledge. Science became a key aspect of a global
intelligence system, which served best the interests of those best placed to receive

---

[36] Nicholas Jardine *et al.,* eds., *Cultures of Natural History* (Cambridge: Cambridge Univ. Press,
1996).

its data. Science became, in turn, both a colonizing ideology and an agency of colonial self-identity. For a colony, the pursuit of science became a license for membership in the community of nations. Today, the processes by which science served colonial expansion are fundamental to understanding science in the modern world. As such, they form the basis of this book.

This volume of *Osiris* presents not only new work, but also a fresh approach to the study of science and its European/colonial interactions. Its coverage is not global. Nor does it focus upon a single discipline, geographical area, or sphere of influence. Instead, it takes up certain themes and examines them from multiple standpoints. Overall, it is concerned to show the variety and complexity of European experience in the encouragement, support, and legitimization of imperial exploration and discovery, and colonial development in the context of metropolitan interests, between the late eighteenth century and the present period of post-cold war uncertainty. Using examples from French, British, Spanish, Portuguese, and Latin American history, it hints at the difficulties in theorizing similarities between imperial science and colonial science in parallel empires. It suggests that studies based on imperial archives mask different local realities. On the other hand, it shows how different colonial traditions can nevertheless produce universal generalizations.

The exercise relies, *au fond,* upon a common valuation of comparative approaches to the development of a political language of science. The scene has shifted considerably from asking whether science was a feature of imperialism (it was, and is), and whether imperialism "advanced" science (likely, if not always possible to prove), to a broader range of questions, some implicit in the Basalla model itself. For example, the "phase" of colonial science, far from being transitional, remains for many countries a condition of apparently permanent existence, with enormous implications for human welfare and economic development. It is only now, at the end of the millennium, that the World Bank has decided to support university education and research in many Third World countries. It is not least for this reason that this volume of *Osiris,* appearing in this equinoctial year, is so timely.

For reasons of method and convenience, this volume is divided into four sections. As we believe readers of *Osiris* to be reasonably familiar with Anglo-American experience, we deliberately invited a range of authors with knowledge of other realms. The essays in Part I are devoted to the "metropolitan perspective," and to what might be called "imperial legacies," giving pride of place to the vision and experience of the Iberian countries. First, Juan Pimentel examines the way in which Spain in the early modern period foresaw the world outside Europe as coming within a larger ecumene, a universal monarchy enlightened by reason, to unite mankind under the banner of religion. The vision transcended science, but defended nature, and in the process established a new conversation between Europe and colonial America. Next, James McClellan and François Regourd enter the stage with a glimpse of France a century later, on the eve of revolution, commandeering natural knowledge in the service of the state to maintain a huge and unwieldy colonial empire, threatened by republicanism, indifference, and England. In the third paper, we look north with Sverker Sörlin, who explores Sweden's less familiar imperial experience—one in which gathering "scientific intelligence" of plants and peoples was to secure presence and authority in the sub-Arctic. It was a dream more than a destiny, but its influence persists. Far to the south, Alberto Elena and Javier Ordóñez remind us how Spain in the early nineteenth century bore the loss of its vast American possessions, and by the end of the century, saw its colonial failures in Africa and the Pacific mirrored in the failures of Spanish science.

That complex history of vision and experiment, success and failure brings us to Part II, where we consider metaphors and meanings in the European colonial enterprise. Visiting "colonial nature," Suzanne Zeller shows how the science of geology became a meeting place for political negotiation in North America, while in South America, Margaret Lopes and Irina Podgorny show how the "museum idea" helped new nations of Latin descent declare their intellectual independence.

With similar effect, Kapil Raj shows how in British India, well into the late nineteenth century, the enterprise of "mapping" not only created a sense of possession, but also accompanied scientific practices that traveled throughout the entire British Empire. The political economy of empire returns as a theme in Michael Osborne's paper on the acclimatization movement, and its role in "civilizing" the unruly domains of French North Africa and Australasia. In France, appeals to science further united the colonial party, which called upon science to rescue colonized peoples from barbarism. In Africa as elsewhere, several decades would elapse before the self-contradictions of "civilization" became apparent, and not until the 1970s was its rhetoric finally dismissed. Its problematic legacy returns in Part IV, and in the postponed history of postcolonial development.

Why this legacy endured for much of the last century becomes clear in Part III, where essays on the "colonial project" examine the culture of local science—as seen from across the Atlantic (by Antonio Lafuente), and as seen on the ground in colonial Brazil (by Silvia Figueirôa and Clarete da Silva). In both cases, science was first cast as an instrument of colonization, but in the second, it became also a matter of colonial self-interest. Colonial conditions could, and did, marginalize the practice of science, but the seeds of colonial nationalism can be traced among these attempts to create a distinctive, local scientific sensibility. From mining to medicine, Part III continues with papers by Harriet Deacon on racism and the professions at South Africa's Cape Colony; and by Michael Worboys on changing conceptions of tropical medicine throughout the British Empire. Both illuminate the role of medicine and science in creating a framework of improvement, even at the cost of creating "differences." With the construction of an imperial medical science comes the image of a global mission, of a war against disease, invoking the weapons of the new bacteriology. It reads as a story as disturbing as any we see in the Third World today.

With Part IV we come full circle back to the European world, to the world system, and to some of the achievements and omissions of colonialism. Wade Chambers and Richard Gillespie skillfully guide us into a newfoundland of place and time where, increasingly, indigenous voices ask to be heard and given priority. Postmodern, postcolonial modalities do not threaten realist preferences, but they challenge our point of departure. Chambers and Gillespie argue that our approach to understanding the pursuit of knowledge and the history of science, must, in some sense, be turned inside out; that the scholarly agenda should privilege local contexts of science, and so ultimately dissolve the lingering distinction between those who write about "science," and those who write about the place of science in colonial and neocolonial history.

In the last three essays, Deepak Kumar, Christophe Bonneuil, and John Merson discuss the role of science as a tool of development. Until the 1960s, the colonial powers viewed economic development as an alternative to decolonization. Colonial economics became the foster parent of development economics in the postcolonial world. But as these authors show, the fragility of independence is everywhere in doubt, even in India, despite its vast resources and stable middle class, its nuclear program, and its investment in information technology.[37] The rhetoric has proven unconvincing in Africa and elsewhere in the tropics where, despite the encouragements of governments and aid agencies, scientific institutions have had limited success; and where the erosion of property rights in local knowledge threatens indigenous peoples and small nations alike. Globalization and commercial "bio-colonialism" may, in their indifference to local interests, stir memories of European colonial expropriation and exploitation. There is no more appropriate moment to recall Santayana's injunction that we would be wise not to forget history, lest we repeat it.

A volume of such global scope must not risk making global promises. Inevitably, there are omissions, some of which a wider conscription (or more volunteers) would have remedied. For example, we conspicuously lack both a Dutch and a German dimension and thus omit regions of the world in which the practices of colonial science were skillfully advanced, both well before and during the nineteenth century. While we may be stronger on the Iberian impulse than is usual in English, there is far too little in our pages on France, and nothing on the United States, a regrettable omission given the role played by American science, technology, and medicine in the Philippines and Cuba, not to mention the postcolonial world. Among the many aspects of European science tributary to our theme, there is far more to be said about the study of race; and it may be observed that the essays that take India and Africa as their foci pay insufficient attention to caste and class. Few of us allude to the rich and extensive exchanges of ideas that took place between Europeans and indigenous peoples, nor do we explore the relations between science and belief, on which so many revisionist accounts of colonial contact are now being constructed. Finally, the history of technology and of medicine—long separate fields of scholarship, but clearly part of the same colonial experience—could each easily justify a volume. For this admittedly poor reason, neither has been given sufficient space in these pages. They surely warrant greater attention in future issues of *Osiris*.

It remains to record a word of acknowledgement. No editor could have been better served by his authors, both those who appear among us, and those who were obliged for various reasons to leave our convoy. In most cases—and in all the Spanish and French cases—the authors kindly allowed me to circulate their drafts among colleagues, so that all might participate in what became a virtual seminar. For the French-subject authors, especially, the seminar became actual, thanks to the hospitality of the Maison Suger in Paris and its director, M. Jean-Luc Lory. To him, and to all others concerned, may I express my warmest thanks.

---

[37] For recent accounts of colonial science in India, see Jim Masselos, "The Discourse for the Other Side: Perceptions of Science and Technology in Western India in the Nineteenth Century," in *Writers, Editors and Reformers: Social and Political Transformations of Maharashtra, 1830–1930,* ed. N. K. Wagle (New Delhi: Manohar, 1999), pp. 114–29; and more generally, David Arnold, *Science, Technology and Medicine* (Cambridge: Cambridge Univ. Press, 2000).

Part I: Imperial Legacies

The Iberian Vision: Science and Empire in the Framework of a Universal Monarchy, 1500–1800

*Juan Pimentel**

ABSTRACT

This essay is devoted to the study of science in a long-successful political structure, the Universal Monarchy. Because the Iberian Empires did not survive into modernity, they have been viewed as incompatible with modern science. This, however, is a matter of perspective. From 1500 to 1800, science was one of the main instruments of Iberian representation in the New World. While it was not expressed in the familiar language of objectivity and was far from experimentalism, a kind of science defined by religious, courtier, and symbolic meanings shaped the dream of a Universal Monarchy. When this political concept became peripheral in the new Western order, Creole cultures reappropiated its practices to mark the identity of their new nations. However, even before colonial emancipation, these new national identities (American, not European; local, not universal) were based firmly in the natural knowledge of the New World regions they represented.

INTRODUCTION

IN A WELL-KNOWN PASSAGE IN *THE WEALTH OF NATIONS* (1776), Adam Smith affirmed that the two most important events in the history of mankind had been the discovery of America and the voyage to India around the Cape of Good Hope.[1] By the end of the Enlightenment, this was a commonplace repeated by G. T. Raynal, William Robertson, and other North European authors, as well as a recognition—often implicit—of the role of the Iberian nations in the expansion of the West. Meanwhile, another theory—in the same academies and among the same *philosophes*—viewed the role that Iberia played on the terrain of experimental science as marginal. The pejorative article on Spain in the *Encyclopédie*, for instance, illustrates a view that was once held by many.[2]

---

* Centro de Estudios Históricos, Consejo Superior de Investigaciones Científicas, Madrid, Spain.
[1] Adam Smith, *An Inquiry into the Nature and Causes of the Wealth of Nations*, ed. A. Cannan (1776; reprint Chicago: Univ. of Chicago Press, 1976), vol. 1, p. 470.
[2] The article on "Spain" that appeared in the *Encyclopédie*, written by Masson de Morvilliers, implied that the country had made no contributions to the increase of knowledge. As expected, it unleashed a resounding reaction in the peninsula, and in fact the echoes of that debate can be traced in the historiography of Spanish science. See Jose M. López Piñero, *Ciencia y técnica en la sociedad española de los siglos XVI y XVII* (Barcelona: Labor, 1979).

Subsequently, stereotypes have circulated around these two truisms and their multiple versions, and have been used to evaluate the historical role of the Iberian monarchies. Their colonial scientific past has given rise to contradictory visions: the heroic, apologetic one maintained by conservative Iberian historians, and the "Black Legend" vision of an ignoble and violent past that is held by French, English, and left-oriented Spanish historians. The peninsula's colonial history has also embodied a central paradox: Iberia's role in European expansion evolved in isolation from another great achievement of the period, the construction of modern science.[3]

Western historiography has always had its centers and peripheries. For a long time, certain products and ways of making science, as well as certain forms of colonial organization, have been highly esteemed, enshrined as precursors of what in those same centers was understood to be "science" and "colonial empire." It is not by chance that a triple perspective (scientific, colonial, and historiographic) from the north of Europe (and by extension, from the Anglo-Saxon world), has tended to consider the Iberian world as episodic and marginal, without a doubt peripheral, and even lacking in interest.[4]

However, any evaluation of science in the Iberian colonial world in the early modern age must consider the transformations our discipline has undergone on two themes that have shaped that perspective: the very notion of "Scientific Revolution" and the phenomenon of knowledge expansion and diffusion.[5] Over the past twenty-five years, the "big picture" of modern science has been gradually fading away. Current historiography is concerned with more than simply boosting the hagiography of the mainstream. New actors and themes have entered the scene. Particularly, the *spatial* and the *local* have come forward as dimensions to be acknowledged when discussing the transmission of scientific theory and practice. The myth of universality has disappeared, and in its place we have a more contingent, more cultural image of what the sciences are, and how they have been constructed. All of this should affect how we look at the scientific aspect of colonial models such as the Iberian. Unquestionably, neither the Iberian colonial model nor its ways of producing knowledge survived modernity well enough to find a place in those Western institutions where not only the present, but fundamentally, the past, are ordered. Nonetheless, beyond the commonplaces and stereotypes forged over time, the Iberian colonial world possessed a scientific dimension that deserves a fresh look.

The purpose of this essay is to offer a synthetic vision of the characteristic features of that colonial scientific past. And although this vision is perhaps too generalized, it expresses something unique, something that distinguishes this case from other

---

[3] Roy MacLeod, "On Visiting the 'Moving Metropolis': Reflections on the Architecture of Imperial Science," in *Scientific Colonialism: A Cross-Cultural Comparison*, eds. Nathan Reingold and Marc Rothenberg (Washington, D.C.: Smithsonian Institution Press, 1987), pp. 217–51, where it is noted how Western science, in general terms, has tended to view its own past in a benevolent, apolitical, and neutral manner. Thus, being outside the contamination of imperial history, Western science's contribution to a politically suspicious enterprise is concealed.

[4] Needless to say, this generalized trend has noteworthy exceptions, including the work of the outstanding Hispanist John H. Elliot and of Anthony Pagden. See also Richard M. Morse, *El Espejo de Próspero: Un Estudio de la dialéctica del Nuevo Mundo* (México: Siglo XXI, 1982).

[5] As an example of this vision of the Scientific Revolution, see Steven Shapin, *The Scientific Revolution* (Chicago: Univ. of Chicago Press, 1996). For a review of the literature on science and empire, see David W. Chambers, "Locality and Science: Myths of Centre and Periphery," in *Mundialización de la ciencia y cultura nacional*, eds. Antonio Lafuente, Alberto Elena, and Maria L. Ortega (Madrid: Doce Calles, 1993), pp. 605–19.

historical ways of acquiring knowledge and building empires. This review can be divided into two sections that coincide more or less with the flourishing and decline of the Iberian Empires. The first section will focus on the scientific products generated from the metropolis and its vision of empire; the second will discuss characteristic ways of performing science in the colonies as an expression of the syncretic character of Creole societies.

ADAM'S LAST WILL AND TESTAMENT AND THE DREAM OF A UNIVERSAL MONARCHY

It is said that when Pope Alexander sanctioned the Treaty of Tordesillas (1494) and the division of the world between the Castile and Portugal monarchies, the French King Francis I approached him and asked to see Adam's last will and testament. Whether fact or fiction, this anecdote reflects Europe's reaction to the Hispano-Portuguese decision to treat the globe as theirs alone.

The 1494 division of the New World by the meridian located 370 leagues west of Cape Verde would be complemented by the 1529 Treaty of Saragossa's establishment of a counter-meridian in the Pacific Ocean.[6] Both events—which provided the origin of the would-be Iberian monopoly over vast unknown regions, and a source of conflict between the two monarchies and the European crowns—also served to express the nature of the Iberian Empires at the time of their gestation. The Iberian vision of the colonial world had always been imbued with a patrimonial and spiritual, almost messianic, conception. Colonization was a symbolic possession, a possession more sought-after than real, indissolubly associated with its Catholic dimension, its ecumenical purpose. Moreover, scientific knowledge of the natural world exercised a decisive role when it came time to implement and represent that possession.

Such agreements to divide the world arose from medieval formulas that the Christian kings of Castile and Portugal used to partition reconquered areas of the Iberian Peninsula. Needless to say, cosmographic information was insufficient to plot the two meridians with precision. David Turnbull has explained the impossibility of achieving a unified image of the world from the Portulan maps, the characteristic manner of representation in Iberia at the end of the fifteenth century.[7] To a certain degree, the same argument applies to all fields of knowledge involved in the recognition of the New World: the stock of knowledge in geography or natural history, for example, circulated within the traditional framework of learning. The incorporation of new lands, new races, and new natural species was the result of a slow, uphill effort to see something different and new from a perspective that was, by definition, obsolete and fragmentary.[8]

---

[6] Concerning Tordesillas and Hispano-Portuguese relations, see the minutes of the International Congress *El Tratado de Tordesillas y su época*, 3 vols. (Madrid: Sociedad V Centenario del Tratado de Tordesillas, 1995).

[7] David Turnbull, "Cartography and Science in Early Modern Europe: Mapping the Construction of Knowledge Spaces," *Imago Mundi*, 1996, 48:5–24.

[8] The incorporation of the New World within the cultural framework of humanism is a vast domain of study. Americanists have more than proven how slowly the image of the New World was transmitted to Europe, and the low-key impact of a process that extended over many decades. See John H. Elliot, *The Old World and the New 1492–1650* (Cambridge: Cambridge Univ. Press, 1970). This explains, for example, the reiterated metaphors—images, myths, and fables of Christian, Greco-Latin, Hebrew, and medieval-chivalrous traditions—with which Iberian travelers fecundated their

Figure 1. A la espada y al compás, a más y más (*"With the sword and the compass, more and more"*). *From* Milicia y descripcíon de las Indias *by Bernardo de Vargas Machuca (Madrid, 1599). (Biblioteca Nacional de Madrid, sig. BN R/8113.)*

On a second plane, there was insufficient time to arrange and assimilate the new facts being brought to light by the many and frequent transoceanic discoveries. Portugal and Spain entered the sixteenth century with an advantage in nautical technology and navigation relative to other European nations. Moreover, Spain evinced clear signs of modernity in royal administration and bureaucratic organization. The speed with which both nations flung themselves at the planet is remarkable. In less

descriptions of the New World. This was the case for El Dorado, the paradise of Adam and Eve, California, and the search for King Solomon's mines. See Juan Gil, *Mitos y utopías del descubrimiento*, 3 vols. (Madrid: Alianza Editorial, 1989).

than forty years, between 1492 and 1529, the Iberians arrived in America, India, and the Philippine Islands. They sailed around the Cape of Good Hope and through the Strait of Magellan; they circumnavigated the globe; and they explored a large part of the Andean and Mesoamerican regions as well as the great Mexican plateau. Their sheer voracity for discovery contrasts markedly with other colonial models.[9]

The scientific aspects of this stage of expansion are marked by certain traits: by the will to make the new exert an influence on the old; by the theoretical difficulty of embracing the one without altering the other; and by the speed with which the idea of the world was expanding and transforming. Among the classical products of Iberian science were the so-called chronicles of the Indies, the genre in which the first great contributions to the scientific knowledge of the New World appeared. One of the most outstanding works was the *Historia general y natural de las Indias* (1535) by Fernández de Oviedo, an encyclopedic, descriptive work based on direct observation. Later, the Jesuit José de Acosta would produce the *Historia natural y moral de las Indias* (1590), a text endowed with an analytical and systematic spirit that marks the intellectual incorporation of the New World. Works of this type constituted the first true geographies and natural histories of the New World. They also put forward theories on the origin of man in America and included the first contribution to modern anthropology, that of the Franciscan Bernardino de Sahagún, thanks to whom a knowledge of pre-Columbian cultures survived the conquest. Outstanding for their work on the pharmacopoeia and on the therapeutic properties of simples are the Sevillian Nicolás Monardes and the Portuguese García de Orta. As did works in other fields, their studies of plants in America and India, respectively, circulated throughout Europe in the editions of other European authors (in this case, those of Clusius).

Although many advances in learning took place outside the scope of the royal administration, being the fruits of missionary zeal or individual erudition, the courts of the Avís and the Spanish monarchs of the House of Austria set up a framework for accumulating and ordering information on their domains. The Casa de Mina in Lisbon, the Casa de Contratación in Seville, and the Academia de Matemáticas and the Consejo de Indias (Council of the Indies) in Madrid were among the first European scientific centers created specifically for such ends.[10]

The Casa de Mina and the Casa de Contratación served as repositories for the cartographic information produced by the discoveries, with the goal of making general maps. Less well known is the so-called Academia de Matemáticas of Philip II, linked to the royal palace El Escorial. There, the monarch made provisions for teaching regular lessons in navigation and cosmography, building nautical instruments, and attempting to solve navigational problems, such as the calculation of longitude.[11] The fact that in 1582 the Portuguese Juan Bautista Labanha was appointed as a professor in that academy illustrates not only the routine movement between subjects of the two crowns, but also that the recent inclusion of the kingdom of Portugal under the monarchy of Spain (1580–1640) had crystallized.

---

[9] Contrast the slow progress of Anglo-Saxon North America from the Atlantic to the Pacific, a process that continued until the twentieth century.

[10] See Ursula Lamb, *Nautical Scientists and their Clients in Iberia (1508–1624): Science from an Imperial Perspective* (Lisbon: Instituto de Investigação Científica Tropical, 1984).

[11] María I. Vicente Maroto and Mariano Esteban Piñero, *Aspectos de la ciencia aplicada en la España del Siglo de Oro* (Salamanca: Junta de Castilla y León, 1991).

However, perhaps the most significant project was the program of *Relaciones geográficas*, initiated in 1570 by the reform of the Council of the Indies. This gave rise to the "cosmographer-chronicler," who was responsible for compiling information on the geography and natural phenomena of the different territories. This information was obtained from questionnaires, the so-called *cuestionarios de Indias*. Over a period of two decades, a set list of fifty questions relating to geographical coordinates, inhabitants and their customs, and plants and animals circulated throughout the kingdoms of America.

It is significant that a similar project was carried out at approximately the same time in the peninsular territories. Thus, America was seen as a place for experimental administrative measures and scientific projects that would subsequently be applied to the peninsula, and vice versa. Among many such examples, the *Relaciones geográficas* indicate how the monarchy considered the American kingdoms as simply further additions to its peninsular kingdoms, with no distinction made in either a legal or a theoretical sense.

The quest for natural knowledge led to organization of what is usually called the first scientific expedition to the New World. In 1570, the crown sent to New Spain the Erasmian physician Francisco Hernández, who explored the territory accompanied by painters, engravers, and a cosmographer. Hernández returned to the court with more than thirty volumes of manuscripts and over two thousand images of the flora, fauna, and soils of New Spain. Publication of these works had to wait for an overdue and incomplete version by the Neapolitan Nardo Antonio Recchi, associated with the Accademia dei Lincei.[12] This is not different from the fate of many other travel accounts and descriptions of the Iberian overseas territories: many originals were not published as complete collections until centuries later, or else parts were circulated in versions by other European authors.[13] Iberian monarchs were not as interested in revealing the secrets of their domains as they were in learning about them. This attitude, often viewed as obscurantist and as indifferent towards the advancement of knowledge, should be seen within the context of international rivalries, the patrimonial spirit of the Iberian Empire, and the functions performed by such descriptions and collections of maps and plants within a courtly society. The Iberian monarchies did not promote trips, natural histories, and geographies in order to reveal to the world a new natural environment, but rather to implement and represent their *empire of the world*.

The example of the Dominican monk Tomasso Campanella illustrates the role of science in a monarchy that had reason, particularly following the annexation of Portugal, to cherish the old imperial dream resuscitated in Europe with the reign

[12] Concerning the mission of Hernández, see Jose M. López Piñero and José Pardo Tomás, *Nuevos materiales y noticias sobre la historia de las plantas de Nueva España, de Francisco Hernández* (Valencia: Consejo Superior de Investigaciones Científicas, 1994). Also Raquel Alvarez and Florentino Fernández, eds, *De Materia Medica Novae Hispanae: Manuscrito de Recchi*, 2 vols. (Madrid: Doce Calles, 1999).

[13] Like the *Diálogos* of García de Orta (spread in Europe by the Clusius version), the *Suma Oriental* of Tomé Pires, known partially in Europe thanks to the travel collection by Ramusio, remained unpublished in its full version until much later—in this case, 1944. In the case of Spain, a good part of the *Crónicas* and the *Relaciones geográficas* were also published in complete editions much later, thanks to scholars such as Jiménez de la Espada in the nineteenth century and others working as late as the second half of the twentieth century. See, for example, Francisco de Solano, *Cuestionarios para la formación de las relaciones geográficas de Indias: Siglos XVI/XIX* (Madrid: Consejo Superior de Investigaciones Científicas, 1988).

of Charles V. In his *Della Monarchia di Spagna* (c. 1600), Campanella calculated the possibilities of the Spanish Crown achieving that imperial dream in relation to its geographical nature and effective Catholicity. Compared with the Ottomans, Spain was at a disadvantage as its domains were not adjacent. However, Spain was far ahead in navigational resources. Its mastery of the seas, possession of a huge fleet, and knowledge of the art of navigation constituted ideal instruments with which the monarchy would fulfill its universal destiny. Campanella developed a program to spread the king's dominion, and advised the king to send technicians of different nationalities to the territories. Belgian and German mathematicians should be sent to all parts of the crown's overseas possessions in order to ascertain the movement of the stars, the position of the oceans, and the flow of the tides. According to Campanella, astrologers were preferably to be German, not only because of their prestige, but also in order to wrest them from a territory infested with heresy. These specialists would study the heavens and describe the stars for the greater glory of the monarchy. The Portuguese and Genoese would take charge of navigational matters. Italians would look after the government administration; transalpines (central Europeans) would provide manufactures, machinery, and fireworks; and Spaniards would be responsible for fortresses, exploration expeditions, and matters of religion.

A Catholic monarchy, thus understood, was a collective enterprise; an organization whose members all contributed to the single goal of a Universal Monarchy; a community where science played both a divine and a human role, as symbolic as it was practical. Campanella concluded by predicting:

> This learning will make the one who possesses it master of the sea, the land and the people, and will illustrate the empire more than any other thing imaginable for making a King great. Because God Himself wants His works to be known and He gives them to he who knows them.[14]

This conception, which in certain aspects brings to mind the ideas of his contemporary, Francis Bacon, tells us of the integrated nature of the program. It intended, within a single plan, to construct a universal sovereignty in which true religion would triumph by means of faith, politics, and learning. Thus, from the Iberian peninsula, the organization and development of scientific activity was marked by a determination to build an empire: a real and symbolic appropriation of the world, whereby knowledge gleaned from nature in the context of a baroque, courtly, and Catholic society would serve the mission that Providence had reserved for the monarchy.

CREOLE CULTURES AND PERIPHERIES OF MODERNITY

Given that this unitary, Catholic, almost cosmic program could survive only under the hegemony of a Universal Monarchy, when the imperial edifice gradually deteriorated during the seventeenth century, the metropolitan scientific institutions also began to decline. This does not mean that science was no longer pursued in the Iberian Peninsula, but rather that the science being promoted remained attached to forms of representation that had lost credibility. The Iberian world was relegated to the periphery of international affairs on many planes. In the scientific sphere, the sources

---

[14] Tomasso Campanella, *Della monarchia di Spagna*, cited in Luis Díez del Corral, *El pensamiento político europeo y la monarquía de España* (Madrid: Alianza Editorial, 1983), p. 318.

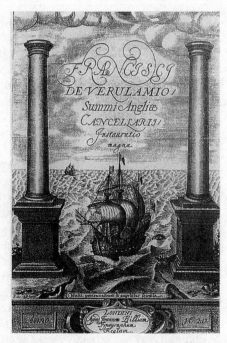

Figure 2. *On the left, the title page of* Regimiento de navegación *by Andrés García de Céspedes (Madrid, 1606). The ship beneath the columns of Hercules and the legend* Plus Ultra *are the national emblems of Spain. Note the striking similarity with the title page, right, of Francis Bacon's* Instauratio Magna *(London, 1620), with its legend* Multi pertransibunt et augebitur Scientia *("Many will cross it and Science will be enlarged"). A splendid coincidence: the image is used in the Anglo-Saxon tradition to represent increase of knowledge (the Baconian peaceful empire of man over nature), while in the Iberian tradition it represents knowledge gained through discovery and conquest of the New World.* (Regimiento de navegación *is in the Biblioteca Nacional de Madrid.*)

of authority shifted to other locations in Europe, to institutions where new forms of learning about the natural world were beginning to take hold. In a sense, with the onset of experimentalism, mechanist rationalism, the mathematicization of natural phenomena, and other ingredients of the "new philosophy," "Adam's last will and testament" was rewritten and the Iberians nations appeared to be left out.

In the Iberian colonial world, the decline of the monarchy led to a situation in which scientific activity began to be performed and utilized by others: the religious orders, the viceregal administrations, and the Creole elite, who capitalized on those syncretic cultures from which new nations would eventually emerge.

Nevertheless, an examination of the institutions and ways of producing science in the colonies requires, in the first place, clarification of the duality inherent in the "Iberian vision." Originally, both for Spain as well as for Portugal, the objective was to discover new routes to the Moluccas and the much-sought-after Oriental spices. However, the most famous calculation error in the entire history of overseas discovery (an incorrect estimate of the earth's perimeter) was responsible for the encounter with the American continent, which decisively altered the imperial projects of both nations. Although the Orient came under Portuguese sovereignty, the greater part of the New World fell to the monarchy of Spain. This division gave rise to two different

Figure 3. *Entry of Joao de Castro in Goa, 1547. (Kunsthistorisches Museum, Vienna.)*

forms of dominion linked to the geographical nature of their respective possessions and to what the two nations sought therein. While Portugal was interested mostly in establishing safe and stable trade routes, Spain set out to conquer the most heavily populated areas of the New World. As a result, the two most important nuclei of the viceroyal power were founded in Mexico and Peru, focal points that were home to two hegemonic cultures at their peak when the conquest took place (the Aztec and Inca societies).

This fact is of capital importance in understanding the formation of two different colonial models. Basically, the Portuguese model did not seek the occupation of territories, but rather the establishment of *feitorias*, mercantile enclaves, and supply ports along the route to the Far East.[15] The Spanish model, on the contrary, tried to reproduce the social structures of the peninsula in the American domains, which led to the rejection of the term "colonies" by traditional Spanish historiography when referring to the kingdoms of America.

This peculiarity of the Spanish imperial model (it was closer to a federated monarchy than to a colonial empire in the nineteenth-century sense) also determined the type of science that developed in the viceroyal societies. The Spanish-American viceroyalties contrasted with Portugal's strongly militarized and restrictive model of scientific culture (e.g., in Brazil and Goa). In 1538, the first American university was founded in Santo Domingo, and a few years later, in 1551, came the universities of Lima and Mexico.[16] Universities, colleges, hospitals, and printing presses, usually connected with religious orders but also emerging from the efforts of the viceroyal administration, were established in almost all major cities of the Spanish Empire. They constituted the nuclei from which scientific culture developed, which was, in many ways, indigenous.

By 1650, Spanish America already had six *tribunales de protomedicato* in Lima, Mexico, Santa Fe de Bogotá, Cartagena, Havana, and Guatemala. There, physicians and surgeons took examinations in order to receive official diplomas. These panels operated independently of the peninsular tribunal, located in Seville. Moreover, hundreds of hospitals run by religious orders operated as charity institutions, and were normally equipped with a pharmacy (or chemist's shop) whose practices made it possible to combine ancient, indigenous herbalist traditions with European therapeutics.

The Spanish colonial colleges deserve separate mention, as they were perhaps the institutions of learning that exerted the greatest influence in the New World. Dominated by the religious orders (Dominicans, Franciscans, and Jesuits), they had their own statutes, privileges, and an academic standing often equivalent to the pontifical and royal universities. Mexico, for instance, had five colleges, all founded before 1600. One of them, the Franciscan Imperial College of Santa Cruz de Tlatelolco, produced treasures such as the *Códice Badiano* and the works of Bernardino Sahagún.[17] But Mexico was not an isolated case: Quito, Mérida, Lima, Cuzco, Panama, Santa Fe de Bogotá, and Havana also boasted a large number of major and minor colleges. The local intelligentsia were educated in these institutions, and the great natural and moral histories of the seventeenth century came to life there. In this sense, it would be appropriate to speak of the *Jesuit vision*, rather than the Iberian

[15] There are classic comprehensive works on the Portuguese colonial world, such as Charles R. Boxer, *The Portuguese Seaborne Empire 1415–1825* (London: Hutchinson, 1969). However, the most recent vision is from Francisco Bethencourt and Kirti Chaudhuri, *História da expansão Portuguesa*, 5 vols. (Navarra: Círculo de Leitores e Autores, 1998).

[16] These were modeled on the universities of Salamanca and Alcalá de Henares. See Agueda Rodríguez Cruz, *Historia de las universidades Hispanoamericanas* (Bogotá: Instituto Caro y Cuervo, 1973).

[17] Well before the rules of grammar first appeared in Dutch (1584) or English (1586), the Spanish friars had already established grammars for the Tarasco (1558), Quechua (1560), Náhuatl (1571), and Zapoteco (1578) languages.

one. The Company of Jesus maintained a network of colleges throughout the His-pano-Portuguese territories with educational programs that included local cultures, languages, and traditions.[18] The Jesuits were also responsible for the major cosmographic programs through the Colegio Imperial of Madrid, which continued the work of the Academia de Matemáticas beginning in the early seventeenth century. They were also eventually the promoters of social experiments of a utopian nature, such as the famous *Reducciones* of Paraguay in the eighteenth century.

By 1700, when the bases of modern science were already established in Europe, the Spanish-American viceroyalties had a flourishing assortment of scientific and academic institutions. The type of natural knowledge they produced possessed the traits of a syncretic learning process. In cosmography, for example, the defense of Catholic dogma meant that, even in that era, the heliocentric theory was accepted merely as an hypothesis. Nevertheless, this did not prevent extensive empirical observations, the reception of European theories, nor attempts to combine them (in an exercise of eclecticism) with Catholic dogma and native traditions. Sigüenza y Góngora, for example, the great Mexican polygraph and astronomer of the seventeenth century, immersed himself in polemics involving the comets and the reconciliation of biblical chronologies with pre-Columbian calendars.

The eighteenth century witnessed the last effort to modernize the imperial structure of the baroque monarchy. The Bourbons in Spain and the Marquis de Pombal in Portugal promulgated a number of measures that are usually grouped under the heading of "reformism" or "enlightened despotism." On the scientific plane, the imperial projects focused to a large extent on a renewed program of travel and scientific expeditions. And it was through these expeditions that Newtonian physics, Linnean botany, and other products of European science reached overseas territories. Once again, the differences between the Portuguese and Spanish colonial worlds are apparent. Although Brazil was the destination of numerous expeditions and explorations, the appearance of institutions on Brazilian soil was a superficial phenomenon, and it was not until the Portuguese monarchy moved to Brazil at the beginning of the nineteenth century that one can properly speak of Brazilian science.[19] The Hispanic viceroyalties, on the other hand, saw a noteworthy flourishing of university chairs, botanical gardens, mining seminars, intellectual circles, periodicals, and other elements that substantiate the fact of a firmly established scientific culture.[20]

The point of coincidence comes, however, with the expulsion of the Company of Jesus, a measure taken by both monarchies in the second half of the century. This act finally sealed the centralist and regalist vocation of Iberian reformism, as well

---

[18] In Portuguese Brazil, the Jesuits founded four colleges in the second half of the sixteenth century (Piratininga, Bahía, Rio de Janeiro, and Olinda), which they maintained up to their expulsion, decreed by the Marquis de Pombal in 1759. In the Spanish-American viceroyalties the list is extensive: it includes, from very early times, dozens of colleges and universities in New Granada, Mexico, Chile, Peru, and Rio de la Plata.

[19] Concerning the Portuguese expeditions, see William J. Simon, *Scientific Expeditions in the Portuguese Overseas Territories (1783–1808) and the Role of Lisbon in the Intellectual-Scientific Community of the Late Eighteenth Century* (Lisbon: Instituto de Investigação Científica Tropical, 1983); Angela Domingues, *Viagens de exploração geográfica na Amazónia em finais do século XVIII: Política, ciência e aventura* (Lisbon: Centro de Estudos de História do Atlántico, 1991).

[20] Although there is an abundant bibliography, two good collections are Lafuente, Elena, and Ortega, *Mundialización de la ciencia* (cit. n. 5); and Juan J. Saldaña, ed., *Historia social de las ciencias en América Latina* (Mexico: Porrúa/Universidad Autónoma de México, 1996).

as its distrust of the followers of St. Ignatius. This event was of decisive significance: the Jesuits had been little less than the intellectual and spiritual arm of the monarchy and its natural connection with the papacy, as well as the teachers of the elite and the nobility par excellence. The Jesuits had also been responsible for much of the geographical exploration overseas and were protagonists of the reconciliation between Catholic dogma and the new science; between scholasticism, the counterreformation, and experimentalism.

The Jesuits also played an important role in the so-called polemic of the New World, the debate that inspired a good part of the intellectual production on American nature. This debate can help us detect the ideological slant that the Creole scientific manifestations acquired in the second half of the eighteenth century. Antonello Gerbi has demonstrated the ubiquity of theories on the inferiority of the American natural environment in authors characteristic of the European Enlightenment—historians of the New World such as G. T. Raynal or William Robertson, and naturalists such as Buffon or Cornelius De Paw. Gerbi also traced the theories' multiple variants, as well as reactions to them from the American continent.[21] It can be said that all of these theories taken together (on the weakness, excessive youth, immature character of the continent and its species, etc.) appeared as a kind of final and summary judgment whereby Europe resolved the doubts and questions raised by the New World ever since its "discovery." The same sentence was also handed down with regard to the intellectual and moral level of its inhabitants.

Both judgments on the deficiencies of the New World were added to the stereotyped vision of the Hispanic world held by northern European *philosophes*. The monarchy's loss of vigor and the Spanish authors' loss of credibility as a source of reliable knowledge led to an extension of this disqualification to the viceregal societies. In fact, a large part of the scientific activity of this period in Latin America can be interpreted in the light of this polemic.[22] The *historias* of Mexico and Chile by the Jesuits Francisco Clavigero and Ignacio Molina, for example, constitute immense frescoes, heirs to the format of the traditional moral and natural histories. However, at this point in the Enlightenment, they signify a patriotic exaltation of the natural environment of America, converting the old category of the "marvelous" into a new cult object related to national identity.[23] Much the same could be said of the works of León y Gama, a New Spain Newtonian known for his astronomical observations, but also for his impressive studies on the *Piedra del Sol* and the *Coatlicue*, perhaps the two most important archeological examples of Aztec culture.[24] Along the same lines are the journalistic enterprises of José A. Alzate in Mexico or Hipólito Unanue in Peru, publications that questioned taxonomical models such as the Linnean sys-

    [21] Antonello Gerbi, *La Disputa del Nuevo Mundo: Historia de una polémica, 1750–1900* (México: Fondo de Cultura Económica, 1982).

    [22] The works of Jorge Cañizares are a good exponent of this perspective, which tends to reassess Gerbi's work and put it in the content of current historiographical issues, such as the credibility and authority of the sources. See, for example, Jorge Cañizares, "Entre Maquiavelo y la jurisprudencia natural: William Robertson y la disputa del Nuevo Mundo," *Quipu*, 1991, *8*, 1:279–91; and *idem*, "Spanish America: From Baroque to Modern Colonial Science," in Roy Porter, ed., *Eighteenth Century Volume of the Cambridge History of Science* (Cambridge: Cambridge Univ. Press, in press).

    [23] Juan I. Molina, *Compendio della storia geografica, naturale e civile del regno del Chile* (Bologna, 1776); Francisco J. Clavigero, *Storia antica del Messico* (Cesena, 1780–1781).

    [24] Antonio León y Gama, *Descripción histórica y cronológica de las dos piedras que con ocasión del nuevo empedrado que se está formando en la plaza principal de México se hallaron en ella en el año de 1790* (Mexico, 1832).

Figure 4. Images of indigenes as gods. From Historia universal de las cosas de la Nueva España, or Códice florentino (1575–1577) *by Bernardino de Sahagún. (Biblioteca Laurentina, Florence.)*

29

tem or chemical methods such as Ignaz Elder von Born's amalgamation of metals.[25] Thus, European forms of learning were deemed useless for classifying or profiting from a diverse natural environment.

Just as science served the imperial ends of a Universal Monarchy, from the late seventeenth century, and particularly in the second half of the eighteenth century, scientific activities in New Spain, Chile, Peru, and New Granada served to vindicate forms of natural knowledge constructed to a large extent as an alternative to the would-be universality of European learning and dominion. By 1800 those scientific cultures, at least in the Hispanic case, were already well established, as reflected not only in the testimonies of Humboldt but in his program of geographical investigation, which owed much to the contributions of his Spanish-American collaborators.

These scientific cultures were understood quite early as a mark of identity, which explains why the pantheon of liberating heroes is overflowing with scientists: Fausto de Elhuyar, Francisco José de Caldas, José A. Alzate, Celestino Mutis, and Humboldt himself.[26] As products of baroque and mestizo societies, these cultures offered approaches to science that preserved (and perhaps still preserve) elements that, from the perspective of Europe, were peripheral in space and obsolete in time. This is understandable from the moment the West became the pacesetter and geographer of modernity and learning. But the West's centrality is debatable, in the end, if one admits at least the possibility that other ways of understanding and practicing natural knowledge have always existed.

[25] We are referring to the *Gacetas de Literatura de México* and to the *Mercurio Peruano*, two of the most representative periodicals on Spanish-American scientific culture in the second half of the eighteenth century.

[26] See Jose L. Peset, *Ciencia y libertad: El Papel del científico ante la independencia Americana* (Madrid: Consejo Superior de Investigaciones Científicas, 1987).

The Colonial Machine: French Science and Colonization in the Ancien Régime

James E. McClellan III and *François Regourd***

ABSTRACT

Although France's colonies were small in number and in size in the eighteenth century, their economic importance made France a major colonial power in the period. The central government, notably the Ministère de la Marine et des Colonies, systematically engaged the elaborate scientific infrastructure of Ancien-Régime France in its colonizing efforts, and French savants provided an essential expertise. This paper examines this bureaucratized scientific arm of France's contemporary "colonial machine" that included the Académie Royale des Sciences, the Académie Royale de Marine, the Observatoire Royal, the Jardin du Roí, the Société Royale de Médecine, the Société Royale d'Agriculture, and the Compagnie des Indes. These institutions and the individuals associated with them undertook coordinated efforts to support and extend contemporary French colonization. Their activities deal with tropical medicine, taxonomic and economic botany, cartography, and a host of related matters. With Paris and Versailles as the hub, by the end of the century an intricate web of institutions and expertise spanned the French colonial world from the Americas to the East Indies. Informal and unofficial colonial networks complemented the official administrative apparatus. The Revolutionary and Napoleonic periods witnessed the destruction of Ancien-Régime colonial structures—scientific and otherwise—yet, the lesson of the utility of science for the creation and maintenance of colonies was not lost on imperial planners who followed.

B EGINNING IN THE SEVENTEENTH CENTURY, FRANCE CARVED out small settlement and exploitation colonies in wild, peripheral sites in the Americas (Quebec, Louisiana, the West Indies, and Guyana); along the slave coasts of Africa; and in the Indian Ocean (Ile de France, Ile Bourbon—today's Mauritius and Réunion—Madagascar, and trading posts on the subcontinent). With its fantastically productive sugar islands—notably Saint Domingue (Haiti)—eighteenth-century France rivaled England as the world's most economically potent colonial power.[1]

---

* Department of Humanities and Social Sciences, Stevens Institute of Technology, Castle Point, Hoboken, New Jersey 07030, e-mail jmcclellan3@compuserve.com.
**Université de Nanterre (Paris X), 200 avénue de la République, 92001 Nanterre, France, e-mail Fregourd@aol.com.
[1] On French colonialism in this period, see Pierre Pluchon, *Histoire de la colonisation française,* vol. 1, *Le Premier empire colonial* (Paris: Fayard, 1991).

Concurrently, French science was institutionally and intellectually the strongest of any nation.

To say that France during the ancien régime offers a preeminent case for exploring connections between science and colonization, or even that contemporary French science provided knowledge useful in establishing and maintaining overseas colonies, merely states the obvious. What is not obvious—and what this essay examines—is that from the time of Louis XIV, the royal administration created and supported an elaborate scientific and technical infrastructure that was not merely tapped on occasion to aid colonization, but which quickly became integral to the process. This scientifico-colonial machine, so to speak, was highly bureaucratized and centralized and, although composed of diverse parts, it functioned in a coordinated way to advance the colonial, national, and dynastic interests of France and the Bourbon monarchy. It was not a static mechanism: the colonial machine existed only in action, and its existence and efficacy derived from the capacity of the administration to mobilize, organize, centralize, and unify material and intellectual resources in France and in the colonies. The lack of a comparable colonial science bureaucracy in contemporary Britain underscores the main point of this paper.[2]

IN SICKNESS AND IN HEALTH

French physicians and the instrumentalities of French medicine came with the first wave of formal colonization in the early seventeenth century. In addition to the church, the army, and the navy, medicine was instrumental in establishing and maintaining overseas colonies.[3] The Marine Royale, headed by the Ministère de la Marine et des Colonies, provided the primary institutional basis for colonial medicine, as for so much else of the scientifico-technical arm of French colonialism. Etienne Taillemite has labeled the contemporary French naval bureaucracy "a prodigious research laboratory."[4] In any event, the Marine Royale had many sick and injured to deal with both in France and in the colonies. As depicted in Figure 1, naval hospitals in Rochefort, Brest, and Toulon—all with origins in the seventeenth century—were the initial institutional manifestations of a colonial medical bureaucracy. The navy needed trained physicians and surgeons, and in the early decades of the eighteenth century, the Ministère de la Marine created naval medical schools in these three cities.[5] These were teaching centers with upwards of two hundred students. A short-

---

[2] John Gascoigne documents the gradual appearance of analogous structures in Britain, but only after the 1780s. See his *Science in the Service of Empire: Joseph Banks, the British State and the Uses of Science in the Age of Revolution* (Cambridge: Cambridge Univ. Press, 1998).

[3] On colonial medicine, see James E. McClellan III, *Colonialism and Science: Saint Domingue in the Old Regime* (Baltimore and London: The Johns Hopkins Univ. Press, 1992), chap. 8; Pierre Pluchon, ed., *Histoire des médecins et pharmaciens de marine et des colonies* (Toulouse: Privat, 1985); Caroline Hannaway, "Distinctive or Derivative? The French Colonial Medical Experience, 1740–1790," in *Mundialización de la ciencia y cultura nacional*, eds. Antonio Lafuente, Alberto Elena, and Mariá Luisa Ortega (Madrid: Doce Calles, 1993), pp. 505–10.

[4] Etienne Taillemite, *Les Archives de la Marine conservées aux Archives Nationales*, 2nd ed., updated by Philippe Henrat (Vincennes: Service Historique de la Marine, 1991), p. 123.

[5] See A. Lefèvre, *Histoire du Service de Santé de la Marine Militaire et des écoles de médecine navale en France, depuis le règne de Louis XIV jusqu'à nos jours, 1666–1876* (Paris: J. B. Gaillière et fils, 1867). The naval medical schools had periods of greater or lesser vitality and were subject to occasional reform; their full story remains to be told. On the Ecole de Chirurgie de la Marine, see

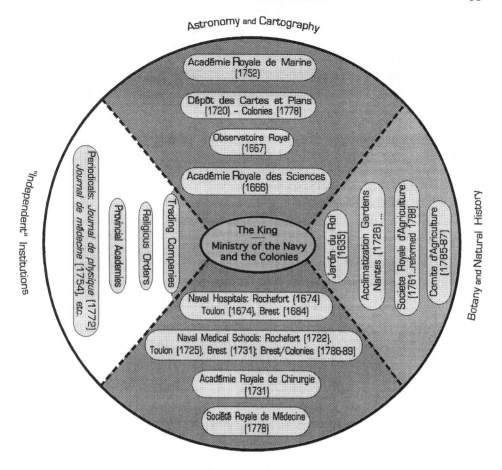

Figure 1. *The metropolitan apparatus.*

lived Ecole Pratique de Médecine, founded in Brest in 1783, capped this educational structure with an institution specifically designed to introduce recent graduates of medical faculties to diseases particular to the colonies and port cities.[6] The bureaucratization of naval medicine culminated in 1763 with the creation of the powerful posts of *inspecteur et directeur général de la médecine de la marine et des colonies*. Pierre-Issac Poissonnier, a major figure in this story, held these positions until 1791, and nominally received reports from all royal physicians—*médecins du roi*—posted in the colonies.[7]

Chartered in 1778, the Parisian Société Royale de Médecine was a well-known,

---

Dr. Hamet, "L'Ecole de Chirurgie de la Marine à Brest (1740–1798)," *Archives de Médecine et de Pharmacie Navale*, 1923, *114:*165–99, 245–59, 321–39, 385–404.

[6] E.-L. Boudet, "Le Corps de Santé de la Marine et le service médical aux colonies au XVIIe et au XVIIIe siècle (1625–1815)," *La Géographie: Terre—Air—Mer*, March–August 1934, 34–6; Hamet, "L'Ecole de Chirurgie," (cit. n. 5), pp. 329–30.

[7] See, for example, the royal warrant of J.-B. Roux (dated 5 April 1766), where one reads that Roux was to "rendre régulièrement compte audit sieur Poissonnier des maladies qu'il aura traitées, de leurs différentes natures, & de la manière dont il les aurat traités des découvertes qu'il pourra

progressive institution that organized physicians and medical science on a national basis. Less well recognized today was its strongly colonial and naval side. The Société elected a number of navy physicians and correspondents from ports and from the colonies. It received regular reports of disease outbreaks in the colonies, published colonial meteorological data in its *Mémoires*, and seriously concerned itself with the health of sailors and conditions in port hospitals. For example, it sponsored prize questions on scurvy, proper rations for sailors, and maintaining troops in hot climates. Along with the Académie Royale de Marine at Brest (1752), the Société became involved in experiments with Parmentier's sea-biscuits, baking bread with sea water, and preserving fresh water at sea. The Société Royale de Médecine also received a steady flow of reports about colonial products and their medical applications, such as cinchona from Saint Domingue, seeds from Sainte Lucie (a.k.a. Saint Lucia), torpedoes (to treat gout), lizards (eaten alive to treat skin diseases) from Guyana, and tisanes from Madagascar. Elephantiasis was such a threat that the *ministre de la marine* himself asked the Société to investigate, which it did in characteristic fashion by preparing a questionnaire and, through official administrative channels, surveying physicians posted in the colonies.[8]

The colonial-medical dimension of the earlier Académie Royale de Chirurgie (1731) was less spectacular.[9] It elected some colonial correspondents and received many reports from the colonies. The Académie Royale Marine at Brest likewise dealt with health issues affecting its personnel. The Académie Royale des Sciences (1666) possessed its colonial-medical side, too, for example, through the election in 1738 of Jean-Baptiste-René Pouppée-Desportes, chief royal physician in Saint Domingue and author of a three-volume *Histoire des maladies de S. Domingue* (1770).[10]

A complementary set of medical structures arose in the colonies themselves, most notably royal hospitals staffed by state-appointed *médecins du roi*. These royal physicians served as local agents for the construction of colonial knowledge, conducting research on the spot. In the most developed case of Saint Domingue, a full-fledged medical community established its independence from the structures of naval and colonial medicine per se, despite resistance from Poissonnier in his role as inspector of colonial medicine. This community included virtual guilds of physicians, surgeons, apothecaries, and veterinarians.[11]

Inoculation against smallpox began in the colonies in the 1740s, decades before

---

faire tant dans la médecine, la pharmacie et la botanique que dans les objets relatifs à l'histoire naturelle." "Séance du Conseil Supérieur de la Martinique 5 mai 1767," Archives Départementales de la Martinique, Fort-de-France, B 12, fol. 47.

[8] See the Procès-Verbaux of the Société Royale de Médecine held in the library of the Académie Nationale de Médecine (hereafter cited as Acad. Nat. Méd. Lib.), Paris; see also Centre des Archives d'Outre-Mer, Aix-en-Provence (hereafter cited as Aix-en-Provence Archives), Colonies C10 C3, dossier 6 (1786).

[9] See Salem Yaqubi, "Contribution à l'histoire de l'Académie Royale de Chirurgie" (Thèse pour le doctorat en médecine [no. 624], Faculté Mixte de Médecine & de Pharmacie de Rennes, Rennes, 1967), and Yves Dordain, "La Chirurgie provinciale française au XVIIIe siècle: Son niveau technique d'après les membres non résidents de l'Académie Royale de Chirurgie" (Thèse pour le doctorat en médecine [no. 281], Faculté Mixte de Médecine & de Pharmacie de Rennes, Rennes, 1962).

[10] McClellan, *Colonialism and Science* (cit. n. 3), p. 137 and note.

[11] See *ibid.*, cit. n. 3, and Isabelle Homer, "Médecins et chirurgiens à Saint-Domingue au XVIIIe siècle," Thèse d'Ecole des Chartes, Paris, 1998.

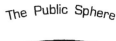

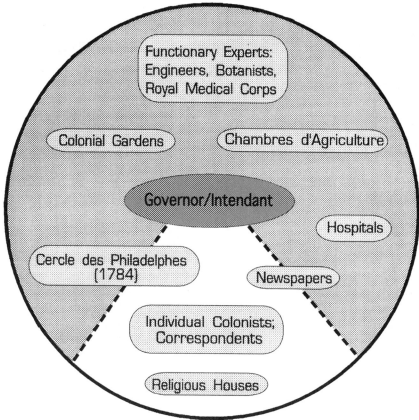

Figure 2. *The Colonial World*

the practice was accepted in France, and a notable subspecialty of eighteenth-century French medicine concerned itself with tropical diseases, their treatment, and related medical issues, such as acclimatizing colonials from France and Africa to local pathologies. A number of monographs published in the colonies and in the metropolis treated colonial medical topics. For example, in Saint Domingue the colonial scientific society, the Cercle des Philadelphes, produced volumes on tetanus and on veterinary medicine (See Figure 2). In France, Poissonnier's brother and deputy *inspecteur général* for colonial medicine, Poissonnier-Desperrières, published a *Fièvres de St. Domingue* (1763) and a *Traité des maladies des gens de mer* (1767). One of Poissonnier's agents, the physician J.-B. Dazille, was especially productive, publishing on illnesses affecting slaves (1776), on tropical diseases in general (1785), and on tetanus (1788).

Science piggy-backed on colonial medicine. Colonial authorities enlisted experts of all sorts, and, as it happened, the medical and the scientific converged in the service of colonial France.

THE CONQUEST OF COLONIAL SPACE

Creating and maintaining colonies in the eighteenth century required a command of the seas. Like its rivals, the French state needed reliable knowledge of the position of its colonies, and mastery of colonial space was seen to depend upon mastery of astronomical space. The authorities turned to a constellation of sciences—astronomy, geodesy, cartography—and established a set of institutions to address, in part at least, this core issue. These included, notably, the Académie des Sciences, the Observatoire Royal (1667), and the Académie de Marine (1752).[12] From this point of view, much of the pure science of eighteenth-century astronomy emerged out of, and was inseparable from, practical problems faced by the colonial machine.

The Académie des Sciences established colonial links through its corresponding members.[13] Michel Sarrasin, a surgeon and royal physician stationed in Canada, was an early correspondent, elected in 1699.[14] In the 1730s, Réaumur in France and his correspondent on the Ile de France, Jean-François Charpentier de Cossigny, published a series of concurrent weather observations.[15] The Académie similarly elected correspondents from Cayenne in South America (e.g., Pierre Barrère in 1725), the West Indies (e.g., Jean-Antoine Peyssonnel in 1723), and the Indian Ocean (e.g., Jean-Baptiste-François Lanux in 1754). The Académie also recruited colonial officialdom, including the Marquis de La Galissonnière (*lieutenant général de la marine* and former *gouverneur-général* in Canada) and the Count de La Luzerne (*ministre de la marine* and former *gouverneur-général* in the Antilles). Resident academicians themselves held posts within the colonial system—Henri-Louis Duhamel du Monceau for many years was *inspecteur-général de la marine*. These connections seem typical of the ties established by the Académie des Sciences with the colonies and with centers of power in France.

The central problem for the seagoing powers was the determination of longitude at sea, and the solution emerged in the 1760s with John Harrison's marine chronometer. In the decades prior, the Académie des Sciences had played a leading role in

[12] On these institutions, see Charles C. Gillispie, *Science and Polity in France at the End of the Old Regime* (Princeton: Princeton Univ. Press, 1980). On the Académie de Marine at Brest, see Alfred Doneaud du Plan, *Histoire de l'Académie de Marine*, 6 vols. (Paris: Bergot-Levrault, 1878–1882); Josef W. Konvitz, *Cartography in France, 1660–1848: Science, Engineering and Statecraft* (Chicago: The Univ. of Chicago Press, 1987), pp. 75–7. The Brest Académie needs a new historical study.

[13] Alfred Lacroix's biographical studies from the 1930s are the most explicit treatments of the colonial face of the Académie des Sciences; see *L'Académie des Sciences et l'étude de la France d'outre-mer de la fin du XVIIe siècle au début du XIXe siècle (Figures des Savants, vols. 3–4)* (Paris: Gauthier-Villars, 1938).

[14] On Sarrasin (or Sarrazin) and his successor, Jean-François Gaultier, see Luc Chartrand, Raymond Duchesne, and Yves Gingras, *Histoire des sciences au Québec* (Montreal: Boréal, 1987), pp. 46–53, 56–8. For Sarrasin, see also *Mémoires de l'Académie Royale des Sciences (hereafter cited as Mém. l'Acad. Roy. Sci.)*, 1704, 48–65; for Gaultier, see also *Mém. l'Acad. Roy. Sci.*, 1747, 466–88 and 1750, 309–10.

[15] On these individuals, see Institut de France, Académie des Sciences, *Index biographique de l'Académie Royale des Sciences* (Paris: Gauthier Villars, 1979). For Cossigny's and Réaumur's papers, see *Mém. l'Acad. Roy. Sci.*, 1733, 417–37; *Mém. l'Acad. Roy. Sci.*, 1734, 553–63; *Mém. l'Acad. Roy. Sci.*, 1735, 545–76; *Mém. l'Acad. Roy. Sci.*, 1738, 387–403.

longitude research by administering a fund of 125,000 *livres* (bequeathed in 1714 by the parlementarian Rouillé de Meslay) for two prizes.[16] The first prize, on motion, has attracted comparatively more attention from historians. The second, to determine "longitude at sea and discoveries useful to navigation and long-distance voyages," is less well studied. The Académie made the first navigation award—a substantial 2000 *livres*—in 1720, and it continued to pose research questions and award the Meslay prize into the 1780s.

As the Académie *Histoire* remarked proudly in 1745, "Points of longitude determined astronomically are to Geography what medals and monuments are to History."[17] Before the chronometer, practical men at sea and in the colonies (who were variously sponsored by the Académie des Sciences, the Académie de Marine, the Observatoire, the Marine Royale, and the Compagnie des Indes) used astronomical techniques to determine the geographical location of French overseas territories. Using simultaneous observations of Jupiter's satellites, eclipses, or occultations of stars, observers in Paris and in a colonial locale could determine the difference in time, and hence the longitudinal separation, between the two stations. For example, in 1685 Jean-Dominique Cassini and Philippe de la Hire determined the longitude of Quebec by observing a lunar eclipse in Paris while Jean Deshayes watched it in Canada.[18] In 1703 and 1704, coordinating with Louis Feuillée, a Minim priest sent to the West Indies for this purpose, Jacques Cassini II used simultaneous observations of the satellites of Jupiter to establish the longitude of Martinique.[19] To the end of the century a steady stream of papers appeared in the *Mémoires* and the *Savants Etrangers* of the Académie des Sciences on colonial longitudes and related matters.

The French expended considerable effort determining the location of their Indian Ocean possessions and developing related navigational aides. The merest, remote specks in that great body of water—Ile de France, Ile Bourbon, and their dependency, Ile Rodrigue—were the key transit stations for French mercantile voyages to and from India and China. In the middle decades of the eighteenth century, J.-B.-N. Denis d'Après de Mannevillette led efforts to chart the region. D'Après rose to become inspector general for the Compagnie des Indes, and was a member of the Académie de Marine and a correspondent of the Académie des Sciences. Sailing in the Indian Ocean in 1740, he worked with the academician Pierre-Charles Le Monnier to determine the longitude of Ile Bourbon. In 1745, with the approbation of the Académie des Sciences, he published his *Neptune oriental* and a navigator's manual (*Routier*) for the Indian Ocean.[20] In 1750–1753, d'Après returned to the region on

[16] See Ernest Maindron, *Les Fondations de prix à l'Académie des Sciences: Les Lauréats de l'Académie, 1714–1880* (Paris: Gauthier-Villars, 1881), pp. 14–22, 44, 46, 48–9. A skilled worker at the end of the ancien régime earned on the order of five hundred *livres* a year; the value of $5–10 U.S. to the *livre* is a helpful rule of thumb.

[17] *Histoire de l'Académie Royale des Sciences* (hereafter cited as *Hist. l'Acad. Roy. Sci.*), 1745, 77.

[18] See Chartrand, Duchesne, and Gingras, *Histoire des sciences au Québec* (cit. n. 14), pp. 25–7; Joseph-Nicolas Delisle reports this observation and further work with Joseph-Bernard de Chabert in 1750 to confirm Quebec's longitude in *Mém. l'Acad. Roy. Sci.*, 1751, 36–9.

[19] Jacques Cassini, *Mém. l'Acad. Roy. Sci.*, 1704, 338–44.

[20] Pierre-Charles Le Monnier, *Mém. l'Acad. Roy. Sci.*, 1742, 347–53; J.-B.-N. Denis d'Après de Mannevillette, *Savants Etrangers*, 1768, 5:90–232; also Andrew Cook, "An Exchange of Letters Between Two Hydrographers; Alexander Dalrymple and Jean-Baptiste d'Après de Mannevillette," in *Les Flottes des Compagnies des Indes 1600–1857*, ed. Philippe Haudrère (Vincennes: Service Historique de la Marine, 1996), pp. 173–82, see p. 174.

the orders of the *ministre de la marine*, the Académie des Sciences, and the Compagnie des Indes with the goal of producing a new set of maps for the area.[21]

Another part of d'Après's assignment in 1750 was to transport the *associé astronome* of the Académie des Sciences, the abbé Nicolas-Louis de Lacaille, to the Cape of Good Hope. With ministerial approval and the material support of the Compagnie des Indes (which Lacaille praised for its "zeal for Navigation which in large measure depends on Astronomy"), Lacaille went to Africa primarily to map the stars of the southern hemisphere.[22] In 1753 the *ministre de la marine* ordered Lacaille to the Iles de France and Bourbon to fix the islands' "exact geographical location." Lacaille had seen d'Après's recent data and thought them satisfactory, but used his time well, reporting longitude and other astronomical observations, performing declination and pendulum experiments, and drawing maps. His meteorological observations included an annotated list of monsoons for the period 1733 to 1754.[23]

The observation of the transit of Venus in 1761 on the Ile Rodrigue by Alexandre-Guy Pingré, an *associé libre* of the Académie, represented the next step in continuing French efforts to chart the Indian Ocean.[24] But it was the 1760–1771 odyssey of Guillaume Legentil de la Galaisière that capped French scientific navigation in the East Indies in the eighteenth century. An *astronome adjoint* of the Académie des Sciences, Legentil went to observe the 1761 Venus transit at Pondicherry, but the entrepôt had fallen to the English. He remained in the Indies for the transit of 1769, returning to France only in 1771.[25] Legentil sailed on the ships of the Compagnie des Indes, sent reports back to the Académie des Sciences from Ile de France, Ile Bourbon, Ile Rodrigue, Pondicherry, Madagascar, and Manilla, and returned to publish his two-volume *Voyage dans les mers d'Inde* in France in 1779–1781.[26] By that point the French had mastered longitudes and techniques for getting to and from the Indian Ocean.

In conducting research in distant locales, investigators did not limit themselves to longitude observations. Informally, they reported various astronomical, cartographical, meteorological, geological, botanical, natural historical, economic, anthropological, and other scientific findings. For example, following Jean Richer's discovery in Cayenne in 1672 that a standard pendulum beats more slowly towards the equator, it became customary to perform pendulum experiments on every expedition, even after the "shape of the world" controversy had been settled. Observations of magnetic variations from true north deserve particular mention because some savants in

[21] See J.-B.-N. Denis d'Après de Mannevillette, *Savants Etrangers*, 1763, 4:399–457; and Pierre-Charles Le Monnier, *Mém. l'Acad. Roy. Sci.*, 1751, 270–2.

[22] Nicolas-Louis de Lacaille, *Mém. l'Acad. Roy. Sci.*, 1751, 519–36.

[23] *Hist. l'Acad. Roy. Sci.*, 1754, 110–16; Lacaille's memoirs, *Mém. l'Acad. Roy. Sci.*, 1754, 44–56; *Mém. l'Acad. Roy. Sci.*, 1754, 94–130; see also his log book, Bibliothèque de l'Observatoire (Paris), MS C 3, 26.

[24] See *Hist. l'Acad. Roy. Sci.*, 1761, 107–111; Alexandre-Guy Pingré, *Mém. l'Acad. Roy. Sci.*, 1761, 413–86. According to Pingré, even the *latitudes* of the Indian Ocean possessions were still not known with sufficient accuracy. Harry Woolf, *The Transits of Venus: A Study of Eighteenth-Century Science* (Princeton: Princeton Univ. Press, 1959) remains the definitive treatment of the transits of Venus.

[25] See entry on Guillaume Legentil by Seymour L. Chapin, *Dictionary of Scientific Biography*, ed. Charles C. Gillispie, vol. 8, pp. 143–5.

[26] See Guillaume Legentil papers, *Mém. l'Acad. Roy. Sci.*, 1768, 237–47; *Mém. l'Acad. Roy. Sci.*, 1771, 247–80.

the 1750s believed that *variations* in the magnetic needle held the possibility of solving the longitude problem.[27] J.-D. Cassini wrote an early series of papers on the subject, and so-called declination data repeatedly show up in the research.[28] Other researchers contributed memoirs to the Académie des Sciences, the Société Royale de Médecine, and other organizations on such practical subjects as gauging the tonnage of ships, ventilating ships' holds, preserving drinking water at sea, rendering fresh water from sea water, and arming ships with lightning rods.[29]

In the long run, the success of France's colonial endeavor depended on the strength of French cartography.[30] Like other European seagoing powers, the French centralized their cartographical and hydrographical efforts, creating a formal Dépôt des Cartes et Plans—a great "calculating" center—within the Navy Ministry in 1720 to collect and consolidate cartographical information.[31] In 1776, the library and collections of D'Après de Mannevillette became a subsidiary Dépôt des Cartes, Plans et Journaux de la Marine in Lorient. A separate Dépôt des Cartes des Colonies was created in 1778, indicating the importance of the colonies in the eyes of officialdom.[32]

The colonies themselves required mapping and surveying. Local surveyors did some of this work for private plantations, and the army and the navy each had an engineering corps charged with building bridges, roads, fortifications, and port installations for colonial defense. In addition, engineer-geographers (*ingénieurs-géographes*), "supersurveyors" in royal service, were deployed in the colonies to draw up local maps.[33]

---

[27] Pierre-Louis Maupertuis, for example, in his *Lettre sur le progrès des sciences* (1752) proposes this solution. See this *Lettre* appended to the Diderot Editeur edition of *La Vénus physique* (Paris: Diderot Multimedia, 1997), pp. 124–6.

[28] See Jean-Dominique Cassini's papers, *Mém. l'Acad. Roy. Sci.*, 1705, 8–13; *Mém. l'Acad. Roy. Sci.*, 1705, 80–2; *Mém. l'Acad. Roy. Sci.*, 1708, 292–6. See also paper by Guillaume Legentil explicitly devoted to declination in the Indian Ocean, *Mém. l'Acad. Roy. Sci.*, 1777, 401–19.

[29] See Dortous de Mairan, "Remarques sur le jaugeage des navires," *Mém. l'Acad. Roy. Sci.*, 1721, 76–107 and his "Instruction abrégée, & methode pour le jaugeage des navires," *Mém. l'Acad. Roy. Sci.*, 1724, 227–41; see also articles by Sébastien-François Bigot de Morogues, "Mémoire sur la corruption de l'air dans les Vaisseaux," *Savants Etrangers*, 1750, *1*:394–410; Jean-François Charpentier de Cossigny, "Expériences réitérées pour s'assurer si les filtrations de l'eau de la mer aux travers des pores du verre sont possibles," *Savants Etrangers*, 1760, *3*:1–18 and his "Essai de la manière de conserver à la mer l'eau potable, dans les voyages de long cours," *Savants Etrangers*, 1774, *6*:94–109; Gabriel de Bory, "Mémoire sur les moyens de purifier l'air dans les vaisseaux," *Mém. l'Acad. Roy. Sci.*, 1781, 111–19; Jean-Baptiste Le Roy, "Mémoire sur un voyage fait dans les ports de guerre de l'océan, pour y établir des paratonnères, et en faire placer sur les vaisseaux," *Mém. l'Acad. Roy. Sci.*, 1790, 472–84.

[30] On the background of contemporary French cartography, see Konvitz, *Cartography in France* (cit. n. 12) and Mary Sponberg Pedley, *Bel et Utile: The Work of the Robert de Vaugondy Family of Mapmakers* (Tring, Herts., England: Map Collector Publications, 1992); Monique Pelletier, *La carte de Cassini: L'extraordinaire aventure de la carte de France* (Paris: Presses de l'Ecole Nationale des Ponts et Chaussées, 1990).

[31] See Bernard Le Guisquet, "Le Dépôt des Cartes, Plans et Journaux de la Marine sous l'Ancien Régime," *Neptunia*, 1999, *214*:3–21; Olivier Chapuis, "Charles-François Beautemps-Beaupré (1766–1854) et la naissance de l'hydrographie moderne" (Thèse de doctorat, Université Paris IV-Sorbonne, 1997) pp. 1391–2; also, Bruno Latour, *Science in Action* (Cambridge, Mass.: Harvard Univ. Press, 1987), pp. 232–47.

[32] Cook, "An Exchange of Letters" (cit. n. 20), p. 180; Taillemite, *Les Archives de la Marine* (cit. n. 4), p. 5.

[33] See Comité des Travaux Historiques et Scientifiques, *La Martinique de Moreau du Temple: La Carte des ingénieurs géographes* (Paris: Comité des Travaux Historiques et Scientifiques, 1998);

While astronomical techniques for determining longitude worked on terra firma, the solution to the problem at sea lay elsewhere. John Harrison solved this more narrow concern.[34] It is less well known that the French, notably the royal clockmaker Ferdinand Berthoud, followed not far behind the English in developing comparable clocks. In the French case, although the chronometric solution emerged from the world of technology, it took astronomers and the Marine Royale to test and certify its utility. Of several French sea trials, two ventured to Saint Domingue in 1769 and 1771.[35] The results of these tests demonstrated the value of chronometers and thereafter French navigators regularly employed them.[36]

The chronometer also became a precision instrument for cartography. Earlier in the century, cartographers usually worked from offices in Paris or Versailles, and even the leading mapmakers, such as Jacques-Nicolas Bellin, came under heavy criticism for imprecision. The chronometer allowed more accurate work to be done in the field, and in 1784–1785 French authorities sent Antoine-Hyacinthe-Anne de Chastenet, Count Puységur to Saint Domingue to take data for a new set of maps of the colony and its waters.[37] Puységur returned to France in 1785, and in 1787 the government published his highly accurate *Pilote de l'Isle Saint-Domingue* and his *Détail sur la navigation aux côtes de Saint-Domingue*.

CULTIVATING NEW EDENS

The colonies were a long way from France and, especially in the early years, were exotic places waiting to be explored and studied. Colonial authorities and the scientific establishment took a keen interest in botany and natural history, and investigating colonial flora and fauna was a defining element of the French colonial enterprise.[38]

Missionaries, especially Dominicans, wrote the first descriptions of tropical nature in the French colonies.[39] Raymond Breton, for example, published a *Dic-*

Jean-Louis Glénisson, "La Cartographie de Saint-Domingue dans la seconde moitié du XVIIIᵉ siècle (de 1763 à la Révolution)" (Thèse d'Ecole des Chartes, 1986); Chartrand, Duchesne, and Gingras, *Histoire des sciences au Québec* (cit. n. 14), pp. 21–8.

[34] See, in particular, the recent collaboration by Dava Sobel and William J. H. Andrewes, *The Illustrated Longitude* (New York: Walker, 1998). For an entrée into the French side of the story, see Catherine Cardinal, *Ferdinand Berthoud (1727–1807), Horloger mécanicien du roi et de la marine* (La Chaux-de-Fonds, Switzerland: Musée International d'Horlogerie, 1984). Note that at roughly the same time Tobias Mayer and others developed an alternative astronomical solution to the problem of longitude that used lunar tables.

[35] McClellan, *Colonialism and Science* (cit. n. 3), pp. 122–3.

[36] The evident end of the longitude story, as least as far as the French were concerned, is plain in papers by Alexandre-Guy Pingré (*Mém. l'Acad. Roy. Sci.*, 1770, 487–513) and the Marquis de Chabert (*Mém. l'Acad. Roy. Sci.*, 1783, 49–66).

[37] François Regourd, "L'Expédition hydrographique de Chastenet de Puységur à Saint-Domingue (1784–1785)," in *Mélanges Paul Butel*, ed. S. Marzagalli (in press). Regarding longitude work in the Indian Ocean, see Philippe Buache, "Observations géographiques sur les Isles de France et de Bourbon, comparées l'une avec l'autre," *Mém. l'Acad. Roy. Sci.*, 1764, 1–6.

[38] For in-depth treatments see McClellan, *Colonialism and Science*, (cit. n. 3), chap. 6; François Regourd, "Maîtriser la nature: Un Enjeu colonial: Botanique et agronomie en Guyane et aux Antilles (XVIIᵉ–XVIIIᵉ siècles)," *Revue Française d'Histoire d'Outre-Mer* (hereafter cited as *Rev. Fran. d'Hist. d'Outre-Mer*),1999, 86:39–63. In the same number of this journal, on pp. 7–38, see also the important overview of this topic by Marie-Noëlle Bourget and Christophe Bonneuil, "De l'inventaire du monde à la mise en valeur du globe: Botanique et colonisation (Fin XVIIᵉ siècle-début XXᵉ siècle)."

[39] See note 38 and P. Fournier, *Voyages et découvertes scientifiques des missionnaires naturalistes français à travers le monde, pendant cinq siècles* (Paris: Paul Lechevalier, 1932).

tionnaire caraïbe in the 1660s that was full of details about Amerindians and Caribbean plants. Jean-Baptiste Du Tertre published a four-volume *Histoire générale des Antilles* (1667–1671) after spending sixteen years in the French West Indies. Jean-Baptiste Labat sojourned in the Antilles for more than a decade, and his *Nouveau voyage aux Isles de l'Amérique* (first published in 1722) contained notable chapters on island botany, as did Du Tertre's. The role of missionaries qua researchers and colonial *rapporteurs* indicates that, to some extent, the emerging colonial machinery was able to co-opt monastic networks. For example, the Jesuit and *académicien honoraire* of the Académie des Sciences, Thomas Gouye (1650–1725), served as a key intermediary between his order and secular scientists, communicating to the Académie numerous reports originating in the Antilles and elsewhere.[40]

The key step from preparing catalogues to exploiting nature was initiated by Jean-Baptiste Colbert, Louis XIV's finance minister and, concurrently, *ministre de la marine*. Motivated by the economic potential of the colonies, in the 1660s and 1670s Colbert began the systematic collection of plants and animals.[41] To further the movement to catalogue and collect colonial flora (Latour's "mobilisation of resources"), Colbert and his successors relied on the Jardin du Roi (1635) and the Académie des Sciences.[42] Under the aegis of these royal scientific institutions, several notable botanists traveled to the colonies. The Minim monk Charles Plumier made three government-financed trips to the West Indies between 1689 and 1697, sending to the Jardin du Roi thousands of drawings and plants that were later used for his *Description des plantes de l'Amérique* (1693) and his eight-volume *Botanicon americanum* (1705).[43] Another Minim monk, Louis Feuillée, went to the West Indies primarily to do astronomy but pursued botany in his spare time, and published an interesting *Histoire des plantes médicales de l'Amérique* (1714).

The great botanists from the Jardin du Roi and the Académie des Sciences—like Antoine and Bernard de Jussieu—occasionally recommended their pupils for scientific appointments in the colonies. Pierre Barrère, for example, held the post of royal physician-botanist for four years in French Guyana, publishing his *Histoire naturelle de la France equinoxiale* (1741). Similarly, Jean-Baptiste-Christophe Fusée-Aublet was posted to the Ile de France in 1753 and then to Guyana in 1762 before publishing his great *Histoire des plantes de la Guiane française* in 1775. Michel Adanson, the talented pupil of Antoine and Bernard de Jussieu, was em-

---

[40] See Académie des Sciences, *Index biographique* (cit. n. 15) and Jacques-Phillippe Maraldi, *Mém. l'Acad. Roy. Sci.*, 1724, 365–79; *Hist. l'Acad. Roy. Sci.*, 1707, 113; *Hist. l'Acad. Roy. Sci.*, 1703, 57; *Hist. l'Acad. Roy. Sci.*, 1704, 42; etc.

[41] For example, on 21 June 1670, Colbert wrote to the director of the Compagnie des Indes Occidentales: "Je désire que vous examiniez bien toutes les fleurs, les fruits et même les bestiaux, s'il y en a de naturels du pays et que nous ne voyons point en Europe, et tout ce qu'il faut observer pour les faire venir. Il faudra m'en envoyer par tous les vaisseaux qui viendront afin que si l'un manque, l'autre puisse réussir; surtout envoyez-moy de l'ananas [for Louis XIV's table!], afin de tenter si l'on en pourra faire venir ici." Pierre Clément, ed., *Lettres, instructions et mémoires de Colbert*, 8 vols. in 10 (Paris, 1861–1882), vol. 3-2, p. 486.

[42] On the Jardin du Roi, see Gillispie, *Science and Polity* (cit. n. 12); Yves Laissus, "Le Jardin du Roi," in *Enseignement et diffusion des sciences en France au XVIIIᵉ siècle*, ed. René Taton (Paris: Hermann, 1964; reprint 1986), pp. 287–341. The colonial dimension of the Jardin du Roi likewise is in need of systematic study. See also Latour, *Science in Action* (cit. n. 31), pp. 224–8.

[43] Jean Lescure, "L'epopée des voyageurs naturalistes aux Antilles et en Guyane," in *Voyage aux Iles d'Amérique*, exhibition catalogue, April–July 1992, l'Hôtel de Rohan (Paris: Archives Nationales, 1992), p. 61.

ployed in Africa by the Compagnie des Indes, published his *Voyage au Sénégal* in 1757, and ended up as a *pensionnaire* of the Académie des Sciences in Paris.

With the passing of time, a notable shift of emphasis occurred. As the colonies became better established, scientific and taxonomic botany increasingly gave way to economic botany and efforts to introduce economically useful plants into production. This step called for implanting institutions and personnel on site, and to this end the French state created official botanical gardens in the colonies and staffed them with salaried *botanistes du roi* and *médecins du roi*. Gardens were created in Saint Domingue, Guadeloupe, Cayenne, Ile de France, and Ile Bourbon. The Cercle des Philadelphes launched two, and other gardens were sponsored by individuals or connected with colonial hospitals. These colonial gardens functioned alongside a special naturalizing garden in Nantes, established at the beginning of the century to receive colonial specimens.

The advent of practical horticulture greatly affected the colonies. The rise of Ile Bourbon, for example, stemmed from the introduction of coffee trees and coffee cultivation in the 1730s. Coffee culture transformed the French West Indies and provided a model the French hoped to imitate with other commodities. The *modus operandi* of these botanical efforts matured in the 1750s. The crown agent and later colonial administrator Pierre Poivre successfully conducted three covert raids in the 1750s on the Dutch East Indies and the Philippines to collect (read: steal) stocks of plants for acclimatization and cultivation on Ile de France and Ile Bourbon. These included pepper, cinnamon, clove, and nutmeg.[44] Poivre returned to the Indian Ocean to direct two more such raids, one in 1769 and another in 1771–1772.

The notable point is that in the 1770s and 1780s, French authorities were mobilizing colonial botanical gardens, their staffs of *botanistes du roi*, and navy transport capabilities to link colonial gardens in the Indian Ocean with those in Cayenne and the West Indies.[45] Shipments of exotic plants from the Indian Ocean to the Caribbean included pepper, cinnamon, mango, mangosteen, and breadfruit trees. While this traffic proceeded on the periphery of the French colonial system, other connections linked the colonies with Paris. In particular, using government channels, the royal botanist in Saint Domingue, Hippolyte Nectoux, and other colonial agents exchanged seeds and specimens with André Thouin, chief gardener at the Jardin du Roi and associate member in botany at the Académie des Sciences.[46]

[44] See Madeleine Ly-Tio-Fane, *Mauritius and the Spice Trade: The Odyssey of Pierre Poivre* (Port-Louis, Mauritius: Mauritius Archives Publication Fund, 1958) and her *Mauritius and the Spice Trade: The Triumph of Jean Nicolas Céré and his Isle Bourbon Collaborators* (Paris-La Haye: Mouton, 1970). The coconut, thought to be an antidote for poisons, was also successfully introduced onto Ile de France; see Pierre Sonnerat, "Description du cocos de l'Ile Praslin," *Savants Etrangers*, 1776, 7:263–6.

[45] For these details see Patrice Bret, "Le Réseau des jardins coloniaux: Hypolite Nectoux (1759–1836) et la botanique tropicale, de la mer des Caraïbes aux bords du Nil," in *Les Naturalistes français en Amérique du Sud XVIe–XIXe siècles*, ed. Yves Laissus (Paris: Comité des Travaux Historiques et Scientifiques, 1995), pp. 185–216; see also his "Des 'Indes' en Méditerranée? L'Utopie tropicale d'un jardinier des Lumières et la maîtrise agricole du territoire," *Rev. Franc. d'Hist. d'Outre-Mer*, 1999, 86:65–89. See also Jean-Louis Fischer, *Le Jardin entre science et représentation* (Paris: Comité des Travaux Historiques et Scientifiques, 1999).

[46] See Yvonne Letouzey, *Le Jardin des Plantes à la croisée des chemins avec A. Thouin (1747–1824)* (Paris: Muséum National d'Histoire Naturelle, 1989), esp. pp. 174–97; Emma Spary, "Making the Natural Order: The Paris Jardin du Roi, 1750–1795," (Ph.D. diss., Univ. of Cambridge, 1993). See also the sizeable contemporary literature about shipping botanical specimens, including works by Académie members Duhamel du Monceau (whose *Avis pour le transport par mer des arbres* . . .

To stimulate colonial agronomy, the monarchy had established colonial Chambres d'Agriculture in Saint Domingue, Martinique, Guadeloupe, and Sainte-Lucie in 1763.[47] These organizations produced some minor dissertations on mills, cotton, and cochineal for their respective colonial administrators, but by and large their contributions were not significant.[48] Local officials sometimes organized competitions on subjects relevant to the development of colonial products. For example, in Sainte-Lucie in the 1780s, they offered prizes worth several thousand *livres* for the best processed sugar, the best rum, and the best cotton mill.[49] The idea of locally based research organized by the royal administration cropped up again at the very end of the ancien régime in a project that Dutrône La Couture defended before the Assemblée Nationale in 1791. In addition to a revitalized network of colonial botanical gardens, Dutrône La Couture proposed a new set of colonial agricultural societies that would mutually correspond under the authority of a renewed Société Royale d'Agriculture.[50]

In Paris, the Société Royale d'Agriculture was called on to play a role in the development of applied botany and agronomy in the colonies.[51] (Although founded in the 1760s, it assumed this role actively only after 1785.) Through its trimestral *Mémoires*, the Parisian agricultural society assured the diffusion of papers sent from the colonies (i.e., on West-Indian cotton, on rice from Ile Bourbon, on sugar cane from Guyana, and on colonial plant and seed transfers).[52] These texts were read in expert circles in France, thereby integrating colonial knowledge into French agronomy. The Société Royale d'Agriculture also encouraged its colonial correspondents by awarding gold medals; one went to Jean-Nicolas Céré and another to his colleague Joseph Martin for their horticultural efforts in the Indian Ocean and West Indian colonies.[53]

Official institutions occasionally recognized individual colonists and their private botanical-agricultural research. For example, A.-T. Broussonnet, secretary of the Société Royale d'Agriculture, recommended a colonial planter from Guadeloupe, one Badier, to the *ministre de la marine*. Badier studied chemistry and botany in Paris and after his return to Guadeloupe, received official encouragement for his experiments on cotton. Following a familar bureaucratic pattern, Broussonnet solicited the advice of the *inspecteurs généraux des manufactures* attached to the Bureau du Commerce before asking the *ministre de la marine* to underwrite Badier's expenses.[54]

---

appeared in 1753) and the Abbé Tessier, whose "Mémoire sur l'importation et le progrès des arbres à l'épicerie dans les colonies françois" appeared in the Académie's *Mémoires* for 1789.

[47] McClellan, *Colonialism and Science* (cit. n. 3), pp. 44–5 ff.

[48] These memoirs are found in the Aix-en-Provence Archives (cit. n. 8), Colonies F3, pp. 124–26.

[49] *Ibid.*, Colonies C10 C3, parts 6 and 10.

[50] See Jacques-François Dutrône la Couture, *Extrait des mémoires que M. Dutrône a présentés à l'Assemblée Nationale au mois de janvier 1791, tiré du "Journal des Colonies"* (Paris: Imprimerie du Cercle social, n.d.), p. 2.

[51] François Regourd, "La Société Royale d'Agriculture de Paris face à l'espace colonial (1761–1793)," *Bulletin du Centre d'Histoire des Espaces Atlantiques* (Université Bordeaux-III), 1998, 8:155–94; Emile Justin, *Les Sociétés Royales d'Agriculture au XVIIIe siècle (1757–1793)* (Saint-Lô [publisher not indicated], 1935).

[52] Regourd, in "La Société Royale d'Agriculture" (cit. n. 51) gives a complete list of colonial memoirs published by the Société.

[53] See Société Royale d'Agriculture de Paris, *Mémoires*, Fall Trimester 1788, vi–vii; and Fall Trimester 1789, xiii.

[54] The Aix-en-Provence Archives (cit. n. 8), Colonies E 14, dossier Badier. These *inspecteurs généraux des manufactures* and their relations with scientific circles have been studied by Philippe Mi-

Further indicative of the agricultural concerns of the colonial machine, an ephemeral Comité Consultatif d'Agriculture met informally under the aegis of the *contrôleur général des finances*. Born after food shortages in the winter of 1784–1785, this Comité consisted of ten top-level scientists and physicians (Lavoisier, Poissonnier, and others) and met about seventy times between 1785 and 1787 before merging with the Société Royale d'Agriculture in 1788. The Comité Consultatif paid serious attention to such matters as Thiery de Menonville's introduction of cochineal in Saint Domingue, Duchemin de l'Etang's call for importing so-called Guinée grass (a fast-growing fodder), naturalizing insect-eating birds from Ile Bourbon and India, and founding a new naturalizing garden in the south of France.[55]

ON THE MARGINS

Thanks to the considerable institutional and human investment made by the monarchy, the scientific bureaucracy of the ancien régime successfully brought its expertise to bear on colonial development, and the world of science became thereby enlarged. Not every scientific act associated with the colonies was necessarily tied—wholly or in part—to overall institutional structures or was bound to submit to norms controlled by the colonial machine. The features that were complementary to "official" colonial science must be kept in perspective. They were relatively unimportant, but a closer examination of sources reveals other modes of circulation, publication, and elaboration of knowledge within the colonies, all of which call for nuancing the monolithic character of the scientifico-colonial machine that we have depicted thus far.

For example, the scientific model exported by the metropolis was not unconditionally accepted in the colonies. In the preface to his *Nouveau voyage aux isles de l'Amérique* (1722), Father Jean-Baptiste Labat proudly claimed the right not to classify plants and animals according to artificial taxonomies as "some persons of consideration" advised.[56] He objected to the "itinerant" science of botanists sent to the West Indies, proposing instead a "field science" born of careful observation and a modest description of nature. Labat repeatedly pointed out—with obvious satisfaction—the mistakes and lapses of Father Charles Plumier, the emissary of the Académie des Sciences.[57]

Similarly, much information transmitted by missionaries escaped the secular apparatus. The Jesuits at the College at Quebec, for example, held official positions as royal hydrographers, but what else did they do for science outside the purview of

---

nard in *La Fortune du Colbertisme: Etat et industrie dans la France des Lumières* (Paris: Fayard, 1998), and in his "Les Savants et l'expertise manufacturière au XVIIIe siècle," in *Histoire et mémoire de l'Académie des Sciences: Guide de recherches*, eds. Eric Brian and Christiane Demeulenaere-Douyère (Paris: Tec & Doc Lavoisier, 1996), pp. 311–18. In addition, Lavoisier was regularly consulted, for example on sugars or cloves; see *Œuvres de Lavoisier* (Paris, 1862–1893), vol. 4, p. 478, vol. 6, p. 74, and ff.

[55] The minutes of this *comité* have been published; see Alfred de Foville and Henri Pigeonneau, eds., *L'Administration de l'agriculture au contrôle général des finances (1785–1787): Procès-verbaux et rapports* (Paris, 1882).

[56] R. P. Jean-Baptiste Labat, *Nouveau voyage aux isles de l'Amérique*, 4 vols. (Martinique: Saint-Joseph, 1979), vol. 1, p. 17.

[57] *Ibid.*, p. 147, where he writes: "C'est à quoi s'exposent ceux qui veulent faire des Relations d'un pays qu'ils ne voyent qu'en passant et comme en courant."

the monarchy?[58] While only a careful examination of monastic records will reveal the knowledge collected and housed in monastic libraries, we do know that Father Mongin sent astronomical observations to his Jesuit superior at the end of the seventeenth century; that the missionary Philippe de Beaumont refers to shipping natural history samples to his order; and that the naturalist Father Nicolson left his important collection from Saint Domingue to his mother house in France (and *not* to the Jardin du Roi).[59]

Religious communities were not alone in doing science on the colonial periphery. Examples abound, particularly in the second half of the eighteenth century, of agricultural, physical, chemical, even electrical or aerostatic experiments conducted by individuals in private, unaffiliated colonial circles.[60] Manuscripts occasionally turn up that exemplify this activity: a compendium of plants from Saint Domingue illustrated with watercolors by an artillery captain; a flora of the West Indies with seventy colored drawings compiled around 1764 by an imprisoned army officer of Guadeloupean origin.[61]

The metropolitan scientific bureaucracy proved flexible in responding to the growing needs of the royal administration and to the increasing specialization of knowledge—while still attracting the goodwill of enlightened amateurs. But in the end the "machine" showed obvious limitations. The distant and elitist character of official institutions and difficulties in publishing through their official outlets presented obstacles for colonials. Similarly, because of the explosive growth of knowledge in the eighteenth century and the growing complexity of science's intellectual and material tools, the scientific effort originating in the colonies increasingly became diverted toward other, second-tier institutions in metropolitan France.

French provincial academies, for example, were more likely to acknowledge the merits of colonial authors—or at least to gratify their vanity. Indeed, many reports that can be called "colonial" are housed today in the archives of provincial academies. For instance, in the archives of the Académie des Sciences, Belles-Lettres et Arts in Bordeaux, one finds reports coming from Sainte-Lucie (by Jean-Baptiste Cassan), Guadeloupe (by Peyssonnel) and Guyana (by Pierre Barrère and Desrivierre-Gers).[62] Other colonial memoirs are waiting to be discovered in the archives of Marseilles, Rouen, and the big Atlantic ports.[63]

---

[58] Chartrand, Duchesne, and Gingras, *Histoire des sciences au Québec* (cit. n. 14), pp. 22–31.

[59] On these examples, see Bibliothèque Municipale de Carcassonne, MS 82, fol. 74–7; Gabriel Debien, "Un Missionnaire auxerrois des Caraïbes: Claude-André Leclerc de Château-du-Bois à la Dominique et à la Guadeloupe (XVIIe siècle)," *Bulletin de la Société des Sciences Historiques et Naturelles de l'Yonne*, 1976, *108*:43; Fournier, *Voyages et découvertes* (cit. n. 39), p. 70.

[60] McClellan, *Colonialism and Science* (cit. n. 3), chap. 10 outlines these for the case of Saint Domingue; see also François Regourd, "Un Médecin des Lumières à Sainte-Lucie: Le Docteur Cassan," Comité des Travaux Historiques et Scientifiques Annual Congress (Fort-de-France, 1998), in press.

[61] For the compendium, see Bibliothèque Municipale de Besançon, MS 446. The flora is an uncatalogued MS in the Bibliothèque Mazarine (Paris), presented in Marcel Chatillon and Jean-Claude Nardin, *De la découverte à l'émancipation: Trois siècles et demi d'histoire antillaise à travers les collections du Docteur Chatillon et de la Bibliothèque Mazarine* (Catalogue of exhibit at the Bibliothèque Mazarine, 2 Nov. 1998 to 29 Jan. 1999) (Paris: Institut de France, 1998), entry 16.

[62] Bibliothèque Municipale de Bordeaux, Fonds Ancien, MS 828 (I–CVI).

[63] For a sense of these archival sources and of provincial academies generally, see the magisterial work by Daniel Roche, *Le Siècle des Lumières en Province: Académies et académiciens provinciaux, 1680–1789* (Paris-La Haye: Mouton, 1978).

Likewise, several free societies and *musées* created in France in the 1780s welcomed colonial correspondents. The Parisian Société d'Histoire Naturelle, for example, had eight colonial associates in 1791.[64] The Bordeaux Musée established contacts with the Cercle des Philadelphes and individuals in Saint Domingue.[65] And the Parisian Musée founded by Court de Gébelin had the famous Moreau de Saint-Méry for its president, a renowned figure in colonial science who was also permanent secretary of the Musée of Pilatre de Rozier.[66]

Metropolitan newspapers were willing to open their columns to contributions from the colonies. The *Journal de Physique*, the *Journal des Savants*, and even more specialized sheets like the *Feuille du Cultivateur* or the *Gazette de Santé* published numerous articles written in the tropics. Many of these articles were, in fact, passed on by distinguished members of the various royal academies, but in many other instances authors addressed editors directly, bypassing official channels. Newspapers published in the colonies were also vehicles for circulating agricultural information and other reports.[67] If some of these reports are attributable to local administrators, most of the planters' submissions were published on the initiative of editors (like the famous Charles Mozard, publisher of Saint Domingue's *Affiches Américaines*), who played a significant role in endowing local scientific statements with a degree of legitimacy.

Some of the activities conducted on the margins of the French colonial machine could be exploited by rival machines, notably England's. Peyssonnel, the Caribbean-based physician and botanist, actually turned to the rival machine and published in the *Philosophical Transactions* when his discovery of the nature of coral reefs as living organisms was rejected by Réaumur.[68] Such behaviors were common practice. Somebody like Cassan, for example, the royal physician in Sainte-Lucie in the late 1780s, was an active correspondent of the Ministère de la Marine, the Société Royale d'Agriculture, the Académie de Marine, and the Académie Royale de Bordeaux, but he did not hesitate to send his report on the Sainte-Lucie volcano to the Swedish *Vetenskapsakademie*.[69] The strictly national logic encouraged by the colonial machine (and by the *ministres de la marine* in particular) was not always upheld by the actors in the field. By the same token, in a spirit of fraternity following the American war, botanical exchanges began between and among British and French gardens in the Caribbean. Nectoux, the superintendent of the French royal garden in Saint

[64] On the Société d'Histoire Naturelle of Paris, see Pascal Duris, *Linné et la France* (Geneva: Droz, 1993), esp. pp. 69–96. For a list of colonial associates, see *Actes de la Société d'Histoire Naturelle de Paris*, 1792, 1:131–32.

[65] McClellan, *Colonialism and Science* (cit. n. 3), pp. 267–8.

[66] Anthony Louis Elicona, *Un Colonial sous la Révolution en France et en Amérique: Moreau de Saint-Méry* (Paris: Jouve, 1934), pp. 19–20.

[67] McClellan, *Colonialism and Science* (cit. n. 3), pp. 97–101; Jean Sgard, ed., *Dictionnaire des journaux 1600–1789* (Paris: Universitas, 1991), passim. See also Jean Fouchard, *Plaisirs de Saint-Domingue: Notes sur la vie sociale, littéraire et artistique* (Port-au-Prince: Imprimerie de l'Etat, 1955; reissued 1988).

[68] For a list of Peyssonnel's papers appearing in the *Philosophical Transactions*, see Auguste Rampal, "Notes biographiques sur J.-A. Peyssonnel," *Bulletin de Géographie Historique et Descriptive*, 1907, 22:32.

[69] Regourd, "Un Médecin des Lumières" (cit. n. 60); Jean-Baptiste Cassan, "Beskrifning om Vulcanen på Sainte-Lucie," in Kongliga svenska vetenskaps-academiens, *Nya Handlingar*, 3rd trimester 1790, pp. 161–78.

Domingue, twice traveled to Jamaica where he and his English counterpart, Dr. Thomas Clarke, exchanged interesting specimens from their gardens, including tea.

While on the geographical periphery of the French colonial empire, the Cercle des Philadelphes in Cap-François, Saint Domingue represented the fullest extension of ancien-régime colonial science, and at the same time evidenced some of the centrifugal forces discussed here. On the one hand, like provincial academies, it began by private initiative (in 1784), and its existence suggests the emergence of autonomous colonial scientific practices. It automatically became *the* regional scientific authority. It cultivated its own botanical garden outside the control of colonial administrators, and published a number of small scientific and medical works in the colony.

But the Cercle des Philadelphes began to receive government subsidies in 1787, and in May of 1789 it became the official Société Royale des Sciences et Arts du Cap François. Compared with the typical French provincial academy, the Cercle des Philadelphes was on a fast track to government recognition and support. Included in that support were gold and silver *jetons de présence*, just like those awarded by the big academies in Paris, and money for prizes. The Cercle des Philadelphes integrated itself within the metropolitan academic network by formalizing a virtually unprecedented association with the Académie Royale des Sciences. It also established links with the Société Royale de Médecine and a few provincial academies and *musées*, and it showed its subservience to the state by electing colonial *gouverneurs-généraux* and *intendants* as honorary members. By the same token, provincial academies in France were notoriously independent, and as a provincial, if tropical, academy, the Cercle des Philadelphes likewise exemplified the desire of colonials for scientific self-determination while maintaining friendly links with the rest of official science. In so doing, the Cercle des Philadelphes legitimated a voice for science originating in and aimed at the colonies.[70]

Finally, parallel forms of knowledge persisted on the periphery of the French colonial machine. We are not referring to the mature scientific traditions of India or China, which the envoys of the Académie des Sciences and the Jesuits identified and collected quickly enough.[71] We have in mind, rather, Amerindian knowledge of nature and that introduced and enriched by African populations brought as slaves to colonial plantations in the West Indies or the Mascareignes. References occur particularly in connection with cures effected by plants best known to Amerindians or slaves. The French colonial machine showed a keen scientific and political interest in this potentially valuable, yet also threatening, indigenous knowledge. Already in 1647, Father Raymond Breton wrote about West Indian natives: "They certainly have wide knowledge and they exploit the rare virtues of many things we have no name for in Europe."[72] A century later the French engineer Fresneau identified and

---

[70] On the Cercle des Philadelphes, see McClellan, *Colonialism and Science* (cit. n. 3), part III; Pierre Pluchon, "Le Cercle des Philadelphes du Cap Français à Saint-Domingue: Seule académie coloniale de l'Ancien Régime," *Mondes et Cultures*, 1985, *45:*157–85.

[71] See for example, Guillaume Legentil's "Remarques et observations sur l'astronomie des Indiens," *Mém. l'Acad. Roy. Sci.*, 1784, 482–501 and related papers in volumes for 1785 and 1788. See also the many reports from the Jesuits in China held in the archives of the Observatoire in Paris.

[72] Père Raymond Breton, *Relations de l'île de la Guadeloupe*, vol. 1 (Basse-Terre: Société d'Histoire de la Guadeloupe, 1978), p. 50. See also Chartrand, Duchesne, and Gingras, *Histoire des sciences au Quebéc* (cit. n. 14), pp. 65–8.

located the rubber tree, thanks to the aid and expertise of the Indians of Guyana.[73] The Société Royale de Médecine received a report about a plant well known in Guadeloupe and in Saint Domingue that was used by slaves supposedly to cure smallpox.[74] In another instance, the *ministre de la marine* queried the Société Royale de Médecine about certain ant nests used by blacks (but not Indians!) in Guyana to staunch hemorrhages.[75] Behind much of this concern loomed the threat of poisonings, which regularly terrorized white colonists. Moreau de Saint-Méry wrote about slaves with knowledge of poisons "giving classes on this repulsive weapon everywhere in the colony."[76] Nightly practices in Saint Domingue that blended *vaudon* rites with Mesmeric principles imported from France attracted the attention of colonial authorities and confirmed—if confirmation were needed—that such underground, badly understood, and possibly threatening knowledge existed and needed policing.[77]

As revealing as they are, secondary centers and modes of activity did not rival the functioning of the central institutions. The great and powerful French colonial machine possessed informants and an allure such that Paris would ultimately retrieve any useful information that appeared in its colonial networks.

CONCLUSION

From its seventeenth-century origins to the end of the ancien régime, the scientifico-technical dimension of French colonialism formed an integral element of a colonial mode of production superintended by the French government. Medicine, astronomy, cartography, botany, and the marine sciences were pressed in every way to produce useful outcomes. French colonial science was heavily institutionalized—witness the Académie Royale des Sciences, the Académie Royale de Marine, the Société Royale de Médecine, the Société Royale d'Agriculture, the Cercle des Philadelphes, botanical gardens, port hospitals, and other organizations. Centered around the Ministère de la Marine, these institutions came to function in a coordinated way, as part of a larger bureaucracy, normalized in the service of the state. Although not without interest, little scientific activity took place outside of this bureaucratized apparatus, and our story involves comparatively little pluralism.

Notable by their near-complete absence from this account are the conventional interests of French commerce. Traders in Nantes or Bordeaux who grew rich from the slave trade, and planters and others in the colonies and in France who profited from sugar and coffee production, expressed comparatively little interest in research

[73] Charles-Marie La Condamine, *Mém. l'Acad. Roy. Sci.*, 1751, 326–8.

[74] Acad. Nat. Méd. Lib. (cit. n. 8), Archives of the Société Royale de Médecine, "Manuscrits #7: Plumitif de la Société R. de Med: Depuis le 1 Xbre 1778 jusques au 30 9bre 1779," p. 141 [Meeting for Friday, 23 April 1779.]

[75] *Ibid.*, "Manuscrits #11: Plumitifs depuis le 10 mars 1786 jusqu'au 24 juillet 1789," fol. 31r [Meeting for 4 July 1786]; also, MS 132, dossier 49.

[76] Louis Médéric Moreau de Saint-Méry, *Description topographique, physique, civile, politique et historique de la partie française de l'isle de Saint-Domingue*, eds. Blanche Maurel and Etienne Taillemite (Paris: Société Française d'Histoire d'Outre Mer, 1984), p. 630. Pierre Pluchon, *Vaudou sorciers empoisonneurs de Saint-Domingue à Haiti* (Paris: Karthala, 1987) is the fundamental and pioneering work in this regard.

[77] See Pluchon, *Vaudou sorciers* (cit. n. 76); McClellan, *Colonialism and Science* (cit. n. 3), p. 146; and Gabriel Debien, "Assemblées nocturnes d'esclaves à Saint-Domingue (La Marmelade, 1786)," *Annales Historiques de la Révolution Française*, 1972, 44:273–84.

science or in using the forces of knowledge to improve production. The economics of the contemporary colonial game simply did not reward long- or even medium-term investment.

French colonial science in the ancien régime is not completely congruent with French exploration in the eighteenth century. Voyages of discovery preceded formal colonization, but in the final analysis the better-known examples of scientific research expeditions undertaken by the French—the missions to Lapland and Peru in the 1730s, the Venus transit observations in the 1760s, and the voyages of Bougainville and La Pérouse later in the century—had little to do with colonization per se, even if the colonial enterprise ultimately benefited. These great expeditions may be said to have been driven by the internal demands of science and an Enlightenment program of measuring the world and documenting its contents.[78]

Why, then, did the monarchy go to such lengths to engage science and expertise in support of colonization? Answers are not hard to find. First, colonial commerce was crucial for the French economy in general and government finances in particular. The economic principles of mercantilism guided Bourbon policy in colonial matters. On the eve of the Revolution, one person in eight in France lived off of colonial commerce, and 60 percent of 632 million *livres* of French foreign trade involved colonial products, with taxes, tolls, and duties (totaling up to 50 percent) bringing millions into state coffers.[79] Second, the colonies served as treasured possessions in contemporary geopolitics, especially vis-à-vis England, and from the 1660s to the French Revolution a driving purpose of the scientifico-technical apparatus of France was to develop the colonies as part of the country's international presence. France's loss of much of its first colonial empire after 1763 and defeat in the Seven Years' War embittered many, especially in the Marine Royale, and gave a new impetus for reform and expansion of the Marine, especially by scientifically inclined officers.[80] Needless to say, this movement found vindication in the victory over the English in the American war. Finally, although arguably first, the sheer glory and magnificence of a great empire was enough of a *raison d'être* for the French state to underwrite costly colonial investments. What we think of as nationalism was only part of the story; the rest, foreign to us in the twenty-first century, concerned the majesty *Très Chrétienne* of the Bourbon kings, a majesty not to be taken lightly.

The revolutionary and Napoleonic periods destroyed the first French colonial empire and, with it, the distinctive ancien-régime style of French colonial science. The French lost Saint Domingue in 1804 to the rebellious slaves of the new country of

---

[78] On French scientific voyages, see especially the recent Ph.D. dissertation by Jordan Kellman, "Discovery and Enlightenment at Sea: Maritime Exploration and Observation in the 18th-Century French Scientific Community" (Princeton University, 1998). See also John Gascoigne, *Joseph Banks and the English Enlightenment: Useful Knowledge and Polite Culture* (Cambridge: Cambridge Univ. Press, 1994). Regarding the Enlightenment and colonialism, see David Wade Chambers, "Period and Process in Colonial and National Science," in *Scientific Colonialism: A Cross-Cultural Comparison*, eds. Nathan Reingold and Marc Rothenberg (Washington, D.C.: Smithsonian Institution Press, 1987), pp. 297–321, here pp. 299, 306–10; McClellan, *Colonialism and Science* (cit. n. 3), p. 293.

[79] See Jean Tarrade, *Le Commerce colonial de la France à la fin de l'Ancien Régime*, 2 vols. (Paris: Presses Universitaires de France, 1972), esp. chap. 19 and table 3, p. 739; Pluchon, *Histoire de la colonisation français* (cit. n. 1), pp. 687–79 and appendices, esp. p. 1018. See also McClellan, *Colonialism and Science* (cit. n. 3), chap. 4, esp. p. 63 and notes, and Paul Butel, *Européens et espaces maritimes (vers 1690–vers 1790)* (Bordeaux: Presses Universitaires de Bordeaux, 1997), chap. 4.

[80] This theme echoes through the volume edited by Ulane Bonnel, *Fleurieu et la Marine de son temps* (Paris: Economica, 1992).

Haiti, and after 1815 France was left with only crumbs of its former empire. The remaining *veilles colonies* declined in importance as France struggled to build a second colonial empire around new, more substantial holdings in Africa and in Southeast Asia.[81] European imperialism in the nineteenth century eclipsed in scale and importance earlier, mercantilist episodes of colonial development. The distinctive "civilizing mission" of the French Republic replaced the pursuit of Bourbon glory. But undoubtedly, as de Tocqueville might have it, the roots of French colonial science reach back to the ancien régime.

[81] On these developments see Denise Bouche, *Histoire de la colonisation française,* vol. II, *Flux et reflux (1815–1962)* (Paris: Fayard, 1991). On French science and colonization in the nineteenth and twentieth centuries, see Lewis Pyenson, *Civilizing Mission: Exact Sciences and French Overseas Expansion, 1830–1940* (Baltimore: The Johns Hopkins Univ. Press, 1993).

Ordering the World for Europe: Science As Intelligence and Information As Seen from the Northern Periphery

Sverker Sörlin*

ABSTRACT

This essay examines science and information networks built by European scientists, often in connection with military and/or commercial enterprises and expeditions. It is focused on Europe's northern peripheral powers, particularly the empires, or post-empires, of Sweden and Denmark. Both Copenhagen and Stockholm/Uppsala served as brokerage centers for knowledge and commodified scientific objects. Both cities also had collections, academies, and other institutions that not only ordered northern Europe for science, but contributed significantly to ordering and mapping the world's natural systems so as to make them accessible for colonialism. Scientific travel is also analyzed here as an important part of the construction of value, both symbolic (collections, scientific prestige) and concrete (mercantilist import substitution).

THE GROWTH OF WESTERN GLOBAL DOMINATION FROM THE Renaissance to the nineteenth century was as much a scientific process as an economic and military one. Historical research has elaborated on the emergence and growth of Western and colonial centers of calculation, and the history of scientific travel abounds with evidence of envoys of European powers that were able to combine the goals of science with those of empire expansion and economic exploitation.

In its fine nuances, this history was never simple or straightforward. It appeared in different shapes and variations depending on which nation and part of the world were involved. This history was also a matter of personal aspirations, motives, and temperaments that constantly changed the agenda of the scientific project. On the face of it, the Victorian traveler Richard Burton in his *Narrative of a Pilgrimage to El Medinah and Meccah* (1855) is the ardent gatherer of scientific evidence, self-consciously refuting the primitive misunderstandings of the Renaissance Mecca-traveler Varthema from Bologna, who in 1503 had colored his Arabian vision with imaginary unicorns. They might, concludes Burton, have been African antelopes,

---

* Department of History of Science and Ideas, Umeå University, S-901 87 Umeå, Sweden.

"but the suspicion of fable remains."[1] On the other hand, Burton and his European fellow travelers in the eighteenth and nineteenth centuries were pulled by many different triggers, ranging from the artistic to the zoological, from crude profit to sexual curiosity. Burton himself was a prime example of the latter, but one would have no problem finding others, both before and after, who were not entirely—or even remotely—scientific in their approach to distant lands.[2]

Artistic representation is another area in which personal, commercial, and scientific motives joined forces successfully, and often had quite complex connections to the imperial project. One could, for evidence, cite the marvelous example of Maria Sibylla Merian, a Dutch flower-painter and divorced mother of two daughters, who in 1699, at the age of fifty-two, traveled to Surinam to collect materials for her superb *Metamorphosis insectorum Surinamensium* (1705), a masterpiece in the iconography of butterflies and caterpillars that has been sought by collectors ever since. True, like most travelers of her generation, she understood the economic potential of the natural marvels and species that she encountered—the *Rothschildia aurota* should, she noted, "bring good silke and make a good profit." She was inspired to take her trip by collections of objects from the Dutch colonies that she saw during the years she lived in the Waltha castle, owned by her Friesland host Cornelis van Aersen van Sommelsdijk, and her trip was sponsored by, among others, the famous director of the Dutch East India Company, Nicolaas Witsen in Amsterdam.[3] Obviously, she was a colonialist artist/scientist. But at the level of her own personal decision and career making, she was a curious, business-minded woman who strove to find her way in an increasingly problematic personal situation.[4]

When we explore and analyze patterns of Western domination in the modern period we should not, therefore, be blinded by its seemingly linear trajectory of constant growth and ever more efficient networks of commerce, science, and military intelligence. This particularly applies to this paper which, to a large extent, will deal with nations in northern Europe that in and of themselves were in an ambiguous position vis-à-vis the major imperial powers of Europe. These northern nations were at the same time colonial, with their own empires, and colonized in the sense that

[1] Richard Burton, *Narrative of a Pilgrimage to El Medinah and Meccah* (London, 1855, new ed. 1893), vol. 2, p. 337, quoted in Jas Elsner and Joan-Pau Rubiés, "Introduction," in *Voyages and Visions: Towards a Cultural History of Travel* (London: Reaktion Books, 1999), pp. 3, 272 n. 5.

[2] See, for example, George S. Rousseau on the connections between colonialism, travel, and sexuality in "Magic, Science, and the 'New World' Origins of Freud's Fetisches," in *Mundialización de la ciencia y cultural nacional,* eds. Antonio Lafuente, Alberto Elena, and María Luisa Ortega (Madrid: Ediciones Doce Calles, 1993), and several essays in his collection *Perilous Enlightenment: Pre- and Post-modern Discourses* (Manchester/New York: Manchester Univ. Press, 1991). See also Alan Bewell, "'On the Banks of the South Sea': Botany and Sexual Controversy in the Late Eighteenth Century," in *Visions of Empire: Voyages, Botany, and Representations of Nature,* eds. David Philip Miller and Peter Hanns Reill (Berkeley/Los Angeles/London: Univ. of California Press, 1996).

[3] On Witsen, see P. J. A. N. Rietbergen, "Nicolaas Witsen (1641–1717) Between the Dutch East India Company and the Republic of Letters," in *All of One Company: The VOC in Biographical Perspective* (Utrecht: HES uitg., 1986).

[4] Elisabeth Rücker and William T. Stearn, eds., *Metamorphosis insectorum Surinamensium,* including: "Maria Sibylla Merian in Surinam: Kommentar zur Faksimil-Ausgabe der Metamorphosis insectorum Surinamensium (Amsterdam 1705) nach den Aquarellen in der Royal Library, Windsor Castle/Commentary to the Facsimile Edition of *Metamorphosis insectorum Surinamensium*" (London: Pion, 1982); Kurt Wettengl, ed., *Maria Sibylla Merian, 1647–1717: Artist and Naturalist* (Amsterdam: Verlag Gerd Hatje, 1997); Tony Rice, *Voyages of Discovery: Three Centuries of Natural History Exploration* (London/Hong Kong: Scriptum Editions, 1999), chap. 2.

within their own boundaries, they had large hinterlands with ethnic groups and minority populations.

Furthermore, we should realize that the European nations had a number of scientific centers within their own borders, each of a particular size and character. International relations were not symmetrical. Some of the centers, like Paris and London, were central on a European scale; others, like Copenhagen or Uppsala, were central mostly in the national arena; still others had a provincial orientation. On the other hand, if special disciplines are considered, a peripheral scientific institution could be central, as was the case for botany at Uppsala, and also for astronomy and mathematics with Leonhard Euler at the Imperial Academy of Sciences in St. Petersburg. Still others, such as Kew in London, served as brokerage institutions within empires.[5] Centrality or peripherality was not primarily a matter of geographical location, but of the combined effect of social, scientific, and—not least—power relations.[6]

This polycentricity of science and of the distribution of scientific institutions should be approached cautiously. Individual savants in different parts of Europe had close and frequent contacts, and studies of correspondence and individual travels clearly demonstrate their far-ranging transnational networks.[7] Naturalists, however, like other people, bore identities; they belonged somewhere, and they were loyal to something or someone. Even more importantly, the daily activities of natural history were carried out in a framework of institutions, agendas, career opportunities, civic and moral principles, financial support, and patronage systems.[8] The decisive components of this framework were nationally defined, and coincided with the parallel growth of the central state.[9] The process was irregular. Nonetheless, science was part

[5] Lucile H. Brockway, *Science and Colonial Expansion: The Role of the British Royal Botanical Gardens* (New York: Academic Press, 1979), p. 78; Ray Desmond, *Kew: The History of the Royal Botanical Gardens* (London: Harvill, 1995).

[6] David Wade Chambers, "Locality and Science: Myths of Centre and Periphery," in Lafuente, Ortega, and Elena, *Mundialización* (cit. n. 2), pp. 605–17; *idem*, "Does Distance Tyrannize Science?" in *International Science and National Scientific Identity*, eds. R. W. Home and Sally Gregory Kohlstedt (Dordrecht/Boston/London: Kluwer Academic Publishers, 1991), esp. p. 32; Roy MacLeod, "On Visiting the 'Moving Metropolis': Reflections on the Architecture of Imperial Science," in *Scientific Colonialism: A Cross-Cultural Comparison*, eds. Nathan Rheingold and Marc Rothenberg (Washington, D.C.: Smithsonian Institution Press, 1987), pp. 217–49.

[7] Biographies of eighteenth-century naturalists provide good examples, from Linnaeus to Lavoisier to Laplace. A study of the correspondence between European astronomers gives detailed evidence of these transnational networks; see Sven Widmalm, "A Commerce of Letters: Astronomical Communication in the 18th Century," *Science Studies*, 1991, *5*, 2:43–58.

[8] The literature on the social embeddedness of scientific practice has grown tremendously. Suffice it here to cite works such as Bruno Latour, *Science in Action: How to Follow Scientists and Engineers through Society* (Cambridge, Mass.: Harvard Univ. Press, 1987); Stephen Turner, "Forms of Patronage," in *Theories of Science in Society*, eds. Susan E. Cozzens and Thomas F. Gieryn (Bloomington/Indianapolis: Indiana Univ. Press, 1990), pp. 185–211; Steven Shapin, *A Social History of Truth: Civility and Science in Seventeenth-Century England* (Chicago/London: The Univ. of Chicago Press, 1994); Nicholas Jardine, James A. Secord, and Emma C. Spary, eds., *Cultures of Natural History* (Cambridge: Cambridge Univ. Press, 1996). For contributions on the social context of international science, see Elisabeth Crawford, Terry Shinn, and Sverker Sörlin, eds., *Denationalizing Science: The Contexts of International Scientific Practice*, Sociology of the Sciences Yearbook *16* (Dordrecht/Boston/London: Kluwer, 1993).

[9] Raymond Williams, *Keywords: A Vocabulary of Culture and Society*, 2nd ed. (London: Fontana/Croom Helm, 1981), p. 178. The literature on nations and nationalism is copious; thematic overviews include John Hutchinson and Anthony D. Smith, eds, *Nationalism* (Oxford/New York: Oxford Univ.

of a general trend towards the recognition of the national. It became gradually defined as a state interest, and consequently the scientists emerged as representatives
of their respective countries. This meant a crucial limitation to the cosmopolitan
ideal of science that had been expressed by Enlightenment philosophers and ideologists.

A SINE QUA NON: THE SCIENTIZATION OF TRAVEL

The gathering of information on distant places—that would later become colonies—
was well developed long before scientific stations were located overseas. The
method was travel, observation, and reportage by letter. Local inhabitants provided
intelligence that was sometimes disregarded and frowned upon but that also, as Stephen Greenblatt has shown, could be meticulously recorded. Such was the case for
the exchange of "speech acts" and insignia in the first encounters of Columbus on
San Salvador in 1492, Jacques Cartier by the St. Lawrence River in 1516, and Martin
Frobisher on Baffin Island in the 1570s.[10] The *cosas maravillosas* (marvelous things)
of overseas territories concerned not only material goods and natural exotica; they
were also a matter of words to describe and define them, articulate their value, and
therefore legitimate their conquest. Talk and text were part of a process of commodification, and speech acts were an early chapter in that dismal chronicle that would
later turn into European imperialism.

Beginning in the seventeenth century, travel underwent a process of scientization
that was linked to the new philosophy of science. Ideologists appeared, formulating
programs and methods. In Francis Bacon's posthumous essay "Of Travel" (1625),
the scientific traveler was advised to keep a journal and observe diligently. In his
earlier *Instauratio Magna* (1620), Bacon had stressed the importance of "autopsy,"
the eye being the most superb of all scientific instruments: "All depends on keeping
the eye steadily fixed upon the facts of nature and so receiving the images simply
as they are." From the middle of the century, the idea of scientific travel as part of the
empirical knowledge program gained increased support. Instructions were formulated by the Royal Navy that British ships should observe and record conditions in
foreign countries. Shortly after its inception the Royal Society created a "committee
for foreign inquiries" and demonstrated an interest in travel literature and collecting
natural specimens and exotica.[11] In the very first volume of the *Philosophical Transactions* (1665), Robert Boyle published a list of questions, "General Heads for a
Natural History of a Countrey, Great or Small." The following year, the *Transactions*
printed an instruction for travelers, "Directions for Seamen Bound on Far Voyages."[12] In the following century, as this process of scientization continued, field
methods, instruments, and practices were developed for work under conditions ranging from Arctic to tropical, and the virtues of documentation and precision grew in

---

Press, 1994), and Geoff Eley and Ronald Grigor Suny, eds., *Becoming National: A Reader* (Oxford:
Oxford Univ. Press, 1996).

[10] Stephen Greenblatt, *Marvelous Possessions: The Wonders of the New World* (Chicago: The Univ.
of Chicago Press, 1991). See also Patricia Seed, *Ceremonies of Possession in Europe's Conquest of
the New World 1492–1640* (Cambridge: Cambridge Univ. Press, 1996).

[11] Michael Hunter, *Establishing the New Science: The Experience of the Early Royal Society*
(Woodbridge: The Boydell Press, 1989), p. 93 and chap. 4.

[12] Royal Society, "Directions for Seamen Bound on Far Voyages," *Philosophical Transactions*,
1665–1666, *1*:140–1.

significance.[13] Promoting scientific travel became common practice at the European scientific academies, and the northern centers quickly followed. In Stockholm, for example, Jacob Faggot, permanent secretary of the Royal Swedish Academy of Sciences (founded in 1739), wrote a document vindicating the importance of research journeys, albeit he concentrated on travels within his native Sweden.[14]

Throughout the seventeenth century, commercial and military objectives of a national character reigned supreme, and science and intelligence were regularly subsumed under the principles of colonial realpolitik. The Spanish and the Portuguese had only superficial scientific motives behind their overseas journeys, carrying out their searches for the southern continent in the belief that colossal riches awaited them. The Spaniards later regarded Captain James Cook's third voyage to the northwestern coastline of America (1776–1778) with great suspicion; they even considered putting Cook in prison should he show up on the shores that Vasco Nunez de Balboa had declared Spanish as early as 1512.[15] The Dutch for a long time acted under similar inducements. The driving force behind the Dutch expeditions to the East and to the Southern Hemisphere was the Dutch East India Company, and their findings were not treated as discoveries to be shared with the scholarly world, but rather as closely guarded trade secrets.[16] What we regard as important discoveries were often spin-offs of commercial and military trips. Englishman John Campbell wrote, as late as the 1740s, that the British should join in the exploration of *Terra Australis* and increase trade in the area, as it would eventually make them a "great, wealthy, powerful and happy people."[17]

However, the strict boundary that was often drawn between history prior to Cook's voyages and the new "scientific" era they initiated is, at best, an oversimplification. One case in point is the use of scientific draughtsmen which, despite repeated claims, did not enter the history of travel with Cook.[18] The German-born Georg Wilhelm Baurenfeind was an artist who accompanied the Danish expedition to Arabia in the 1760s, just to mention one counterexample, and scientific iconography associated with travel goes back at least as far as Konrad Gesner's mid-sixteenth century natural histories.[19] Instead of fixed time limits, it would be wiser to talk about a process of scientization, of which the Cook voyages undeniably form an

[13] See Michael T. Bravo, "Precision and Curiosity in Scientific Travel: James Rennell and the Orientalist Geography of the New Imperial Age," in Elsner and Rubiés, *Voyages and Visions* (cit. n. 1), pp. 162–83. Bravo's endnotes provide good examples of the historiography of precision in scientific travel.

[14] Jacob Faggot, "Tankar om fäderneslandets känning och beskrifwande" (Thoughts on the knowledge and description of the fatherland), *Kungliga Vetenskapsakademiens Handlingar* (Proceedings of the Royal Swedish Academy of Sciences), 1741.

[15] Christian I. Archer, "The Spanish Reaction to Cook's Third Voyage," in *Captain James Cook and His Times,* eds. Robin Fisher and Hugh Johnston (Vancouver: Douglas and Macintyre, 1979), p. 106.

[16] Günther Schilder, "New Holland: The Dutch Discoveries," in *Terra Australis to Australia,* eds. Glyndwr Williams and Alan Frost (Melbourne: Oxford Univ. Press, 1988), p. 83.

[17] John Campbell, ed., *Navigantium atque itinerantium bibliotheca,* 2 vols. (London, 1744–1748), vol. 1, p. 332; quoted in Alan Frost and Glyndwr Williams, "Terra Australis: Theory and Speculation," in *Terra Australis* (cit. n. 16), p. 28.

[18] See, for example, Lynn Withey, *Voyages of Discovery: Captain Cook and the Exploration of the Pacific,* 2nd ed. (Berkeley/Los Angeles: Univ. of California Press, 1989), p. 10.

[19] See the numerous examples in Rice, *Voyages of Discovery* (cit. n. 4). See also William T. Stearn, "Botanical Exploration to the Time of Linnaeus," *Proceedings of the Linnean Society of London,* 1958, *169,* and further examples in Allan Ellenius, ed., *The Natural Sciences and the Arts: Aspects of Interaction from the Renaissance to the 20th Century: An International Symposium* (Uppsala/ Stockholm: Almqvist & Wiksell, 1985).

important part. After them, as well as before, national interests dominated in cross-boundary science, with few exceptions of unselfish internationalism until the nineteenth century. Nations were formed by numerous forces, and one of those was the scientific community. Even with regard to the oft-idealized eighteenth-century ideology of the transnational *république des lettres,* we may have to ask ourselves whether scientists themselves did not contribute to the complex formation of nations, especially given their intellectual status and profile.[20]

Previous research into the history of exploration and scientific travel has noted a national or patriotic dimension, as well as the cosmopolitan character of a small number of remarkable projects, like the observations of the first transit of Venus in 1761. This major undertaking was primarily organized by the French, but it involved large numbers of scientists in many parts of the world. Instead of looking at it as an international enterprise, though, one might better understand it as one in a series of steps towards establishing extensive networks for the exchange of goods and intelligence.[21]

SCIENCE IN A NORTHERN EMPIRE: SWEDEN IN THE EIGHTEENTH CENTURY

As early as the seventeenth century, Swedish naturalists, inspired by gothic myths of northern wealth and glory, began formulating a program of northerly directed research. This endeavor was bombastically patriotic, hailing the riches of Europe's distant north that was invariably called "Lapland," although its boundaries in practice extended to the entire northern part of Sweden and Finland. Both of these countries were then part of the sizeable Swedish Empire that reached from the German provinces in the south and the Danish border in the west to Ingermanland in the east and Finnish Lapland in the north, the farthest corner of which was almost on the Arctic Sea. The program was advocated not least by astronomers and naturalists who had themselves traveled in the region. Urban Hiärne, the chemist, made a tour of southern Norrland in the 1680s, carefully noting everything of economic importance. Olof Rudbeck, Jr., the naturalist son of Uppsala University's famous polymath and chancellor of the same name, made a journey (1695) to Lapland and returned to publish a volume, *Nora Samolad* (1701). He traveled with support from the state, in the official hope that he would return with such things as could "lend to our fatherland glory and honour." He praised the region's rich resources and the cold of winter that makes people happy and their lives longer. There was freedom for everyone in the north, and no reason to be afraid of thieves and crooks.[22]

---

[20] Generally on the idea of the republic of letters, see *Res Publica Litteraria: Die Institutionen der Gelehrsamkeit in der Frühen Neuzeit,* Wolfenbüttler Arbeiten zur Barockforschung 14, eds. Sebastian Neumeister and Conrad Wiedemann, 2 vols. (Wiesbaden: In Kommission bei O. Harrasowitz, 1987). On its problems: Lorraine Daston, "The Ideal and Reality of the Republic of Letters in the Enlightenment," *Science in Context,* 1991, *4.*

[21] A. Hunter Dupree, "Nationalism and Science—Sir Joseph Banks and the Wars with France," in *A Festschrift for Theodore B. Artz,* eds. David H. Pinkney and Theodore Ropp (Durham, N.C.: Duke Univ. Press, 1964), pp. 41–2. See also the argument for continental and imperial rivalry in Alan Frost, "Science for Political Purposes: European Explorations of the Pacific Ocean, 1764–1806," in *Nature in Its Greatest Extent: Western Science in the Pacific,* eds. Roy MacLeod and Philip F. Rehbock (Honolulu: Univ. of Hawaii Press, 1988), pp. 27–44.

[22] Olof Rudbeck, Jr., *Nora Samolad Eller Uplyste Lapland* (Northern Sami-land or Enlightened Lapland), (Uppsala, 1701). On the expedition: Carl-Otto von Sydow, "Rudbeck d.y.:s dagbok från lapplandsresan 1695: Med inledning och anmärkningar I," (Rudbeck, Jr.'s journal from the Lapland

Map 1. *Norden in 1696 drawn by Uppsala astronomer Anders Spole. It is presented here reproduced from Erik Dahlberg's copious patriotic work* Suecia antiqua et hodierna *(first published in 1714). The map shows the extension in the late 17th century of Sweden's Baltic Empire, including the present-day regions of Sweden, Finland, and northern Germany, as well as the Baltic provinces and Pomerania. To the west of Sweden was the Danish Empire including Norway and Atlantic islands, to the east the Russian Empire.*

A similar kind of exoticism is present in Carolus Linnaeus (1707–1778), who started out from Uppsala on horse in the spring of 1732, when he was twenty-five years old. He praised the beauty of Lapland's nature, the calm and healthy life of the Sami: "you sleep here under your reindeer hide, free from all troubles."[23] Linnaeus constantly returned to the blessings of the Sami fish-and-water diet, and to the virtues of the Nordic winters.[24] He also observed the area's natural resources, but was sad when he realized they were not utilized to their full advantage. Years later, in 1754, he would propose to the Royal Swedish Academy of Sciences that plantations be established on the Lapland mountains above the tree line.[25]

---

journey of 1695: With introduction and commentary I), *Svenska Linnésällskapets årsskrift* (Yearbook of the Swedish Linnaean Society), 1968–1969.

[23] From *Flora Lapponica* (quotation translated by the present author). A survey in English of Linnaeus and his different activities, including his travels in Lapland, is presented in Tore Frängsmyr, ed., *Linnaeus: The Man and His Work* (Berkeley: Univ. of California Press, 1985).

[24] Carolus Linnaeus, *Diaeta naturalis* (1733) was directly inspired by his travels in Lapland.

[25] *Caroli Linnaei Iter Lapponicum Dei gratia institutum 1732*, first publ. 1889; Carolus Linnaeus, "Tankar om nyttiga växters planterande på de Lappska Fjällen" (Thoughts on the cultivation of useful plants in the Lapland mountains), *Kungliga Vetenskapsakademiens Handlingar*, 1754.

In 1710 a scientific society, Vetenskapssocieteten, had been formed in Uppsala. In 1739, aristocratic circles in Stockholm joined with leading naturalists to establish the Kungliga Svenska Vetenskapsakademien (The Royal Swedish Academy of Sciences). Both organizations contributed to a new utilitarian program, which had strong patriotic overtones, particularly in the early decades of the 1700s. The research program that the societies fostered emphasized what was, in essence, a Swedish hinterland, the north. The leading Swedish naturalists at the time unanimously agreed that the north was a splendid and unique scientific resource. Emanuel Swedenborg suggested that the cold Swedish climate be researched, and in 1717 he proposed to Charles XII that an observatory be built in the country because of the outstanding qualities of the "Hyperborean" skies.[26]

As early as 1739, Mårten Triewald, one of the five men who founded the Academy of Sciences, suggested that it send researchers out to ascertain the natural histories of the Swedish provinces.[27] Even though little happened at first, Triewald would eventually see his suggestion met. In 1780 J. D. Lundmark, another student of Linnaeus, traveled to Lapland with the academy's support to collect plant and animal life.[28] He was accompanied by Clas Fredrik Hornstedt, a zoologist (who shortly thereafter sailed to China on an academy commission), and the young Olof Swartz, later famous for his studies of West Indian flora, which culminated in the three-volume *Flora Indiae occidentalis* (1797–1806).

It was no coincidence that Lapland excursions received state support. The regions of the far north were the subject of a scientific exoticism that in certain respects is reminiscent of scientific curiosity about distant continents. In 1736–1737, a French survey headed by Pierre-Louis Moreau de Maupertuis that was intent on determining the shape of the earth once and for all traveled to northern Scandinavia ("Laponie"). Geodesist Jöns Svanberg repeated this experiment around 1800, reporting his findings to the academy.[29] Also worth noting are the Lapland travels of natural historian Göran Wahlenberg, who covered large expanses of the Arctic region, wrote a dissertation on the environs of Kemi in Finland, studied flora and glaciers, and made height and temperature observations of the mountain regions in latitude 67° north.[30] Lars Levi Lestadius, botanist and renowned preacher, also enjoyed academy support

[26] Sven Widmalm, "Mellan kartan och verkligheten: Geodesi och kartläggning, 1695–1860" (Between map and reality: Geodesy and surveying, 1695–1860) (Ph.D. diss., Department of History of Ideas and Learning, Uppsala University, 1990), p. 43.

[27] Royal Swedish Academy of Sciences, minutes, 15 Dec. 1739.

[28] Royal Swedish Academy of Sciences, minutes, 3 May 1780.

[29] Jöns Svanberg, "Berättelse öfver resan til Pello, företagen på Kongl. Vetenskaps Academiens bekostnad . . ." (Account on the journey to Pello, undertaken at the cost of the Royal Academy of Sciences . . .), *Kungliga Vetenskapsakademiens Handlingar*, 1799; idem, "Historisk öfversigt af problemet om Jordens figur, jemte anledningarne til den nya Lappske Gradmätningen och definitiva resultaterna däraf . . ." (Historical review of the problem of the figure of the earth, with the motives for the new grade measurement in Lapland and the definitive results of it . . .), *Kungliga Vetenskapsakademiens Handlingar*, 1804.

[30] Göran Wahlenberg, *Geografisk och ekonomisk beskrifning om Kemi lappmark i Vesterbottens höfdingedöme* (Geographical and economic description on Kemi Lap district in Vesterbotten county) (Stockholm, 1804), p. 4. *Idem,* "Anmärkningar om Lappska vegetationen, med beskrifning om *Myosotis deflexa* eller ett nytt förgätmig-ej från Lappland" (Remarks on Lapland vegetation, with a description on *Myosotis deflexa* or a new forget-me-not from Lapland), *Kungliga Vetenskapsakademiens Handlingar*, 1810, pp. 106–14; idem, *Berättelse om mätningar och observationer för att bestämma lappska fjällens höjd och temperatur vid 67 graders polhöjd* (Account on measurements and observations to decide the elevation and temperature of the Lapland mountains at 67° north) (Stockholm, 1808); on the academy's support, p. 5.

when he studied the plant geography of the most northerly regions of Sweden in 1824.[31]

Lapland enjoyed special treatment, but most of the Swedish provinces eventually received visits by inquisitive scholars. Several trips won the support of the academy, and travel accounts were published in the *Proceedings* even if undertaken without academy support.[32] An increased interest in regional distinctiveness came in the early nineteenth century, partly because of continued feverish activity in botany, and partly (as we reach the middle of the century) apropos of the surveys and charting of Swedish geology.[33]

One remarkable feature of the Academy of Sciences in Stockholm was its persistent build-up of networks for communication and collecting. Not only did it support individual traveling scientists, but it also sustained local collectors and rapporteurs, and it gradually established a scientific presence at particular sites. Some of these became more or less permanent, and later turned into full-fledged research stations. An epistemological and methodological debate persisted for decades in the academy about the preferred method and style of scientific travel: was it the "lone-traveler model," the "correspondent model," or the "station model"? No clear answer emerged, but there was a gradual evolution from the single and mobile to the plural and permanent.[34] The way the Swedes proceeded in their domestic hinterland north provided a model for extension into the Arctic and North Atlantic waters even further afield. In the nineteenth century, what seemed to be a retreat to the Swedish and the provincial from the merry global travels of the Linnaean disciples in the eighteenth century, was, instead, a geographical reorientation.[35]

Establishing scientific activities in the north complemented other efforts to integrate the hinterland regions and make them productive. This was, in essence, a colonial program, even more so since a significant portion of the northern population was native Sami, an ethnic group reminiscent of non-Christian natives elsewhere in the world. Their long-standing relations to the land made them skeptical of, and of-

---

[31] Lars Levi Laestadius, "Beskrifning öfver några sällsyntare växter från norra delarne af Sverige jemte anmärkningar i växtgeografien" (Description of some rare plants from the northern parts of Sweden and remarks on the geography of plants), *Kungliga Vetenskapsakademiens Handlingar*, 1824, p. 160; continued, *ibid.*, 1826.

[32] E.g., Carl Johan Hartman, "Physiographiska observationer under en resa genom vestliga delarne af Gestrikland, Helsingland och Jämtland" (Physiographic observations during a journey through the western parts of Gestrikland, Helsingland and Jämtland), *Kungliga Vetenskapsakademiens Handlingar*, 1818, pp. 121–60. The majority of Swedish provinces were paid at least some attention, even though the references to Lapland are unusually numerous; A. J. Ståhl, *Register öfver Kongl. Vetenskaps-academiens handlingar ifrån deras början år 1739 till och med år 1825* (Register of the Proceedings of the Royal Swedish Academy of Sciences from their beginning in 1739 through 1825) (Stockholm: Kungliga Vetenskapsakademien, 1831).

[33] For an overview, see Elof Colliander's bibliography, *Kungl. Svenska Vetenskapsakademiens skrifter 1826–1917* (Stockholm/Uppsala: Almqvist & Wiksell, 1917), esp. pp. 261 ff. (botany), pp. 227 ff. (geology).

[34] Pär Eliasson, "Platsens blick: Vetenskapsakademien och den naturalhistoriska resan, 1790–1840" ([With a summary in English] The place's glance: The Royal Swedish Academy of Sciences and scientific travel, 1790–1840) (Ph.D. diss., Department of History of Science and Ideas, Umeå Univ., 1998, publications [skrifter] no. 29).

[35] A good overview of all Swedish expeditions to the Arctic is Gösta H. Liljequist, *High Latitudes: A History of Swedish Polar Travels and Research* (Stockholm: The Swedish Polar Research Secretariat and Streiffert Förlag AB, 1993). For a critical study based on current theories in the history of science, see Urban Wråkberg, *Vetenskapens vikingatåg: Perspektiv på svensk polarforskning 1860–1930* (The Viking raid of science: Perspectives on Swedish polar research 1860–1930) (Stockholm: Center for History of Science, Royal Swedish Academy of Sciences, 1999).

ten victims of, Swedish state policies. The colonizing efforts were spearheaded by parish priests who worked to Christianize the Sami and to establish centers of learning and teaching in the sparsely populated north. Their intelligence activities and substantial topographical reports became, with time, a well-rehearsed *instrumentarium* of Swedish economic and ideological dominance of its territory. A similar strategy was used by the Danes in Greenland.[36]

Sweden, therefore, had much colonial experience when she entered the overseas intelligence business in the 1740s, almost simultaneously with the founding of the Academy of Sciences in Stockholm. This was no coincidence, since the academy quickly became an arena for scientific diplomacy in which Sweden negotiated with foreign powers. After the Nordic Wars, ending in 1718, Sweden was no longer the great European power it had once been, so the academy's forays into scientific exploration did not represent a political or military threat to other nations.

Naturalists, sea captains, politicians, and businessmen all contributed to the scientific intelligence-gathering effort—not least within the Swedish East India Company—as did Swedish diplomats and emissaries abroad. London had a sizeable contingent of Swedes in the eighteenth century—Daniel Solander among them[37]—as did leading Dutch and German cities.[38] Linnaeus was an important actor in all this, both as reigning prince of the botanical stage and as the entrepreneurial genius organizing complex information networks in a peripheral European power.

This was truly a global collection enterprise. The vast itineraries of Linneaus's twenty-some hand-picked students (the "apostles," as he called them, eight of whom died in his service), traveling in known continents over a forty-year period from the mid-1740s, may at first seem like a true manifestation of internationalism. This impression is reinforced by the fact that the Swedish scientists, often young and unmarried, traveled with ships and expeditions dispatched by foreign powers: Dutch (Carl Peter Thunberg), Russian (Johan Peter Falk), English (Daniel Solander and Anders Sparrman, who accompanied Cook's first and second circumnavigations, respectively), and Spanish (Pehr Löfling). The scientific hitchhiking was made possible by two circumstances. First, these young men, as disciples of Linnaeus, were considered estimable collaborators. Second, given Sweden's diminished military powers, granting the disciples space on board implied no security risk.[39] But it was a sensitive task to be in the service of two crowns. For example, twenty-two-year-

[36] Michael Harbsmeier, "Bodies and Voices from Ultima Thule: Inuit Explorations of the Kablunat from Christian IV to Knud Rasmussen" (The Northern Space, Working Paper Series 11, Umeå University, Department of History of Science and Ideas, 1999), and literature quoted therein.

[37] Edward Duyker, *Nature's Argonaut: Daniel Solander 1733–1782: Naturalist and Voyager With Cook and Banks* (Melbourne: The Miegunyah Press, 1998).

[38] Sven Rydberg, *Svenska studieresor till England under frihetstiden* (Swedish study-travels to England in the eighteenth century), Lychnos-Bibliotek 12 (Uppsala: Almqvist & Wiksell, 1951). Instrument makers contributed to scientific traveling, making "industrial espionage" part of the technology transfer from a center, such as London, to Sweden, largely a consequence of the British "tools act" that restricted the distribution of professional intelligence. See Olov Amelin, "Medaljens baksida: Instrumentmakaren Daniel Ekström och hans efterföljare i 1700-talets Sverige" ([With a summary in English] The reverse of the medal: The mathematical instrument-maker Daniel Ekström and his followers in eighteenth-century Sweden) (Ph.D. diss., Uppsala University, 1999), esp. pp. 54–5. See also Duyker, *Nature's Argonaut* (cit. n. 37), chap. 7, on Solander's involvement in industrial intelligence gathering and on his attempts to recruit skilled British artisans to Sweden.

[39] On Linnaeus and his traveling apostles, see Sverker Sörlin, "Scientific Travel: The Linnean Tradition," in *Science in Sweden: The Royal Swedish Academy of Sciences 1739–1989*, ed. Tore Frängsmyr (Canton, Mass.: Science History Publications, 1989), pp. 96–123, and literature quoted therein.

old Pehr Löfling went from Sweden to Madrid in 1751 and later followed José de Iturragias's expedition to South America (1754–1756). Löfling's correspondence typifies the interplay between scientific and national motivations in the Linnean enterprise. In his letters to Linnaeus, Löfling repeatedly stresses his duties as a disciple, a "fellow countryman" and "servant" of his country. He cannot, however, in the service of Spain, promise to send all his findings and specimens directly to Linnaeus. Instead, he will try and send duplicates to Spain in the hope that the Spaniards (as they had indicated) would prove willing to trade them for specimens that Linnaeus could send from Sweden.[40] In other words, scientific results were valuable goods and nothing in their international trade was to be taken for granted.

THE LINNAEAN CONSTRUCTION OF VALUE: SYMBOLIC AND MERCANTILE GOODS

One may wonder what this ideologizing of travel represented, above and beyond Linnaeus's own obvious interest in improving his system of sexual classification and confirming its validity on a global scale. One answer lies in Linnaeus's constant attempts to carve out a special niche for Sweden in the world of science, a status that could substitute for the country's military setbacks. A 1741 speech, followed up by his 1759 dissertation on travel in foreign countries, maintained that Sweden's success as a scientific nation must be built upon the distinctive character of the country—evidently its natural resources, but also its way of pursuing scientific study, which was different from the European way. Swedes could never hope to measure up to Europe: "Where else . . . are there such splendid and more numerous *hospitals* than in London? Where else are more beautiful *surgical operations* performed than in Paris? Where are tidier *anatomical preparations* exhibited than in Leyden? Where are there more *botanical collections* than in Oxford?"[41] Linnaeus's patriotic emphasis on the Swedish way was furthermore based on a fundamental skepticism towards Europe, where his ideas were not catching on as fast as he thought warranted.[42] Linnaeus's sympathy for cosmopolitanism was modest. Instead, his rhetoric underscored the impression that the Swedish nation could gain upon the great scientific powers of the day by following his patriotic research program and utilizing the great resources that Providence had laid down in the Swedish soil.[43]

Another explanation of Linnaeus's project points to its economic significance. As has been convincingly argued of Joseph Banks (1743–1820) as the "autocrat of the philosophers," "presiding genius of exploration," and "imperial impresario," the eco-

[40] Löfling to Linnaeus, 27 Aug. and 1 Oct. 1753, cited in Stig Ryden, *Pehr Löfling: En linnélärjunge i Spanien och Venezuela 1751–1756* (Pehr Löfling: A Linnaean disciple in Spain and Venezuela 1751–1756) (Stockholm: Almqvist & Wiksell, 1965), pp. 160–1.

[41] Carl von Linné, "Om nödvändigheten af forskningsresor inom fäderneslandet" (On the necessity of research travel within the fatherland), Swedish translation of the Latin original, in *Skrifter af Carl von Linné* (Writings by Carl von Linné), 5 vols. (Uppsala: Kungl. Vetenskapsakademien, 1905–1913), vol. 2 (1906), p. 71.

[42] Frans A. Stafleu, *Linnaeus and the Linnaeans: The Spreading of Their Ideas in Systematic Botany, 1735–1789* (Utrecht: A Oosthoek's Uitgeversmaatschappij N. V., and the International Association for Plant Taxonomy, 1971), chaps. 8 and 9.

[43] Among the many examples: Linnaeus to Carl Peter Thunberg, 27 Apr. 1775, Bergianska brevsamlingen, vol. 19, Archives of the Royal Swedish Academy of Sciences, Stockholm (on the glory to be won from a journey to Japan). Pehr Löfling in Spain, about to sail on José de Iturriagas's expedition to South America, is likewise approached by Linneaus: "Dearest, always please me, as the opportunity presents itself, with something remarkable to entertain Europe with . . ." Linnaeus to Löfling, undated, quoted in Ryden, *Pehr Löfling* (cit. n. 40), p. 237.

nomic benefits accruing to Britain were the motivation for the scientific projects that Banks orchestrated in the field and, in particular, from his own "centre of calculation" at 32 Soho Square in London.[44] "His imperial vision," writes David Mackay, "was a profoundly imperialist one," and it governed his carefully organized collecting activities that included more than one hundred collectors from different walks of life and in different regions of the world, and also inspired his grandiose replantation schemes for bread fruit, cotton, tea, and other colonial products.[45] Banks loved new specimens and was profoundly fond of his collections, although he did not always have time to work on them himself. He was passionately engaged in the progress of science as long-term president of the Royal Society. The main quest of his life was for a seamless combination of knowledge and power; the pursuit of science with the eager search for economic utility.[46] Banks did indeed take cues from the Swedish naturalist when he carved out his role in British and global scientific networking.

Linnaeus's career was also marked by a preoccupation with economic utility.[47] The same goes for his program of global scientific data collecting, and he would not restrict himself to a limited repertoire. His interest in utility, and in his own favorite idea of plant transfer, is apparent from the first item on his list of instructions (1745) to the first of his traveling disciples, Christoffer Tärnström, who went to China: "1. To get a Tea bush in a pot, or at least seed thereof, so preserved, as he has been by my given oral instruction." He then went on to the Chinese Mulberry tree, fishes, seeds, et cetera, not to forget "13. Live golden fish for Her Royal Majesty."[48]

For Pehr Löfling, Linneaus formulated a twenty-seven-point instruction that covered mammals, birds, amphibia, fishes, insects, worms, trees, plants, grasses, mosses, soils, cultivation, practical utility, and local nomenclature. The young naturalist was furthermore instructed to study economy, geology, illnesses, and household medicines, to recruit a few Spanish students, and to deliver a "herbarium with

[44] Epithets from Hector Charles Cameron, David Mackay, and others are quoted in David Miller, "Joseph Banks, Empire, and 'Centers of Calculation' in Late Hanoverian London," in Miller and Reill, *Visions of Empire* (cit. n. 2), p. 22.

[45] David Mackay has emphasized this aspect of Banks's work for a long time. See his "Banks, Bligh, and Breadfruit," *New Zealand Journal of History*, 1974, 8; *idem*, "A Presiding Genius of Exploration: Banks, Cook, and Empire, 1767–1805," in Fisher and Johnston, *Captain James Cook and His Times* (cit. n. 15); *idem*, *In the Wake of Cook: Exploration, Science, and Empire, 1780–1801* (London: Croom Helm, 1985); and *idem*, "Agents of Empire: The Banksian Collectors and Evaluation of New Lands," in Miller and Reill, *Visions of Empire* (cit. n. 2), pp. 38–57.

[46] For a general account of Banks's life, work, and significance, see Harold B. Carter, *Sir Joseph Banks, 1743–1820* (London: British Museum of Natural History, 1988). For the contemporary cultural and social context of his work see also John Gascoigne, *Joseph Banks and the English Enlightenment: Useful Knowledge and Polite Culture* (Cambridge: Cambridge Univ. Press, 1994).

[47] Observations on the economical background to Linnaeus's provincial journeys were made at an early stage by economic historian Eli F. Heckscher in "Linnés resor—den ekonomiska bakgrunden" (Linnaeus's travels—The economic background), *Svenska Linnésällskapets Årsskrift*, 1942. Heckscher was followed by his colleague Karl-Gustav Hildebrand: "The Economic Background of Linnaeus: Sweden in the Eighteenth Century," *Svenska Linnésällskapets Årsskrift*, 1978. For a general description of the mercantilist debate in Sweden during the eighteenth century, see Lars Magnusson, *Merkantilism: Ett ekonomiskt tänkande formuleras* (Stockholm: SNS Förlag, 1999), chap. 9; Magnusson's preceding eight chapters are available in English as *Mercantilism: The Shaping of an Economic Language* (London: Routledge, 1994). For a recent contribution, see Lisbet Koerner, "Purposes of Linnaean Travel: A Preliminary Research Report," in Miller and Reill, *Visions of Empire* (cit. n. 2), pp. 117–52.

[48] Carl von Linné, *Bref och skrifvelser af och till Carl von Linné* (Letters and documents of and to Carl von Linné), 2 pts. (Stockholm: Ljus, 1907–1943), pt. 1, vol. 2, pp. 53–4.

all plants of Spain," as well as to achieve a "complete Flora and Fauna of Spain." The most common words in the instruction are "all" and "every." Löfling is supposed to study *all* plants, *all* animals, *every* kind of soil, *every* kind of stone. And as if that were not enough, Linnaeus finally adds, "And ask the reason for everything!"[49]

Löfling was not an adventurer, nor was he a hero. Like the other Linnean disciples, he was a civil servant, obedient to the will of his master. He worked in the tradition of the *Beamten,* the early mandarins of the north European universities and academies. His loyalties were Swedish, not Spanish, but he realized that he had to conform to the will of his employers. He was, in fact, caught up in a diplomatic and scientific contest in which demarcations between Spanish, Portuguese, and Dutch spheres of interests were to be settled. Apart from his commercial instructions, Löfling was also given the task of collecting seeds and plants for the newly founded Botanical Gardens in Madrid and "curiosities" for the planned Royal Cabinet of Natural History.

Only a few years earlier, Pehr Kalm, another young disciple of Linnaeus, had sailed for America, where he spent from September 1748 to February 1751 exploring Delaware, New York, Pennsylvania, and New Jersey. Two northern excursions took him to Quebec and to Ontario, where he saw Niagara Falls. Kalm's trip, organized by Linnaeus and the Royal Swedish Academy of Sciences, was an extensive operation over many years; the four-volume travelogue was not finally published until 1761.[50] Again, mercantilist policies were clearly visible. At the behest of the academy, Linnaeus drew up a prospectus in which he listed all the invaluable plants Kalm might be able to import from the New World: oaks, walnuts, grass seed, species of pharmacological interest, mulberry trees for sericulture. That these specimens would flourish in the Swedish climate was beyond doubt; *Radix ninsi,* a Canadian medical herb, "could thrive as easily as the common columbine in our gardens."[51] Linnaeus repeatedly argued for a northern route, involving Canada and the Hudson's Bay area; indeed, for some time it was planned that Kalm would actually go to Siberia, until America was finally chosen.[52] The reason is obvious: plants that had grown in northern areas could be successfully transplanted to Swedish soil, Linnaeus thought. However, nothing came of these efforts. Only a few ornamental

[49] The instruction is printed in Rydén, *Pehr Löfling* (cit. n. 40). For a detailed presentation of Löfling and his Latin American expedition, see Francisco Pelayo López, ed., *Pehr Löfling y la expedición al Orinoco: 1754–1761* (Madrid: Colección Encuentros, 1990). For further information on Löfling, see Stig Rydén, *Pedro Loefling in Venezuela (1754–1756)* (Göteborg and Madrid: Instituto Ibero-Americano, 1957). An English version of Linnaeus's posthumously printed collection of the papers Löfling left behind, *Iter Hispanicum* (Stockholm, 1758), was published by Jean Bernhard Bossu, "An Abstract of the Most Useful and Necessary Articles Mentioned by Peter Loefling," in *Travels Through That Part of North America Formerly Called Louisiana* (London, 1771), vol. 2, pp. 60–422.

[50] A critical edition in Swedish has been published in four carefully edited volumes (Helsinki, 1966–1988). The classic English translation by John Reinhold Forster has been reissued with an introduction by Ralph M. Sargent: *Travels into North America* (Barre, Mass.: Imprint Society, 1972). See also Adolph B. Benson, ed., *Pehr Kalm's Travels in North America: The English Version of 1770,* 2 vols. (1937; reprint, New York: Dover, 1966).

[51] Carl von Linné, 10 Jan. 1746, in Linné, *Bref och skrifvelser* (cit. n. 48), pt. 1, vol. 2, pp. 58 ff.

[52] Carl Skottsberg, *Pehr Kalm: Levnadsteckning* (Pehr Kalm: Biography), Levnadsteckningar över K. Svenska Vetenskapsakademiens ledamöter 139 (Stockholm, 1951), pp. 306 ff. In January 1745 Linnaeus himself was, according to a letter by Kalm, jumping with delight at the thought of having Kalm go to Siberia.

plants, such as the Virginia creeper, adapted to the northern European climate and remain as lasting reminders of these transplantation attempts.[53]

Kalm also networked on Linnaeus's behalf. He made contacts with several American botanists, among them Cadwallader Colden, who was instrumental in introducing of the Linnean system of classification into America, and who provided hundreds of American plant descriptions to Linnaeus in Uppsala. Another contact helpful to Kalm was John Mitchell, the English-born physician and botanist.[54] Kalm's *Travels* reveal the Linnean style of travel writing: facts first. The traveler was supposed to register anything of importance—and everything seemed equally important. The boring repetitiveness of Kalm's *Travels* made the academic world groan. On the other hand, the factual nature of Kalm's and the other disciples' writings explains their success. The strictly scientific style of their work (by contemporary standards) and their careful professionalism made the disciples diplomatically respectable, apart from giving them credibility in the scientific arena. Their colonial mission was—in contrast to that of the colonial powers such as France, Holland, and England—not to demand land, nor to stake claims on natural resources, but to use science as a substitute for mercantilist import.

Kalm and Löfling, together with all the other Linnean apostles, were humble subordinates on a holy mission with two closely related aims: to promote Linnaeus's system of classification, and to increase Swedish economic prowess. Obviously, there were material objectives, but there was also a symbolic dimension. Science served an important role in the social construction of value. This was manifest in the rhetoric of the day, which stressed the elevated role of the *républiques des lettres*, and in the museums and cabinets filled with treasures from the colonies.[55]

The national and colonial dimensions of eighteenth-century Swedish scientific travel can be better understood when put into this context of symbolic social construction of value. Linnaeus realized this in his intuitive way. American territorial conquest was nothing to him, nor indeed to Sweden. But America could be explored and exploited scientifically and symbolically. Her resources took on the form of specimens and fabulous facts and findings. These represented hard cash on a new market—a virtual stock-exchange of scientific goods—which only gradually appeared as such, veiled as it was by the rhetoric of the aloof and disinterested scientist, member of the republic of letters. Linnaeus sought, in essence, to establish Sweden in general, and Uppsala in particular, as a center of world science, thus bringing Swedish glory to unsurpassed heights, higher even than her great-power status of the seventeenth century. This was the logic of Swedish scientific expansionism in the eighteenth century.

Linnaeus's patriotic and scientific cameralism should also be understood as a key element in Sweden's policy of import substitution. Tea, cochineal, quinine, saffron— all sorts of foreign products could be transferred and adapted to Sweden, Linneaus

[53] Sten Lindroth, *Kungl. Svenska Vetenskapsakademiens historia 1739–1818* (The history of the Royal Swedish Academy of Sciences 1739–1818), 2 pts. (Stockholm, 1967), pt. 1, vol. 2, pp. 50 ff.

[54] The social connections Kalm made can be followed in his extensive travel diary: Pehr Kalm, *Resejournal över resan till norra Amerika*, eds. Martti Kerkkonen and Harry Krogerus, in *Skrifter utgivna av Svenska litteratursällskapet i Finland*, 4 vols. (Helsinki, 1966–1988); on Mitchell, see vol. 1, pp. 277–306, and later volumes *passim*; Colden is described the first time in vol. 2, p. 218.

[55] Oliver Impey and Arthur Macgregor, eds., *The Origins of Museums: The Cabinet of Curiosities in Sixteenth- and Seventeenth-Century Europe* (Oxford: Clarendon Press, 1985). Lorraine Daston, "The Factual Sensibility," *Isis*, 1988, 79:452–67.

believed. This view almost became a national gospel, spread by his disciples. Pehr Kalm, for example, writing from London in June 1748 to his master, demonstrated that he had learned his lesson well: "I do know that *Historia Naturalis* is the base for all Economics, Commerce, Manufactures . . . because to want to progress far in Economics without mature or sufficient insight into Natural History is to want to act as a dancing master with only one leg."[56]

THE COMMAND OF FACTS AND GOODS: THE DANISH EXPEDITION TO ARABIA FELIX

The complex relationship between nationalism and internationalism, science and mercantilism in constructing of colonial scientific networks is also detected in the Danish expedition to the Middle East, 1761–1767. Compared to the large British and French South Seas expeditions, this was modest in both scope and size. The party was small, with only five scientists and a servant, and after an initial journey by sea they traveled mainly over land. Inspiration came from the Dane F. L. Norden's expedition to the Nile Valley in 1737, published as *Voyage d'Egypte et de Nubie* (1755). The brains behind the Arabian expedition, the German philologist and Orientalist Johan David Michaëlis of Göttingen, said of Norden's journey, "What a superb gift from the Danish nation to science!"

Michaëlis had reasons to advocate a Danish expedition. He had identified a number of unresolved scientific problems in Arabia relating to the natural sciences, biblical history, and linguistics. He also had good contacts in the Nordic countries. Aware of the scientific interests of the Danish King Frederik V, he approached the Danish Crown with his idea: an expedition to *Arabia Felix*. He had a clear conception of what he wanted to find, what kind of manuscripts to buy, what places to visit, what questions to ask, and so on; he even had ideas about who should take part. All this he made clear to Johan Hartvig Ernst Bernsdorff, foreign minister to the Danish king.[57]

However, the king did not feel totally secure with only the German professor's word and asked four Danish professors, Peder Ascanius, Johann Christian Kall, Carl Georg Kratzenstein, and Georg Claus Oeder, to comment on Michaëlis's proposal. They accepted it, but with certain telling amendments and clarifications. Kall repeatedly returned to the issue of Danish prestige, an aspect of the journey that was on the mind of Michaëlis as well, who claimed it was "reasonable, that the nation, the King of which himself has arranged this expedition, could herself harvest the fame it produces." Michaëlis also noted that His Majesty could count on receiving "glory in all future school books" for his contributions to this cornerstone of Western knowledge on Arabia.

For Kall, this glory depended on the acquisition of manuscripts. The effort should by no means be hampered by a restricted budget, he declared, lest it not lend the

---

[56] Kalm to Linné, 3 June 1748, in Linné, *Bref och skrifvelser* (cit. n. 48), pt. 1, vol. 8, p. 27.

[57] Documents relating to the expedition, some of which have not previously been published, have been edited and compiled in a magnificent volume: Stig T. Rasmussen, ed., *Den Arabiske rejse 1761–1767: En dansk ekspedition set i videnskabshistorisk perspektiv* (The journey to Arabia 1761–1767: A Danish expedition from the perspective of history of science) (Copenhagen: Munksgaard, 1990). All quotations hereafter refer to this work, unless otherwise stated; translations from the Danish are by the present author. A firsthand account of the expedition is Petrus Forsskål, *Resa till lycklige Arabien: Petrus Forsskåls dagbok 1761–1763* (Journey to Arabia: The journal of Peter Forsskål), ed. Arvid Hj. Uggla (Uppsala: Svenska Linné-Sällskapet, 1950).

Royal Library of Copenhagen "significant new splendour and glory." He also asked if it would not be wise to prohibit the expedition members from sending manuscripts to any place but the Royal Library, the risk being that the Royal Mission at Tranquebar might send the haul to Halle, which, he claimed, had happened before. Since the king of Denmark was the chief financial contributor, His Majesty should also have "the right to the entire result of the journey and to the fame that comes with it, and that the travellers should consequently . . . send all their answers [to questions that would be provided to them by other scientists and scholars, as well as their diaries] . . . to His Excellency Count von Moltke, whereafter His Excellency can forward them to their proper destinations. . . ." Documents from the expedition should be stored in the Royal Library where they "will not get lost, as could otherwise be the case, with a private person in Germany or elsewhere."

These remarks by Kall were emphasized in the final instruction, signed by His Majesty and dated 15 December 1760. This is a remarkably modern, tolerant, and enlightened piece of work, leaning heavily, as one would expect, on Michaëlis and his Danish commentators. The king outlined his egalitarian ideals: all five scientists should enjoy equal status, so that in case of dissent each of their votes would carry equal weight. No formal leader was appointed. It is therefore an error reverberating through the literature on the subject to call it the Niebuhr expedition. The only special mission Carsten J. Niebuhr was given (apart from his scientific duties as mathematician and geographer) was to keep the expedition's purse!

The king was eager to meet the demands of the scientific community. He asked for independently kept journals because "in Europe . . . it is considered most reliable that which is confirmed by several witnesses." That was meant to stress the international character of the enterprise, but what counted was Denmark's standing in the world of science. There should be no doubt that this was a Danish expedition. It was the Danish ambassador at the Sublime Porte in Constantinople who was to undertake local arrangements, issue passports, and so on. Copies of the travelers' journals were to be sent to Europe, addressed to the above-mentioned "Knight of Dannebro, Adam Gottlob Count Moltke." Originals were to be kept with the travelers. As for manuscripts, the king argued along the same lines as Michaëlis and, even more so, Kall. Expanding on the original budget proposed by Michaëlis, the king specifically requested texts in natural history, geography, and history, as well as codices of the Hebraic and Greek bibles. All manuscripts were to be sent to "Our Royal Library in Copenhagen," or to Count Moltke, or be delivered upon the expedition's return to Denmark.

The expedition was to receive and answer questions from academies and learned institutions throughout Europe; explicitly mentioned was the "Académie des Inscriptions et des belles Lettres" of Paris. The procedures for handling answers reveal, however, the national foundations of this seemingly international effort. The answers were to be sent to Copenhagen "*sub sigillo volante*," i.e., unsealed, so that copies of them could be made there. This rule also applied to all sketches, drawings, charts, documents, natural specimens, etc., that were produced alongside the journals. In sum, all new knowledge culled by the expedition, as well as all items collected, must first be brought to Copenhagen; as the instruction put it, "to no other place than right there."

This state of affairs can be interpreted in the following way. There obviously was an international scientific community that was eager to partake in the expedition's results. It was equally obvious that these first-rate scientist-scholars-travelers were cosmopolitan figures. Niebuhr was German. Petrus Forsskål was from Swedish Finland, born in Helsinki, and had studied in Uppsala with Linnaeus and in Göttingen under Michaëlis. Two Danes accompanied the expedition: Frederik Christian von Haven, who had also studied with Michaëlis, and Christian Carl Kramer, a young medical doctor still completing his education. Georg Wilhelm Baurenfeind, draughtsman and artist, was born in Nuremberg but educated in Copenhagen.

There was an apparent risk that the universally valid results might be published or, in the case of the natural objects, stored elsewhere in Europe. Copenhagen was hardly a major scientific center in mid-eighteenth-century Europe, and more attractive fora for scientific work, publication, and exhibition were easy to find. King Frederik V's decision to centralize the collections in the Danish capital was a typical expression of nationalistic thinking. One of science's functions was its value as a showpiece. Science conferred status in the international arena, but such status was first to be acquired on a national level. If not, its function would be dubious in the eyes of kings and princes.

CONCLUSION: THE PATRONS OF NATIONS

As has been noted by numerous authors writing on the history of museums, scientific collections served as tools in the social construction of national consciousness.[58] They also contributed to the construction of value. That which was rare and foreign was treated as valuable, and was lionized and glamorized in museums and other collections.[59] Traditionally, value has been identified primarily with a social elite, the aristocracy, or the business community of overseas trade. William Eisler has demonstrated how art and natural objects from faraway lands acquired a function as gifts and trophies in an intricate European system of values. Hanging on the walls of a director of the Dutch East India Company, axes, shields, or *jakaranda* shrines meant much more than could ever be factually explained. The same thing went for scientific specimens. A plant, a stuffed bird, or a pickled lizard from exotic lands manifested the owner's rich contacts and elevated interests, his importance, wealth, and status. Such collectors were often found in the vicinity of science. Dutch examples include East India director George Clifford, who supported naturalists and artists such as Maria Sibylla Merian, and whose garden Linnaeus described in *Hor-*

---

[58] See, e.g., Karl E. Mayer, *The Art Museum: Power, Money, Ethics* (New York: William Morrow, 1979). See also the contributions in John Gillis, ed., *Commemorations: The Politics of National Identity* (Princeton: Princeton Univ. Press, 1994), particularly Eric Davis, "The Museum and the Politics of Social Control in Modern Iraq," pp. 90–104. For Scandinavia, see Sverker Sörlin, "'Att skapa traditioner som aldrig övergifvas': Artur Hazelius och det nationella arvet under 1800-talet," ('To create traditions that will never be abandoned': Artur Hazelius and the national heritage during the nineteenth century), in *Nordiska museet under 125 år* (The Nordic Museum during 125 years), eds. Hans Medelius, Bengt Nyström, and Elisabet Stavenow-Hidemark (Stockholm: Nordiska museets förlag, 1998), pp. 17–39; Christer Nordlund, "Organising Geology for Display: The Museum of the Swedish Geological Survey 1871–1915," in *Nordisk Museologi* (Nordic Museology), 1999, 7:111–130; Jenny Beckman, *Naturens palats: Nybyggnad, vetenskap och utställning vid Naturhistoriska riksmuseet 1866–1925* (With a summary in English: Nature's palace: Buildings, science and exhibition at the Swedish Museum of Natural History 1866–1925) (Stockholm: Atlantis, 1999).

[59] Daston, "The Factual Sensibility" (cit. n. 55). The theme recurs constantly in Renaissance and early modern cabinets and *Wunderkammern*; see the essays in Impey and Macgregor, *The Origins of Museums* (cit. n. 55). See also Ludmilla Jordanova, "Objects of Knowledge: A Historical Perspective on Museums," in *The New Museum*, ed. Peter Vergo (London: Reaktion Books, 1989).

tus Cliffortianus (1737); and Clifford's colleague Nicolaas Witsen, who for his private collection received some early specimens of the marsupial, soon to become the most popular symbol of the southern continent.[60]

Over time, the status and symbolic value of collections became more closely intertwined with the cultivation of national prestige—both royal and dynastic, but with an emerging sense of the territorial state. Likewise, scientific expeditions and international research programs were means by which a nation and its rulers could acquire acknowledgment and glory. The Danish king wanted Copenhagen to be a center in that sense: internationally renowned but essentially supporting national interests. Recognition was an international matter, not least for the Danish, being more marginal and more dependent on foreign scientific centers for acknowledgment. This explains why Frederik V so strongly urged his travelers to help the international scientific community as much as they could. They were told to stay accurate and zealous, and to adopt all modern methods of science. The king directed Forsskål, the natural historian, to observe precisely the principles outlined by Linnaeus in his dissertation *Instructiones peregrinatoris,* defended by one of his disciples in Uppsala the previous year (1759). As was the case with their Dutch, British, and Swedish counterparts, the Danish expedition could use the East India Company for transportation. The commander of the military ship on which the expedition was traveling to Arabia was instructed to give all possible help to the pursuit of science. National resources—commercial, military, and political—were mobilized in order to further Danish national interests. The *arena* for these efforts, however, was international.

Science was not an activity set above commercial, military, or other interests. However, the rhetorical image of science as neutral ground provided an arena for national performance, as the Dane's Arabia expedition demonstrates. The exploratory quest for knowledge for the sake of mankind became a commonplace in contemporary eloquence, and Cook was hailed (not unlike Newton earlier in the century) as the incarnation of disinterested, self-sacrificing zeal. As eulogized by French astronomer Jean-Sylvain Bailly, "England weeps for a great man; France demands his panegyric. One weeps for him in Tahiti . . . and this sadness is the most beautiful encomium that virtue and genius has ever received."[61] It may be true (though we have noted the exceptions) that the veil of secrecy was being torn off the results of scientific voyages; the transists of Venus are the oft-repeated example of this.[62] But further examples are not quite as frequently given. And while we can show a large

[60] William Eisler, "Terra Australis: Art and Exploration 1500–1768," in *Terra Australis: The Furthest Shore* (Sydney: International Cultural Corp. of Australia, 1988), pp. 15–34, esp. pp. 25–6.

[61] Jean-Sylvain Bailly, "Éloge du Capitaine Cook," in *Discours et mémoires, par l'auteur de l'histoire de l'astronomie* (Paris: de Bure, 1790), vol. 1, p. 348. Barbara Maria Stafford, *Voyage into Substance: Art, Science, and the Illustrated Travel Account, 1760–1840* (Cambridge, Mass.: The MIT Press, 1984), p. 28.

[62] Numa Broc, *La Géographie des philosophes: Géographes et voyageurs français au XVIIIe siècle* (Paris: Ophrys, 1975), pp. 284 ff. However, it is not difficult to find evidence to the contrary. The skeptical French Baron de Gonneville declared in 1783, during French preparations for a follow-up to Cook's voyages, "[The English] . . . pretend to have crossed this region in all directions and, in order to discourage other nations, want to make us believe that everything has been done . . . but let us beware, for this language is practically that of the Dutch, who had good reasons to conceal their possessions near the Southern Lands for their own interest and their own trade, which they wished to extend to the exclusion of other Europeans." The baron feared something like this from the English, and he advocated demarcation lines in order to separate spheres of interest between the European great powers; see John Dunmore, *French Explorers in the Pacific,* 2 vols. (Oxford: Clarendon Press, 1965–1969), vol. 1, p. 252.

number of international scientific networks in operation—between individual scientists and between academies of science, mostly maintained by means of correspondence—we still find emergent nation-state contexts into which results, applications, and glory were fit and funneled.

Finally, in the case of Frederik V, we can also observe the changing role of the patron. As Richard Westfall found in his study of Galileo, there was no ready-to-use system of patronage in the city-state of Florence at the time: "It was a set of dyadic relations betweens patrons and clients, each of them unique."[63] With King Frederik, the patron coincided with the nation's highest representative. This was not always the case, but in countries like France, England, Prussia, and Sweden we nonetheless see central authorities appearing as important patrons of science. In France the state provided the Académie des Sciences with resources; the same was true for the Preussische Akademie der Wissenschaften in Berlin.[64] In Sweden the Academy of Sciences was attributed royal status in 1741 and was given, in 1747, exclusive rights to produce and sell the Swedish almanac, a rich source of income for more than two hundred years. In England and France, rapidly growing military and naval powers were used for scientific purposes. One result of this change (albeit gradual) was the growing independence of science from its old patrons, the churches and courts. Alliance with the emerging nation-state and its institutions meant a new loyalty, but one infinitely more rewarding for science.

There was also a close connection between scientific travel and the commercial capitalism extending around the globe. In Sweden and Holland especially, the East India Companies acted as patrons of science, encouraging scientists to avail themselves of their ships and urging their personnel to help collect specimens and make observations. That was the role of Witsen in Amsterdam, or, for that matter, the directors of the Swedish East India Company in Göteborg. We should note that science in this particular kind of collaboration became interwoven in the same networks as commerce. Science and commerce formed synergetic powers in the process leading towards Western world hegemony.

The fruits of these scientific activities, like the commercial ones, were brought home to the patron's country of origin. In Amsterdam and Uppsala, in London and Copenhagen, scientific objects were transformed into trophies and symbols of glory and enlightenment. Many rose in prestige from this trade: patrons, scientists, centers of learning, even cities that could boast of impressive collections of exotica. But, on an increasing scale, it was also evident that these actors and beneficiaries were all integral components of another, greater quantity: the nation-state, which to a growing extent realized the potentials of science and intelligence as instruments of colonial power.

[63] Richard Westfall, "Science and Patronage: Galileo and the Telescope," *Isis*, 1985, *76*:29. Generally on this topic, see Guy Fitch Lytle and Stephen Orgel, eds., *Patronage in the Renaissance* (Princeton: Princeton Univ. Press, 1981). For a theoretical discussion, see Stephen Turner, "Forms of Patronage," in Cozzens and Gieryn, *Theories of Science in Society* (cit. n. 8), pp. 185–211.

[64] Roger Hahn, *The Anatomy of a Scientific Institution: The Paris Academy of Sciences, 1666–1803* (Berkeley/Los Angeles: Univ. of California Press, 1971), pp. 45–6. "In France . . . there could be no debates about the relationship of state and culture. Knowledge was clearly designed to be the ornament and tool of authority, and new institutions, if they were to survive, had to serve the nation rather than form it." See also James E. McClellan III, *Science Reorganized: Scientific Societies in the Eighteenth Century* (New York: Columbia Univ. Press, 1985).

Science, Technology, and the Spanish Colonial Experience in the Nineteenth Century

Alberto Elena* and Javier Ordóñez**

ABSTRACT

Nineteenth-century Spain—using the resources that remained of its vast empire—struggled to maintain its place as an international power. Following the loss of its colonies on the American continent, however, it could assume only a modest imperial presence. This loss occurred at precisely the time that the country, lagging behind other European powers, was taking its first tentative steps toward industrialization and modernization. The delay in modernizing, along with Spain's still quite modest scientific and technological capacities, made it impossible for the country to become anything more than a spectator during the age of great imperial adventures. As the century closed, Spain, disillusioned, faced the crisis of 1898.

It is true that Spain, convalescing from an illness which had lasted for four centuries, could not now consider providing beings for new societies . . . , because it would first have to colonize itself rather than other unknown and wild regions. Yet, if it is at present unable to found great societies, it should at least endeavor to sow the seeds for them or renounce its hope of immortality . . . ; it should establish here a trading post, there a small farming settlement, set up a military camp in Borneo, in Jolo, on the Red Sea, the islands of the Pacific, the Gulf of Guinea, along the Slave Coast and the Barbary and Saharian coasts, in the Rif, so that, under its present slow development, these outposts might serve as a reserve and pave the way for solving a problem which will be facing us once more in a few years' time.[1]
—Joaquín Costa

* Departmento de Logica, Facultad de Filosophia, Universidad Autónoma de Madrid, 28049 Madrid, Spain.
**Departmento Linguistica y Logica, Facultad de Filosophia y Letras, Universidad Autonoma de Madrid, 28049 Madrid, Spain.

[1] Cierto que España, convaleciente de una enfermedad de cuatro siglos, no puede pensar ahora en dar el ser a nuevas sociedades . . . , que antes debe atender a colonizarse a sí propia que a colonizar regiones ignotas y salvajes. Pero si de momento no puede fundar grandes sociedades, cuando menos debe sembrarlas o renunciar a vivir en la posteridad. . . . Debe establecer ora factorías comerciales, ora pequeños núcleos de población agrícola, ora estaciones militares en Borneo, en Joló, en el mar Rojo, en las islas del Pacífico, en el golfo de Guinea, en la costa de los Esclavos, en las de Berbería y del Sáhara, en el Rif, para que, desarrollándose ahora lentamente, sirvan de reserva y preparen la solución del problema tal como volveremos a plantearlo dentro de pocos años.

Joaquín Costa, *Marina española o la cuestión de la escuadra* (Madrid: Biblioteca Económica, 1913), p. 40.

INTRODUCTION

DURING THE SECOND HALF OF THE NINETEENTH CENTURY, THE industrialized nations of Europe, led by Britain, Prussia, and France, were enjoying enormous economic growth and trading their manufactured goods both at home and throughout the world. After the 1870s, they became the leaders in an aggressive campaign to acquire new colonies, particularly in Africa, which they assumed would make them even more prosperous. Spain, demoralized and impoverished by the loss of most of its former empire, wished to share in this prosperity.

In order to do so, it seemed obvious to many in early nineteenth-century Spain that two changes were essential: (1) Spain must finally catch up with modern science, improve its transportation and communications systems, develop a modern industrial base, and enhance its own productivity and that of the colonies it still possessed; (2) Spain must, like the other great powers, acquire new colonies. The first goal was to be deeply undermined by a trading/manufacturing mentality that Spain had inherited from its earlier empire. The second goal was flawed in itself, since in this era colonies seldom brought the great European powers the expected profits. Wealth came rather from industry and from trading manufactured products widely both at home and abroad. The Spanish, who had gained great wealth from their South American conquests in the sixteenth century, could hardly be expected to anticipate that colonies were now a dubious investment.

ACQUIRING SCIENCE AND TECHNOLOGY

It has been well established that Spain was very slow to embrace modern science and technology, far behind the leading European nations.[2] Spain's leaders were not, however, as has often been asserted, completely indifferent to the need to develop a modern industrial base at home and in the colonies and to enjoy the benefits of modern technology. The Spanish leadership understood that modern science and technology were essential if Spain were to enjoy the prosperity and easier lifestyle that the United States and many nations in Europe possessed by this time.

A traditional Spanish attitude towards trade with the colonies and industry at home, which we might term the *"rentier* mentality," undermined these efforts, as well as the strenuous efforts that were made after 1850 to introduce modern technology, communications, and industry in Spain. In its first colonial experience, Spain had grown used to importing wealth wholesale from the colonies, in a fairly uncomplicated manner: large quantities of silver and other valuable natural products from South America; luxury items such as tobacco, to enhance the lifestyle of the elite; cash in the form of rents, taxes, duties, and fees; foodstuffs; and products that could be used by established industries in Spain—all of these simply came to the metropolis without requiring much cultural change at home to maintain the flow. As for exports, Spain could export many of its agricultural or manufactured products to the colonies, sometimes imposing them through monopolistic trade restrictions. To gain wealth from this economy, Spain had not had to develop many new technologies,

---

[2] See especially Jordi Nadal, *El Fracaso de la revolución industrial en España, 1814–1913* (Barcelona: Ariel, 1975), together with his later reassessment in "Il Fallimento della rivoluzione industriale in Spagna: Un Bilancio storiografico," in Peter Mathias, David S. Landes, S. Berrick Saul *et al., La Rivoluzione industriale tra il Settecento e l'Ottocento* (Milan: Mondadori, 1984), pp. 209–32.

revolutionize its industries, or aggressively seek new markets beyond the colonies themselves. The habit of expecting good economic things to come into Spain from abroad was to have an ironic result. In attempting to modernize its communications and industries between 1850 and 1900 and to develop a scientific and technological infrastructure, Spain and some of its colonies were, in effect, colonized by scientifically advanced countries and failed to develop a firm scientific, technological, or industrial base of its own.

The introduction of the railways in Spain is a good example of the reverse colonization that the country experienced. Enthusiasm for the railways was, if anything, too great in Spain. Prosper Merimée, in a letter to François Arago on 2 February 1859, observed, "Everything has changed in Spain and it has become prosaic and French. People only speak about the railways and industry."[3] The investment of Spanish capital in the first phase of railway construction (1848–1866) was huge; indeed economic historians such as Gabriel Tortella have attributed the failure of the entire project of industrialization in Spain to excessive investment in the railways, which absorbed capital that could have been used to develop other sectors.[4] In his novel *Journey to the Moon* (1865), Jules Verne commented revealingly when he explained Spain's small contribution to the scientific project described in the book: "As for Spain, it was impossible to get together more than one hundred and ten *reales*[,] under the pretext that they had to complete the laying of their railway network." "The truth is that science is not given very much attention in that country. It is still a little backward."[5] The enormous investment made by Spain in railways in the 1860s proved barely profitable and caused the sector itself to go into crisis.[6]

In principle, the railways seemed an irreproachable investment. At first railways had been seen by the industrialized powers primarily as a means for improving regional transportation, but it was not long before their usefulness for ensuring the exploitation of raw materials from the colonies and for penetrating foreign markets was understood.[7] The most recent economic expansion in industrialized countries such as the United States or Prussia had clearly been fueled with wealth generated by the railways.[8] In both cases, however, importing the basic railway technology had required an educational infrastructure to assimilate ideas arriving from Britain and (later) France, together with an industrial and commercial base to supply material for the railways and manufactured goods for commerce.

However, the Spanish case had little in common with the experience of the young United States or Prussia. According to the leading historian of science Juan Vernet, Spain did not even begin to join the wave of scientific and technological develop-

[3] Quoted in M. C. Lécuyer and C. Serrano, *La Guerre d'Afrique et ses répercussions en Espagne: Idéologies et colonialisme en Espagne, 1859–1904* (Paris: Presses Universitaires de France, 1976), p. 21.
[4] Gabriel Tortella, *Los Orígenes del capitalismo en España: Banca, industria y ferrocarriles en el siglo XIX* (Madrid: Tecnos, 1973), pp. 163–200 and 338–40.
[5] Jules Verne, *De la terre à la lune* (Geneva: Fanot, 1979), chap. 12, "Urbi et orbi," pp. 109–10.
[6] Nadal, *El Fracaso de la revolución industrial*, (cit. n. 2), p. 50.
[7] See, for example, Daniel R. Headrick, *The Tools of Empire: Technology and European Imperialism in the Nineteenth Century* (New York: Oxford Univ. Press, 1981), pp. 180–203, and *idem, The Tentacles of Progress: Technology Transfer in the Age of Imperialism, 1850–1940* (New York: Oxford Univ. Press, 1988), pp. 49–96.
[8] See Darwin Stapleton, *The Transfer of Early Industrial Technologies to America* (Philadelphia: American Philosophical Society, 1987), especially pp. 127 ff., and Wolfhard Weber, "Preussische Transferpolitik, 1780 bis 1829," *Technikgeschichte*, 1983, *50*:181–96.

ment until the 1850s.[9] The educational infrastructure was barely established. Although between 1834 and 1855, under Isabel II, engineering colleges had been founded (or refounded, in the case of the Escuela de Caminos, Canales y Puertos [School of Public Works]), it was still too early for these to bear fruit. Spain would be largely dependent upon foreign technologists for building and operating the railways.

Spain's slowness to industrialize meant that the manufactured products required for the railways also had to be purchased abroad. Spain produced little steel and few steel manufactured products. As a result, the building of the railways depended initially on huge purchases of *matériel* from abroad. The fact that no steam locomotive was manufactured in Spain until 1884 is a good indication of the situation. By the same token, the international trade in manufactured products that had so enriched Prussia and the United States with the aid of the railways did not occur in Spain; the country had very few manufactured products to sell.

After the initial railways boom, a shortage of capital occurred, and Spain would have been unable to continue to develop the railways without the aid of foreign investors. After 1866, 60 percent of all the capital invested in the railways was French, and the profits from this investment went to France.[10] This financing did, of course, help to improve the transport network in mainland Spain, but no consideration was given at any time by the Spanish as to whether the network was ever going to be really profitable to them. Instead they gave themselves over to "railways fever" rather as though it were just another luxury.[11] Railways were adopted primarily as a token of progress. The desire for profit was always subordinate to the desire for both internal and international political legitimation of Spain as a modern state.

The modernization of Spain's mining industry is another example of reverse colonization. Mining was one of Spain's traditional sources of wealth, but modernization of the industry in the late nineteenth century enriched other European countries rather than Spain. In their legal modification of the mining industry in 1849 and 1859, as the economic historian Jordi Nadal has pointed out,[12] Spanish legislators were prompted by a desire to make the technologically outmoded industry profitable, but because Spain could not itself provide the necessary technological exper-

---

[9] Juan Vernet, *Historia de la ciencia española* (Madrid: Instituto de España, 1975), p. 284.

[10] See Iván T. Berend and György Ránki, *The European Periphery and Industrialization, 1780–1914* (Cambridge / Paris / Budapest: Cambridge Univ. Press / Editions de la Maison des Sciences de l'Homme / Akadémiai Kiadó, 1982), p. 81. Far from being exceptional, this foreign hegemony likewise influenced other vital sectors such as the chemical industry: see the works of Gabriel Tortella, "La Primera Empresa química española: La Sociedad Española de Dinamita (1872–1896)," in *Historia económica y pensamiento social: Estudios en homenaje a Diego Mateo del Peral*, eds. Gonzalo Anes, Luis Angel Rojo, and Pedro Tedde (Madrid: Alianza, 1983); and Jordi Nadal, "La Debilidad de la industria química española en el siglo XIX: Un Problema de demanda," *Moneda y Crédito*, 1986, *176*, reproduced in *idem, Moler, tejer y fundir: Estudios de historia industrial* (Barcelona: Ariel, 1992), pp. 273–305.

[11] In relation to the building and exploitation of the railways in Spain, see Miguel Artola, ed., *Los Ferrocarriles en España, 1844–1943*, 2 vols. (Madrid: Servicio de Estudios del Banco de España, 1978); Antonio Gómez Mendoza, *Ferrocarril, Industria y Mercado en la modernización de España* (Madrid: Espasa-Calpe, 1989); Francisco Comín Comín, Pablo Martín Aceña, Miguel Muñoz Rubio *et al.*, *Ciento cincuenta años de historia de los ferrocarriles españoles* (Madrid: Fundación de los Ferrocarriles Españoles / Anaya, 1998), vol. 1; *La Era de las concesiones a las compañías privadas*; and Francisco Cayón García, Esperanza Frax Rosales, María Jesús Matilla Quiza *et al.*, *Vías paralelas: Invención y ferrocarril en España (1826–1936)* (Madrid: Fundación de los Ferrocarriles Españoles, 1998).

[12] Nadal, *El Fracaso de la revolución industrial* (cit. n. 2), pp. 90–3.

tise, *matériel,* or capital, it was forced to import all of these from abroad. During the following years Spain indeed experienced a spectacular increase in domestic production of some mineralogical products—especially iron ore but also coal, copper, and mercury—but with little benefit to the economy.

By the beginning of the twentieth century, Spain had become a world leader in the export of iron ore, but at a time when a nation's strength was measured partly by the size of its steel industry, Spain still had only one steel mill, at Vizcaya, in the Basque country. The mill, founded in the early 1880s, produced only a small quantity of steel. The industries that might have required steel for manufactured products did not yet exist, so most of the iron ore was shipped to Spain's European rivals.

Many of the mines were both owned and operated by foreign companies. By 1913, there were more than 130 such companies in Spain, and their combined capital made up more than 50 percent of all of the capital invested in mining in the country. During the last decades of the nineteenth century, the flow of foreign capital ensured that the increase in production fundamentally responded to the need for raw materials of the main colonizing countries (especially Britain, France, and Belgium) rather than to the economic needs of Spain itself.[13] Ironically, Spain, which sought to colonize others, was itself in effect being colonized by the great powers, because of its desire to acquire science and technology.[14]

SCIENCE AND TECHNOLOGY IN THE SPANISH COLONIES

By the mid-nineteenth century, Spain's once-great colonial empire had dwindled to a few widely scattered colonies. Close to home, Spain held the Canary Islands and some territories on the northern shore of Africa. In the New World, only a few Caribbean islands were left; all of the mainland colonies in South and Central America had achieved independence. In Asia, only the Philippines and the Mariana Islands remained, and until 1869, when the Suez Canal was opened, ships could reach these colonies only by making the long journey around the Cape of Good Hope, through seas ruled by other European powers. Communications with these Asian colonies were poor, but Spain's relations with the Caribbean colonies remained close. Not only were they more accessible, but the ruling elite of the colonies, frightened by the successful slave rebellion in Santo Domingo at the end of the eighteenth century, desired the Spanish presence. Independence movements did not gain the active support of the *criollo* population here, as they had in the mainland colonies of South America.[15]

During the late nineteenth century, the colonial empires of the industrialized Euro-

---

[13] See *ibid.,* pp. 87–187; and Rafael Dobado, "La Minería estatal española, 1748–1873," in *Historia de la empresa pública en España,* eds. Francisco Comín and Pablo Martín Aceña (Madrid: Espasa-Calpe, 1991), pp. 89–138.

[14] This echoes the usual concept of Spain as a scientific semi-periphery, a topic first developed by José Sala Catalá, "La Communauté scientifique espagnole au XIXe siècle, et ses relations avec la France et l'Amérique Latine," in *Naissance et développement de la science-monde: Production et reproduction des communautés scientifiques en Europe et en Amérique Latine,* ed. Xavier Polanco (Paris: La Découverte / Conseil de l'Europe / UNESCO, 1990), pp. 122–47.

[15] There is no complete or rigorous history of Spanish colonialism. However, for the underlying ideology, see Elena Hernández Sandoica, *Pensamiento burgués y problemas coloniales en la España de la Restauración, 1857–1887* (Madrid: Univ. Complutense, 1982), and Roberto Mesa, *El Colonialismo en la crisis del XIX español* (Madrid: Ciencia Nueva, 1967), and *idem, La Idea colonial en España* (Valencia: Fernando Torres, 1976).

pean powers grew dynamically, serving as markets for manufactured products and as sources for the raw material required by industry. Wealth flowed in both directions, to and from the metropolis, and it appeared to the great powers, and to most observers, that the colonies were enriching them. This seemed to make investments in science and technology in the colonies not only worthwhile but essential. Improvements in steam navigation and telegraph networks were basic priorities in such colonies in the second half of the nineteenth century.[16]

Because of its technological shortcomings and dependence on foreign technology, Spain was behind in steam navigation and could not send merchant fleets to the Asian colonies. For the same reasons, it had not succeeded in modernizing communications. A contemporary economist wrote, "We must confess, although with profound regret, that Spain is one of the most backward countries in Europe when it comes to communications."[17] For decades, the Madrid government had failed to organize communications within the empire, unable to integrate them into a new and more flexible network.

In order to take a closer look at Spain's efforts to introduce science and technology into its old colonies in the late nineteenth century, two of the most important colonies, the Philippines and Cuba, may serve as examples.

The Philippines

It is surprising how little we know about science and technology in the Philippines during this period, which has not yet received much scholarly attention.[18] As we have seen, the metropolis had few scientists or technologists to send to the colonies, and getting there at all was not easy. Of the few Spanish colonists in the Philippines, many were members of religious orders.[19] Although some teaching orders did help to import scientific and technological knowledge, that was obviously not their first priority.

The Spanish government made limited efforts to increase scientific knowledge in the Philippines. Several scientific institutions were founded during the nineteenth century—the Escuela de Nautica (Nautical College) in 1820, the Escuela de Botanica y Agricultura (College of Botany and Agriculture) in 1858, and the Escuela de Medicina y Farmacia (Medicine and Pharmaceutical College) in 1875—but because of the colony's poor educational system none of them gave either intellectual or material benefits to the Philippines, and scientific practice remained mainly in the hands of Spanish missionaries.[20]

---

[16] See Headrick, *Tentacles of Progress* (cit. n. 7), pp. 18–48 and 97–144.

[17] Brigadier Ramírez Arcas, *Anuario económico-estadístico* (Madrid, 1859), p. 58.

[18] A pioneering and still very useful work is Benito Fernandez Legarda, Jr., "Foreign Trade, Economic Change and Entrepreneurship in the Nineteenth-Century Philippines" (Ph.D. diss., Harvard Univ., 1955). See also Ana María Calavera Vayá, "Inversiones españolas en Filipinas durante el siglo XIX: Estado de la cuestión," in *Extremo Oriente Ibérico: Investigaciones históricas; metodología y estado de la cuestión*, eds. Francisco de Solano, Florentino Rodao, and Luis E. Togores (Madrid: Consejo Superior de Investigaciones Cientificas, 1989), pp. 499–507, and Josep M. Fradera, *Filipinas, la colonia más peculiat: La Hacienda pública en la definición de la política Colonial, 1762–1868* (Madrid: Consejo Superior de Investigaciones Cientificas, 1999).

[19] See Evergisto Bazaco, *History of Education in the Philippines: Spanish Period, 1565–1898* (Manila: Univ. of Santo Tomas Press, 1953).

[20] See Antoni Marimon, *La Política colonial d'Antoni Maura: Les Colònies espanyoles de Cuba, Puerto Rico i les Filipines a finals del segle XIX* (Palma de Mallorca: Documenta Balear, 1994), pp. 97–8.

Although primary education was made compulsory in the 1860s and the level of literacy in the native population was reasonably high, the elementary school system was deficient.[21] Little or nothing was achieved in secondary, and therefore in higher, education. Two church-owned universities, the University of San Ignacio and the University of Santo Tomas, had existed since the seventeenth century, but the former had to shut down in 1768 when the Jesuits were expelled from the colony. A third, the short-lived University of San Felipe, was established in 1707 but closed only two decades later for lack of students. The university curriculum emphasized the humanities, and science was almost completely neglected.

As for communications, the only developments worthy of mention that affected the Philippines were achieved by other nations: the opening of the Suez Canal in 1869 (a French initiative), and the laying of a telegraph cable from Hong Kong by the British in 1880. Throughout the century, the Philippines remained in poor communication with the metropolis.

The only example of scientific excellence during the nineteenth century was the meteorological observatory in Manila, established by the Jesuits in 1865 to help forecast the dreaded *baguios* (typhoons).[22] The observatory was soon receiving requests for exchange of information with others in Asia, especially the one in Hong Kong but also those in Zikawei (in China) and Djakarta. However, even this worthy project became entangled in beaurocratic red tape when the Spanish civil servants tried to charge the observatory the ordinary telegram rate for the transmission of the meteorological information. Although the problem was solved once the Spanish authorities were consulted and the Manila Observatory was officially recognized, official indifference continued to plague its operations.[23]

Industry in the Philippines was increasingly concentrated in the hands of the Chinese, especially towards the end of the century, and the effect of investment from North America was also beginning to be felt.[24] Trade between Spain and the Philippines had little to do with science or technology. Spain was content with the valuable tobacco trade.[25] Export to the colony was quite limited. As a grieving witness wrote

[21] Even so, only in the late nineteenth-century administration of Valeriano Weyler (governor general of the Philippines 1888–1891) was a certain sustained effort noted to encourage the creation of elementary schools, provide them with teaching materials, and efficiently control the salaries of teachers. See Marimon, *La Política colonial d'Antoni Maura* (cit. n. 20), p. 110, and W. E. Retana, *Mando de Weyler en Filipinas* (Madrid: Viuda de M. Minuesa, 1896).

[22] See Miguel Saderra Massó, *Historia del Observatorio de Manila (1865–1915)* (Manila: McCullough, 1915); William C. Repetti, *The Manila Observatory (Manila, Philippines)* (Washington D.C., n. p., 1948); James J. Hennessey, "The Manila Observatory," *Philippine Studies*, 1960, 8:99–120; and John N. Schumacher, "One Hundred Years of Jesuit Scientists: The Manila Observatory, 1865–1965," *Philippine Studies*, 1965, 13:258–86.

[23] See Saderra Massó, *Historia del Observatorio de Manila* (cit. n. 22), pp. 63 ff.

[24] See Edgar Wickberg, "Early Chinese Economic Influence in the Philippines, 1850–1898," *Pacific Affairs*, 1962, 275–85; *idem, The Chinese in Philippine Life, 1850–1898* (New Haven: Yale Univ. Press, 1965); Norman G. Owen, *Prosperity without Progress: Manila Hemp and Material Life in the Colonial Philippines* (Quezon City, Philippines: Ateneo de Manila Univ. Press, 1984); and Benito Fernandez Legarda, Jr., "American Entrepreneurs in the Nineteenth-Century Philippines," *Bulletin of the American Historical Collection*, 1972, 1, 25–52.

[25] See Edilberto C. de Jesus, *The Tobacco Monopoly in the Philippines: Bureaucratic Enterprise and Social Change, 1766–1880* (Quezon City, Philippines: Ateneo de Manila Univ. Press, 1980), especially pp. 161–66; Miquel Izard, "Dependencia y colonialismo: La Compañía General de Tabacos de Filipinas," *Moneda y Crédito*, 1974, 130, 47–89; E. Giralt Raventós, *La Compañía General de Tabacos de Filipinas, 1881–1981* (Barcelona: Compañía General de Tabacos de Filipinas, 1981), pp. 9–99; and Javier López Linage and Juan Hernández Andreu, *Una Historia del tabaco en España*

in 1879, the only business that had increased between Spain and the Philippines in recent times was the export from Spain of playing cards.[26]

Cuba

Cuba was Spain's best-loved colony. Because Cuba wanted Spanish protection and Spanish ships could easily reach Cuba, communications between the metropolis and the colony were good. Although Cuba was much more advanced than the Philippines and far better able to afford modern technology, the *rentier* mentality prevented Spain from helping the colony to modernize.

The primary impetus for acquiring modern technology came from the local elites in Cuba, not from Spain. Thanks to its slave economy, Cuba produced large quantities of sugar, coffee, and tobacco to sell abroad. The capital from the sale of these products produced a buoyant economy and a *criollo* bourgeoisie that was not only anxious to acquire new technology but could afford it. The first railway in Latin America was built between Havana and Güines in 1837, eleven years before the first railway opened in Spain.[27] For help the *criollo* turned first to the British and later to American engineers. By the time the line between Barcelona and Mataro opened in 1848, more than 600 kilometers of rails had already been laid in the colony. The bourgeoisie also pioneered the introduction of steam-engine technology into sugar production.[28]

Spain had little to sell to the colony. In principle, Cuba absorbed excess wheat from the peasant workers of Castille, and from this point of view it could be said that it gave money to Spanish agriculture. But in manufactured goods, only the Catalan textile industry benefited from the Cuban market, thanks to a monopoly imposed

---

(Madrid: Ministerio de Agricultura, Pesca y Alimentación, 1990), especially pp. 86–101 and 150–77. In 1870 the Spanish overseas minister himself, Segismundo Moret, openly revealed to Parliament his frustration over the fact that trade with the Philippines brought Spain no profit, except for the tobacco trade: "The Philippines give nothing, or virtually nothing, to Spain, if we discount the excellent tobacco products sent from their factories. And while this is the case . . . it can be seen that Spain's colonization does not advance, trade does not prosper, and wealth is not developed; in a word, Spanish civilisation would not seem even to have taken possession of those lands." Quoted in Mesa, *La Idea colonial en España* (cit. n. 15), p. 93.

[26] Carlos Recur, *Filipinas: Estudios administrativos y comerciales* (Madrid: Ramón Moreno & Ricardo Rojas, 1879), p. 111.

[27] See Berta Alfonso Ballol, Mercedes Herrera Sorzano, Eduardo Moyano *et al., El Camino de hierro de La Habana a Güines, primer ferrocarril de Iberoamérica* (Madrid: Fundación de los Ferrocarriles Españoles, 1987). See also Oscar Zanetti and Alejandro García, *Caminos para el azúcar* (Havana: Editorial de Ciencias Sociales, 1987), and Eduardo L. Moyano Bazzani, *La Nueva Frontera del azúcar: El Ferrocarril y la economía cubana del siglo XIX* (Madrid: Consejo Superior de Investigaciones Científicas, 1991).

[28] Concerning the cultivation and trading of sugar in Cuba, together with its significant economic importance in the relationship with metropolitan Spain, see the works of Manuel Moreno Fraginals, *El Ingenio: Complejo económico social cubano del azúcar* (Havana: Editorial de Ciencias Sociales, 1978), and *Cuba / España, España / Cuba: Historia común* (Barcelona: Crítica, 1995), as well as those by Manuel Martín Rodríguez, *Azúcar y descolonización* (Granada: Univ. de Granada, 1982), *idem,* "Del Trapiche a la fábrica de azúcar, 1779–1904," in *La Cara oculta de la industrialización española,* eds. Jordi Nadal and Jordi Catalán (Madrid: Alianza, 1990), pp. 43–97, and *idem,* "El Azúcar y la política colonial española (1860–1898)," in *Economía y colonias en la España del 98,* ed. Pedro Tedde (Madrid: Síntesis / Fundación Duques de Soria, 1999), pp. 161–77. For Puerto Rico, see Andrés Ramos Mattei, *La Sociedad del azúcar en Puerto Rico (1870–1910)* (Río Piedras: Univ. de Puerto Rico, 1988).

on the colony: when this monopoly ended after Spanish–American War, Catalonia suffered a serious crisis from which it would take years to recover.[29]

Despite the early introduction of railways, the same shortsightedness that had limited the value of new technology in Spain was evident in Cuba. Technology was used to support traditional agriculture rather than to develop other modern industries or trade, and much of the wealth derived from sales of the major agricultural products was sent abroad to purchase technology and manufactured products. Even the sale of sugar sent capital outside the Spanish Empire. The largest customer for Cuban sugar was not Spain but the United States, which consumed 75 percent of the total output by 1860 and 85 percent by 1890, making the island economically dependent upon its neighbor.[30]

As a result, Cuba was unable to engage in the kind of profitable domestic and international trade that might have brought industrial and economic development on the island and encouraged it in the metropolis. Science and technology made only limited advances in Cuba during the nineteenth century, and because of its dependence on the sale of agricultural products, Cuba became an economic colony of the United States.

SCIENCE AND TECHNOLOGY IN THE SECOND COLONIAL VENTURE

From the 1870s the industrialized nations competed to acquire colonies in Africa. Spain, ambitious to share in the prosperity of its neighbors, did not wish to be left out. As the Spanish invasion of Morocco began in 1859, the newspaper *La Discusión* enthusiastically proclaimed, "The future of our fatherland is on the beaches of Africa."[31]

Spain had long had close ties with northern Africa but, from the fifteenth century, had felt it sufficient to hold a few outposts on the northern coast. These were valued not only for strategic reasons but because they helped to ensure safe passage through the western Mediterranean to the Canary Islands, vital to trade with the New World. Leaving aside Manuel Iradier's incursions farther south in the Gulf of Guinea territories (1875 and 1886), in the late nineteenth century Spanish colonization in Africa was confined to the northern coast—Morocco and the western Sahara.[32]

Public rhetoric in Spain in 1859 presented the invasion of Morocco as a selfless act, a mission to bring civilization to a benighted country. Both parliamentary speeches and more mundane articles in the press were full of fervent patriotic statements justifying the invasion. Naturally, however, the war between Spain and Mo-

[29] See Jordi Maluquer de Motes, "El Mercado colonial antillano en el siglo XIX," in *Agricultura, comercio colonial y crecimiento económico en la España contemporánea*, eds. Jordi Nadal and Gabriel Tortella (Barcelona: Ariel, 1974), pp. 322–56.

[30] Martín Rodríguez, "El Azúcar" (cit. n. 28), pp. 166–8.

[31] *La Discusión*, 6 Dec. 1859. Quoted in Lécuyer and Serrano, *La Guerre d'Afrique* (cit. n. 3), p. 62.

[32] Manuel Iradier Bulfy, *Africa. Viajes y trabajos de la Asociación Euskara "La Exploradora" (reconocimiento de la Zona Ecuatorial de África en las costas de occidente: sus montañas, sus ríos, sus habitantes, clima, producciones y porvenir de estos países tropicales. Posesiones españolas del Golfo de Guinea. Adquisición para España de la nueva provincia del Muni)* (Vitoria: Asociación Euskara para la Exploración y Civilización del Africa Central "La Exploradora," 1887). No recent studies have replaced those of José María Cordero Torres, *Iradier* (Madrid: Instituto de Estudios Políticos, 1944), and Ricardo Majó Framis, *Las Generosas y primitivas empresas de Manuel Iradier Bulfy* (Madrid: Consejo Superior de Investigaciones Científicas, 1954), however ideologically biased they are.

rocco can in no way be understood to have been, as Raymond Carr claimed in his influential *Spain, 1808–1939* (1966), "a classic example of a war of honour unsupported by economic interest."[33] Although economic intentions were not definitive at the outset, the relationship between Spain and Morocco over the last four decades of the nineteenth century was unquestionably marked by economic interests. Spain saw in the kingdom of Morocco a vast, potential market and excellent possibilities for investment. When the experience failed, it served to emphasize the manifest economic supremacy of France and Britain, countries that would exert a shared influence in the region.[34]

Whatever the reasons, the campaign of 1859 marked the beginning of Spain's presence in northern Africa, a presence that would last a century, if we consider that Morocco became independent with the end of the protectorate in 1956. (It lasted even longer if we include Equatorial Guinea and the Sahara.)[35] For those who thought they were on the threshold of a second great Spanish Empire, the enterprise was frustrating: "a large war and a small peace" was the usual lament. For the head of government in Morocco, General Leopoldo O'Donnell, economic factors were decisively important. One of his most ambitious projects consisted of building roads between Ceuta and Tangiers and between Ceuta and Tetúan, paid for by the Moroccan treasury but built under the supervision of Spanish engineers.[36] However, O'Donnell's ideas brought no profit to Spain. When peace was signed, the great beneficiary was Britain. It was the British who financed the war debt and who strengthened their presence in the Straits (a presence that gradually increased throughout the nineteenth century as the strategic value of Gibraltar grew with the opening of the Suez Canal). A British steamship line was inaugurated between Gibraltar and Tangiers in 1857, in open competition with the French ships that had been sailing from Marseilles since 1852, and British textiles were introduced into the kingdom of Morocco. This influence continued throughout the century; in 1875, Sultan Muley Hassan, ruler of Morocco, authorized the laying of a telegraph cable between Gibraltar and Tangiers by the British Eastern Telegraph Company.

Spanish euphoria gave way to the realization that Spain's influence over Morocco could not be sustained by the force of arms alone. Following the treaty of 1860, the presence of civil servants, soldiers, and the clergy in Morocco contributed to a greater cultural and social interest in the country, and many publications appeared on the Moroccan languages, culture, and geography. As a result of this interest, a

---

[33] Raymond Carr, *Spain, 1808–1939* (Oxford: Clarendon, 1966), p. 261.

[34] With regard to European penetration in Morocco, the locus classicus is still Jean-Louis Miège, *Le Maroc et l'Europe*, 4 vols. (Paris: PUF, 1961–1963).

[35] On the Spanish protectorate in Morocco, see the works of Victor Morales Lezcano, *El Colonialismo hispanofrancés en Marruecos (1898–1927)* (Madrid: Siglo XXI, 1976), pp. 21–150, and *idem, España y el Norte de Africa: El Protectorado en Marruecos (1912–1956)* (Madrid: Univ. Nacional de Educación a Distancia, 1984); on the Spanish Sahara, see John Mercer, *Spanish Sahara* (London: George Allen & Unwin, 1976); Juan Bautista Villar, *El Sahara Español: Historia de una aventura colonial* (Madrid: Sedmay, 1977); José Ramón Diego Aguirre, *Historia del Sahara Español* (Madrid: Kaydeda, 1988); and Javier Morillas, *Sahara Occidental: Desarrollo y subdesarrollo* (Madrid: Prensa y Ediciones Iberoamericanas, 1988); on Equatorial Guinea, see Mariano L. de Castro and María Luisa de la Calle, *Origen de la colonización española en Guinea Ecuatorial (1777–1860)* (Valladolid: Univ. de Valladolid, 1992); and Mariano L. de Castro and Donato Ndongo-Bidgoyo, *España en Guinea: Construcción del desencuentro, 1778–1968* (Madrid: Sequitur, 1998).

[36] See Manuel Fernández Rodríguez, *España y Marruecos en los primeros años de la Restauración (1875–1894)* (Madrid: Consejo Superior de Investigaciones Científicas, 1985), p. 14.

pro-Africanist movement arose in Spain, which was consolidated in 1883 by the founding of the Sociedad de Africanistas y Colonistas (Society of Africanists and Colonists) with the support of the Sociedad Geográfica de Madrid (Geographical Society of Madrid).[37] Iradier's journeys to Guinea in 1884, and the journeys of Alvarez Pérez, Cervera, Quiroga, and Rizzo to the Sahara in 1886, were undertaken with the support of the Sociedad de Africanistas.

Free from militaristic overtones, the Africanist programme formed a part of the contemporary European movement that looked upon colonization as a civilizing mission. Despite this fragile commitment, Joaquín Costa and Francisco Coello (the politicians who were the main ideologues of colonialism) were highly critical of the Spanish bourgeoisie's inability to imitate the great colonial powers, Britain and France. The Africanists were aware of the fact that after the war with Morocco, Spanish industry did not follow the soldiers, and warships were never replaced by merchant ships from Andalusia and Catalonia.[38] As a result of the Africanist movement, and with the initiative of the Spanish minister of state, Segismundo Moret, chambers of commerce were opened in Morocco in 1885, creating a framework for economic activity in the country. The Catalan bourgeoisie, the main beneficiary of commerce with Spain's American colonies, also looked for ways to improve economic penetration in northern Africa, imitating not only the policies of other countries but also their products: British textiles were unquestionably the best example, but others were not afraid to imitate the sparkling wines of Champagne or Cuban rum. In 1886, in an attempt to keep up with the British and French, the Compañía Transatlántica (Transatlantic Company) established a regular shipping line between Spain and Morocco to ensure the steady flow of Catalan products to the country.[39] However, the civil wars and the economic crisis in Spain that lasted from 1873 to 1896 substantially decreased the efficiency of Catalan trade with Morocco, and it became restricted, finally, to the Caribbean. In the end, Spain would be one of the few European countries whose trade balance with Morocco was, during the last decades of the nineteenth century, always unfavorable.

It was probably in northern Africa that Spain felt most keenly its inability to carry out a modernizing colonial policy. The French–British hegemony aborted the few Spanish commercial initiatives that existed, and some of its most ambitious projects. This was the case, for example, with the projected Tangiers–Tetúan railway, which was to have been built by Spain under the guarantee of the Sublime Porte. The project came up in 1882, with a proposal by the Spanish ambassador in Constantinople, Juan Antonio Rascón, who emphasized with typically Africanist rhetoric the traditional ties between Spain and Morocco. Rascón was careful to point out that the most powerful reason for supporting the building of the railway was not to favor Spanish penetration in the area, but merely to safeguard Morocco's interests against the ambitions of countries that were traditionally strangers in the area (France, Britain, and Italy). The railway was never built. Large civil works were traditionally carried out on the basis of prior agreement between the powers involved, and the

---

[37] See José Antonio Rodríguez Esteban, *Geografía y colonialismo: La Sociedad Geográfica de Madrid (1876–1936)* (Madrid: Univ. Autónoma de Madrid, 1996), pp. 85–93.

[38] See, for example, the significant speech given by the economist Gabriel Rodríguez in the Teatro Alhambra, Madrid, on 30 Mar. 1884, published in the volume *Intereses de España en Marruecos* (Madrid: Instituto de Estudios Africanos / Consejo Superior de Investigaciones Científicas, 1951), p. 39.

[39] See Lécuyer and Serrano, *La Guerre d'Afrique* (cit. n. 3), pp. 273–5.

government in Madrid declined to promote an initiative that might upset the great powers. In the building of the Tangiers–Tetouan railway, Spain would have openly depended on British technology and was not able to ensure the maintenance of the line.[40]

CONCLUSIONS

The optimistic expectations of economic expansion that had motivated the Spanish to introduce science and technology and to acquire new colonies after 1850 ended badly with Spain's defeat in the Spanish–American War. Once again Spain became demoralized and pessimistic about the future.[41] Many Spaniards pointed to the country's patent scientific and technical backwardness as an essential key to the disaster. On 23 June 1899, deputy Eduardo Vincenti, commenting on the defeat by the United States, told the Parliament,

> I shall not tire of repeating that leaving on one side false patriotism, we should take inspiration from the example given us by the United States. The United States won not only because it was the stronger, but also because it was better instructed and educated, yet under no circumstances was it more valiant. No Yankees bared their breasts before our troops; our army had to confront a machine invented by some electrician or some mechanic. There was no fight. We were beaten in the laboratory and the offices, but not on the sea or on dry land.[42]

The same might have been said about Spain's general failure to achieve the prosperity enjoyed by the industrialized nations or to bring that prosperity to its colonies. Crippled by the *rentier* mentality inherited from its first empire, Spain could not fully integrate science and technology into its own culture, and it also became incapable of modernizing the colonies.

As a final irony, it now seems probable that trade with colonies played little role in the prosperity of the imperialistic nations that Spain wished to emulate. The great powers themselves believed that the colonies were bringing them wealth, and Spain,

[40] See Fernández Rodríguez, *España y Marruecos* (cit. n. 36), pp. 102–5. An illuminating comparison can be made by looking at the strictly contemporary case of the building of the railway between Algeria and Tunisia, recently studied by Mohamed Lazhar Gharbi, *Impérialisme et réformisme au Maghreb: Histoire d'un chemin de fer algéro-tunisien* (Tunis: Cérès, 1994).

[41] The recent commemoration of the first centenary of the crisis has produced some interesting publications on this crucial period of Spanish history. Among the most relevant for our purposes are Juan Pan-Montojo, ed., *Más se perdió en Cuba: España, 1898 y la crisis de fin de siglo* (Madrid: Alianza, 1998); Pedro Laín Entralgo and Carlos Seco Serrano, eds., *España en 1898: Las Claves del desastre* (Barcelona: Galaxia Gutenberg / Círculo de Lectores, 1998); Santos Juliá, ed., *Debates en torno al 98: Estado, sociedad y política* (Madrid: Comunidad de Madrid, 1998); Roberto Mesa, ed., *Tiempos del 98* (Seville: Fundación El Monte, 1998); Raymond Carr, María Dolores Elizalde, Carlos Malamud et al., *Imágenes y ensayos del 98* (Valencia: Fundación Cañada Blanch, 1998); and the excellent monographic issue of *Revista de Occidente*, Mar. 1998, 202–3, devoted to the subject under the heading *1898: ¿Desastre nacional o impulso modernizador?* Of particular interest for our purpose is Tedde, *Economía y colonias en la España del 98* (cit. n. 28).

[42] Quoted by Antonio Moreno and José Manuel Sánchez Ron, "La Ciencia española contemporánea: Del Optimismo regeneracionista a la exaltación patriótica," in *Mundialización de la ciencia y cultura nacional*, eds. Antonio Lafuente, Alberto Elena, and María Luisa Ortega (Aranjuez / Madrid: Doce Calles / Univ. Autónoma de Madrid, 1993), p. 392. For a fine overview of Spanish backwardness in science at the time, see José Manuel Sánchez Ron, "Física, matemáticas y la derrota de 1898," *Arbor*, 1998, *630*:279–94.

perhaps in part because of its earlier colonial experience, shared that belief.[43] The same view was held by authoritative commentators on nineteenth-century imperialism, such as Lenin and J. A. Hobson. In the twentieth century, the accusation that in losing the colonies the government had destroyed the Spanish economy made effective propaganda for the dictators Primo de Rivera and Franco.[44] Until rather recently, the theory that colonies were profitable was also the accepted historical view.

Economic historians who have restudied the issue, however, have demonstrated that during the last third of the nineteenth century colonies were seldom profitable for their possessors.[45] Britain was perhaps the only power that truly managed to obtain a substantial profit from its colonies.[46] If that is the case, then the emphasis on trading with colonies simply distracted Spain's attention from a much more important source of prosperity: domestic and international trade in manufactured goods. Spain's *rentier* mentality and its failure to grasp the crucial relationship between science, technology, and capitalism made the hope of participating in the prosperity of its European neighbors in the second half of the nineteenth century an impossible dream.

[43] There being still no comprehensive reinterpretation of the Spanish colonial model of the nineteenth century, see Candelaria Sáiz Pastor, "El Modelo colonial español durante el siglo XIX: Un Debate abierto," *Estudios de Historia Social*, 1988, *44–7*:651–5; Dolores Elizalde, "Modelos de imperio," in *Viejos y nuevos imperios: España y Gran Bretaña, siglos XVII–XX*, eds. Isabel Burdiel and Roy Church (Valencia: Episteme, 1998), pp. 83–93; and Josep M. Fradera, "La Política colonial española del siglo XIX: Una Reflexión sobre los precedentes de la crisis de fin de siglo," *Revista de Occidente*, 1998, *202–3*:183–99.

[44] An important contribution is, however, that made by Jordi Maluquer de Motes, "El Impacto de las guerras coloniales de fin de siglo sobre la economía española," in Tedde, *Economía y colonias en la España del 98* (cit. n. 28), pp. 101–21, where the author, disagreeing with the ideologies of Primo de Rivera and Franco, convincingly argues that the loss of colonial markets did no serious damage to the Spanish economy.

[45] See, among others, John Gallagher and Ronald Robinson, "The Imperialism of Free Trade," *Economic History Review*, 1953, *6*:1–15; David K. Fieldhouse, *Economics and Empire, 1830–1914* (London: Weidenfeld & Nicolson, 1973); Paul Bairoch, "Le Bilan économique du colonialisme: Mythes et réalités," *Itinerario*, 1980, *1*:29–41; Patrick O'Brien, "European Economic Development: The Contribution of the Periphery," *Economic History Review*, 1982, *35*: 1–18; A. M. Ekstein, "Is There a Hobson–Lenin Thesis on Late Nineteenth-Century Colonial Expansion?" *Economic History Review*, 1991, *44*:297–318; Josep M. Fradera, "La Experiencia colonial europea del siglo XIX: Una Aproximación al debate sobre los costes y beneficios del colonialismo europeo," in *Europa en su historia*, ed. Pedro Ruiz Torres (Valencia: Univ. de Valencia, 1993).

[46] The classic quantitative estimates are still those of Lance E. Davis and Robert A. Huttenback, *Mammon and the Pursuit of Empire: The Political Economy of British Imperialism, 1860–1912* (Cambridge: Cambridge Univ. Press, 1986), and Patrick O'Brien, "The Costs and Benefits of British Imperialism, 1846–1914," *Past and Present*, 1988, *120*:163–210.

Part II: Milieux and Metaphor

The Colonial World as Geological Metaphor: Strata(gems) of Empire in Victorian Canada

*Suzanne Zeller**

ABSTRACT

This essay highlights the experience of British North America during the Victorian age as a case study in the complex relationship between science and empire. It analyzes the development of geology as a Victorian (and highly imperialistic) science in colonial Canada, in three chronological phases. During each, the study of geology helped to structure an imperial-colonial dialogue that reflected changing mutual perceptions and relationships. As colonists undertook geological exploration and interpretation, they modified imperial institutions to suit their goals. They also absorbed the means by which to colonize other peoples and regions. In this sense, the quintessential Victorian science exerted powerful cultural influences, transforming new landscapes into readable texts that redefined the future.

INTRODUCTION

THEORETICAL MODELS, LIKE CINDERELLA'S GLASS SLIPPER, can represent exclusive, even hegemonic, constructions. In the generation since George Basalla and others set about codifying the modern diffusion of European science, their work has stimulated important responses and refinements.[1] For example, their model and its variants did not effectively distinguish welcome from unwelcome imperial incursions, or shared values from imposed ideologies. Ques-

* Wilfrid Laurier University (WLU), Waterloo, Ontario, Canada N2L 3C5.

The author acknowledges research funding from WLU Operating Funds and the Social Sciences and Humanities Research Council of Canada Institutional Grant, and thanks Roy MacLeod, Hugh Torrens, Ian Brookes, John Warkentin, and Graeme Wynn for valuable comments.

[1] See I. Bernard Cohen, "The New World as a Source of Science for Europe," *Actes du IXe Congrès International d'Histoire des Sciences* (Barcelona-Madrid, 1959), pp. 96–130; Donald Fleming, "Science in Australia, Canada, and the United States," *Actes du Xe Congrès International d'Histoire des Sciences*, vol. I (Paris: Hermann, 1964), pp. 179–96; George Basalla, "The Spread of Western Science," *Science*, 1967, *156:*611–22; Raymond Duchesne, "Science et société coloniale: Les Naturalistes du Canada français et leurs correspondants scientifiques (1860–1900)," *History of Science and Technology in Canada Bulletin*, 1981, *5:*99–139; Roy MacLeod, "On Visiting the 'Moving Metropolis': Reflections on the Architecture of Imperial Science," *Historical Records of Australian Science*, 1982, *5:*1–16; Ian Inkster, "Scientific Enterprise and the Colonial 'Model': Observations on Australian Experience in Historical Context," *Social Studies of Science*, 1985, *15:*677–704; Nathan Reingold and Marc Rothenberg, eds., *Scientific Colonialism: A Cross-Cultural Comparison* (Washington: Smithsonian Institution, 1987); Rod W. Home, ed., *Australian Science in the Making* (Cambridge: Cambridge Univ. Press, 1988).

tionable inferences resulted when historical transitions from "colonial" to "independent" science failed to materialize along the path that Basalla's American perspective
had suggested was natural. The nineteenth-century history of geology in British
North America offers a case in point.

In rejecting the American Revolution and adapting British Loyalism to North
American exigencies, the colonial societies of Upper and Lower Canada (later Ontario and Quebec), Nova Scotia, New Brunswick, Prince Edward Island, and Newfoundland developed as complex cultural amalgams. Within them, Anglophone,
Francophone, and Aboriginal peoples manifested sharply contrasting perceptions of
their own status as British subjects. An English-speaking majority outside of Lower
Canada (and a minority inside) sought to share the material progress—with its links
to science, including geology—attained by Britain and the United States. By the
1820s and 1830s, the issue of public funding for mineralogical surveys in British
colonies keen to assess their chances of an industrial future straddled a growing
divide between colonial Tory and Reform politics. Meanwhile, a French-speaking
majority of conquered people in Lower Canada also recoiled from outside threats
to their community. For these heirs to the flourishing scientific tradition that had
valued minerals among New France's natural resources, proposals for public surveys
under the British regime smacked at first of foreign, capitalistic interference. Some
bourgeois *Canadiens* nevertheless came to admire Victorian culture, practicing geology as amateur naturalists.[2] Meanwhile, neither Anglophone nor Francophone societies gave native peoples much reason to embrace European imperialism in any of its
available forms. Native and Métis guides routinely assisted geological explorers
across British North America in ways untold in written documents.

Colonial orientations towards British imperialism, with geology among them,
fanned out widely. While London remained the political metropolis, as well as the
home of the Geological Society of London (GSL), until the 1830s many Anglophone
British North Americans also looked to Edinburgh for geological direction. Even as
London's scientific shadow lengthened with the creation of the Geological Survey
of Great Britain in 1835, American surveys of longer standing were studying—and
naming—contiguous formations, offering alternatives to colonial dependence upon
the mother country.[3] By the 1850s, a growing Canadian expansionist movement had
implicated geology more deeply in the fray that produced Canadian Confederation
in 1867. While some colonial geologists saw themselves as national, imperial, and
North Atlantic linchpins, others quite purposefully retained their regional emphases.

As the "moving metropolis" played out dynamic colonial effects, geology itself
was changing. During the nineteenth century, three phases of geological thought
overlapped political developments in British North America. The first, an explor-

---

[2] Michael D. Stephens and Gordon W. Roderick, "Science, Self-Improvement and the First Industrial Revolution," *Annals of Science*, 1974, *31*:463–70; Luc Chartrand, Raymond Duchesne, and Yves
Gingras, *Histoire des sciences au Québec* (Montreal: Boréal, 1989), pp. 42, 56–8, 146–7; Suzanne
Zeller, *Inventing Canada: Early Victorian Science and the Idea of a Transcontinental Nation* (Toronto: Univ. of Toronto Press, 1987), chaps. 1–2.

[3] Ian Inkster and Jack Morrell, eds., *Metropolis and Province: Science in British Culture, 1780–
1850* (London: Hutchinson, 1983), pp. 11–54, 151–78, 91–119; Suzanne Zeller, "Nature's Gullivers
and Crusoes: The Scientific Exploration of British North America, 1800–1870," in *North American
Exploration*, ed. John Logan Allen, 3 vols. (Lincoln, Ne.: Univ. of Nebraska Press, 1998), vol. 3: *A
Continent Comprehended*, pp. 194–5; and William E. Eagan, "The Canadian Geological Survey:
Hinterland Between Two Metropolises," *Earth Sciences History*, 1993, *12*:99–106.

atory phase from 1815, brought the colonies into Europe's geological sights. This approach proceeded at its peril, as New World evidence first confirmed and then transformed Old World assumptions. The second, an organizational phase after 1830, inducted colonial geologists into the social circles of a dynamic, expansive, increasingly entrepreneurial geological imperialism. They learned their lessons well. The third, a consolidative phase after 1855, saw these lessons applied to broaden Canadian political as well as scientific horizons. Confederation's aftermath saw a second generation of professional geologists, trained as modern scientists, deployed to master an exponentially enlarged transcontinental territory. While these phases outline Victorian geology's particular historical path in Canada, their pattern implies no necessary trajectory from colonial to independent science. Instead, it elucidates the rise of secondary imperialisms, emulations shaped in their turn, too, by collaboration and resistance.

NEPTUNE'S PLAYGROUND

During the nineteenth century, British North America edged increasingly into Europe's geological gaze. In the east, Moravian missionaries sent Labradorean rocks to Britain for analysis. In the west, fur traders assembled a continental overview in exploring Rupert's Land—the enormous territory drained by Hudson Bay—and regions beyond for the Hudson's Bay Company (HBC) and its rivals. While this areal tradition welcomed mineral reports, a growing theoretical synthesis brought useful stratigraphical and palaeontological tools.[4] Recent scholarship also reconfirms the place of A. G. Werner (1750–1817), of the Freiberg School of Mines, in the cornerstone of geology's conceptual foundations. Werner's "geognosy" studied the earth's crust systematically, encouraging active, organized field investigations. As Rachel Laudan explains, Wernerian approaches varied causally, accentuating minerals' external characteristics; and historically, keying "formations" to ordered strata and their fossils. Drawing from chemical cosmogony, Werner privileged Neptunist interpretations of water's mechanical/chemical roles in depositing geological strata over Vulcanist emphases upon fire in the earth's interior.[5]

Werner's teachings reached British North America indirectly, echoing debates be-

---

[4] Rev. [Henry] Steinhauer, "Notice Relative to the Geology of the Coast of Labrador," *Geological Society of London Transactions*, 1814, 2:488–94; Hugh S. Torrens, "The Transmission of Ideas in the Use of Fossils in Stratigraphic Analysis from England to America 1800–1840," *Earth Sci. Hist.*, 1990, 9:108–17; Trevor H. Levere, *Science and the Canadian Arctic* (Cambridge: Cambridge Univ. Press, 1993) pp. 99–102; John Warkentin, ed., *The Western Interior of Canada* (Toronto: McClelland and Stewart Ltd., 1964), chap. 3; Barry Gough, *First Across the Continent* (Toronto: McClelland and Stewart Ltd., 1997); Joseph Burr Tyrrell, ed., *David Thompson's Narrative of His Explorations in Western America 1784–1812* (Toronto: Champlain Society, 1916); John Warkentin, "David Thompson's Geology: A Document," *Journal of the West*, 1967, 6:468–90.

[5] Rachel Laudan, *From Mineralogy to Geology* (Chicago: Univ. of Chicago Press, 1987), chaps. 5, 7–8; Hugh S. Torrens, "Geology in Peace Time: An English Visit to Study German Mineralogy and Geology in 1816," *Algorismus*, 1998, 23:147–75; Dennis R. Dean, *James Hutton and the History of Geology* (Ithaca, N.Y.: Cornell Univ. Press, 1992), chap. 7. See also Alexander M. Ospovat, "Romanticism and German Geology: Five Students of Abraham Gottlob Werner," *Eighteenth-Century Life*, 1982, 7:105–17; "The Importance of Regional Geology in the Geological Theories of Abraham Gottlob Werner: A Contrary Opinion," *Annals of Science*, 1980, 37:433–40; and "The Work and Influence of Abraham Gottlob Werner: A Re-evaluation," *Proceedings of the XIIIth International Congress of the History of Science* (Moscow: Naouka, 1974), sec. 8: *History of Earth Sciences*, pp. 123–31.

tween Neptunist followers of Robert Jameson (1774–1854), the University of Edinburgh's professor of natural history from 1804; and their Vulcanist rivals led by
James Hutton (1726–1797). In particular, Jameson, Werner's student, converted British explorers, surveyors, military officers, and emigrants into an extraordinary corps
of overseas observers. Encouraged by economic conditions, many ventured abroad,
primed to corroborate Werner's Primitive, Transition, Secondary, and Alluvial formations as universal and to report to Jameson's Wernerian Natural History Society.
More easily than did the United States, British North America accommodated European representations of a Neptunist theater accessed almost exclusively through vast
hydrological systems linking three oceans.[6]

After 1815, retreating arctic ice joined Napoleon's defeat in reviving British appetites for northern exploration. Jameson, who formulated specimen collectors' official instructions, also analyzed reports from the arctic expeditions of John Ross in
1818 and of William Parry in 1819–1820, 1821–1823, and 1824–1825, which for
him confirmed Wernerian tenets of a dynamic earth history with even coal found in
colder regions. Jameson's student, the whaling captain William Scoresby, Jr., expanded the Wernerian cabinet with specimens from Norway's Spitzbergen Island;
his remarkable *Account of the Arctic Regions* (2 vols., 1820) extended water's dynamic agency to its icy "denominations."[7]

Arctic geology gained focus through John Franklin's treks along the Mackenzie
and Coppermine Rivers in 1819–1822 and 1825–1827. Franklin's surgeon-naturalist,
the zoologist Dr. John Richardson (1787–1865), another Jameson student, pleased
his mentor with a "luminous sketch" of Wernerian findings, despite harrowing conditions in the Barren Lands. Broader preparations for Franklin's second voyage included lectures by William Fitton (1780–1861), the GSL's secretary and a Jameson
student increasingly inclined toward Huttonian insights. Ensuring the GSL a share of
Franklin's findings, Fitton's portable reference collection supplemented Richardson's
own winter lectures at Fort Franklin. George Back, R. N. and Peter Warren Dease,
a Hudson's Bay Company officer, joined this expedition; consultations with Richardson for future arctic voyages disseminated an evolving Wernerian framework along
formal and informal lines.[8]

[6] Joan M. Eyles, "Jameson, Robert," in *Dictionary of Scientific Biography*, ed. Charles C. Gillispie
(New York: Charles Scribner's Sons, 1971), vol. 7, p. 70; Anand C. Chitnis, "The University of
Edinburgh's Natural History Museum and the Huttonian-Wernerian Debate," *Ann. Sci.*, 1970,
26:85–94; L. J., "Biographical Memoir of the Late Professor Jameson," *The Edinburgh New Philosophical Journal*, 1854, 57:2–3; Laudan, *Mineralogy to Geology* (cit. n. 5), p. 109; Roy Porter, *The
Making of Geology* (Cambridge: Cambridge Univ. Press, 1977), pp. 149–56; Dean, *James Hutton*
(cit. n. 5), chaps. 4, 7; Robert Jameson, *The Wernerian Theory of the Neptunian Origin of Rocks*
(1808; reprint, New York: Hafner Press, 1976) pp. 100, 145, 153, 206; Sally Newcomb, "The Ideas
of A.G. Werner and J. Hutton in America," *Earth Sci. Hist.*, 1990, 7:104.

[7] "Literary and Scientific Intelligence," *The Edinburgh Magazine*, 1817, 1:367–9; Robert Jameson
in William Parry, *Journal of a Third Voyage for the Discovery of a North-West Passage* (1826; reprint,
New Haven: Research Publications, 1977), app., pp. 132–51; Robert Jameson, "General Observations on the Former and Present Geological Condition of the Countries Discovered by Captains Parry
and Ross," *Edinburgh New Phil. Jour.*, 1827, 2:105–6; William Scoresby, Jr., *An Account of the Arctic
Regions*, 2 vols. (Edinburgh, 1820), vol. 1, chap. 4 and app. Jessie M. Sweet, "Robert Jameson and
the Explorers," *Ann. Sci.*, 1974, 31:21–47; *idem*, "Instructions to Collectors: John Walker (1793) and
Robert Jameson (1817)," *Ann. Sci.*, 1972, 29:397–414; Levere, *Science and the Canadian Arctic* (cit.
n. 3), chap. 2; Zeller, "Gullivers and Crusoes," (cit. n. 3), pp. 198–203.

[8] John Richardson, "Geognostical Observations," in John Franklin, *Narrative of a Journey to the
Shores of the Polar Sea* (1823; reprint, New York, Greenwood Press, 1969), app. 1, pp. 497–538;
Robert Jameson in John Richardson, "General View of the Geognostical Structure of the Country

Richardson's Wernerianism diverged from the "modified Volcanic theory" that Fitton promoted, as William Smith's influential *Geological Map of England* (1815) and W. D. Conybeare's and W. Phillips's *Outlines of the Geology of England and Wales* (1822) inspired a distinctive English stratigraphical tradition. Yet Richardson's colleagues shared his heavy reliance, encouraged by Jameson, upon visual comparisons with European geological/geomorphological features. This classic search for similarities, situated by Michel Foucault "at the border of knowledge," resonated deeply in Western culture. Its limitations, suggests I. S. Maclaren, included difficulty discerning uniqueness and difference in a British North American landscape profoundly shaped, for example, by glaciation.[9]

Richardson also judged exploratory travel too cursory to support even much-sought-after geological sketches of the vast terrain he covered. He held out for more information, especially from fur traders on the scene. Along with the geophysicists Edward Sabine and J. H. Lefroy, he admired Alexander von Humboldt (1769–1859), also Werner's student, whose synoptic approach meant relating British North America's geology, flora, fauna, and climate to global distribution patterns. Serving this "cosmic" purpose, Richardson's *Arctic Searching Expedition* (2 vols., 1851) transcended Wernerian foundations by combining firsthand data with classical components of his liberal education. The work reduced British North America to its starkest geometrical frame, revealing in rivers, lakes, and mountain chains an arresting series of interrelated lines, angles, and circles; it was accompanied, finally, by a transcontinental geological map (Map 1).[10]

Extending from Hudson's Bay to the Shores of the Polar Sea," *Edinburgh Philosophical Journal*, 1823, *9:*372–6. John Richardson, "Intelligence from the Land Arctic Expedition," *Edinburgh New Phil. J.*, 1826, *1:*161–6; *idem*, "Overland Arctic Expedition," *Edinburgh New Phil. J.*, 1827, *2:*161–3; *idem*, "Account of the Expedition Under Captain Franklin," *Edinburgh New Phil. J.*, 1827, *6:*107–17; *idem*, "Topographical and Geological Notices," in John Franklin, *Narrative of a Second Expedition to the Shores of the Polar Sea* (1829; reprint, Rutland, Vt.: Charles E. Tuttle Company, 1971), app. 1, pp. i–lviii. Gordon L. Davies, *The Earth in Decay* (London: Macdonald Technical and Scientific, 1969), pp. 266–7; Joan M. Eyles, "Fitton, William Henry," in Gillispie, *Dictionary Scientific Biography* (cit. n. 6), vol. 5, p. 13; John Warkentin, *Geological Lectures by Dr. John Richardson, 1825–26*, Syllogeus No. 22 (Ottawa: National Museum of Natural Sciences, 1979); William O. Kupsch, "John Richardson's Geological Field Work," in *Arctic Ordeal*, ed. C. Stuart Houston (Kingston and Montreal: McGill-Queen's Univ. Press, 1984), app. F, pp. 273–316; Levere, *Science and the Canadian Arctic* (cit. n. 3), chap. 3; Zeller, "Gullivers and Crusoes," (cit. n. 3), pp. 203–7, 220. See also William Henry Fitton, "Geological Notice of the New Country Passed Over by Captain Back During His Late Expedition," in George Back, *Narrative of the Arctic Land Expedition to the Mouth of the Great Fish River* (1836; reprint, Edmonton: M.G. Hurtig, 1970), app. 4, pp. 543–62; and Peter Warren Dease and Thomas Simpson, "An Account of Arctic Discovery on the Northern Shore of America in the Summer of 1838," *Journal of the Royal Geographical Society*, 1839, *9:*325–30.

[9] William H. Fitton, "Presidential Address," *Geological Society of London Proceedings*, 1827–28, *6:*55; and 1828–29, n.s. *1:*125, 133–4; *idem*, "Geological Notice" (cit. n. 8), pp. 549, 552. Marcel de Serres, "Remarks on the Question, does the observation made in coal-mines of Canada and of Baffin's Bay . . . announce a change in the inclination of the ecliptic?" *Edinburgh New Phil. J.*, 1835, *19:*64–71; Richardson, "Geognostical Observations" (cit. n. 8), p. 538. Michel Foucault, *The Order of Things* (New York: Vintage Books, 1970), p. 68; Ian S. Maclaren, "The Aesthetic Mapping of Nature in the Second Franklin Expedition," *Journal of Canadian Studies*, 1985, *20:*39–57; Suzanne Zeller, "Classical Codes: Biogeographical Assessments of Environment in Victorian Canada," *Journal of Historical Geography*, 1998, *24:*20–35.

[10] Jameson's successor at Edinburgh in 1854, the philosophical naturalist Edward Forbes (1815–1854), had published the groundbreaking essay "On the Connexion between the Distribution of the Existing Fauna and Flora of the British Isles and the Geological Changes which have Affected Them," *Memoirs*, Geological Survey of England and Wales, 1846, *1:*336–432; see also Philip Rehbock, *The Philosophical Naturalists* (Madison: Univ. of Wisconsin Press, 1983). John Richardson to Sir William Jackson Hooker, 8 July 1828 and 18 Dec. 1832, Sir William Jackson Hooker Papers, British

Map 1. *"British North America," in Sir John Richardson,* Arctic Searching Expedition, *2 vols. (1851; reprint, New York: Greenwood Press, 1969), vol. 1. Richardson's long-awaited first full geological map of the British North American colonies, showing "Volcanoes; Metamorphic or Primitive Rocks; and Fossiliferous Rocks from the Silurian Strata Upwards."*

Meanwhile, a more mainstream Wernerianism guided geologists in British North America. J. J. Bigsby (1792–1881), an Edinburgh graduate appointed to the International Boundary Commission in 1820, authored the colonies' first geological publications, tracing Lakes Huron and Superior to "a flood of waters and floating sub-

North American Letters, Royal Botanic Gardens, Kew. Richardson, "Overland Arctic Expedition" (cit. n. 8), p. 161; Zeller, "Gullivers and Crusoes" (cit. n. 3), pp. 207–8, 212; Robert Kerr, "For the Royal Scottish Museum," *The Beaver*, 1953, pp. 32–5; M. Arnett MacLeod and Roderick Glover, "Franklin's First Expedition as Seen by the Fur Traders," *The Polar Record*, 1971, *15:*669–82. Richardson, "General View of the Geognostical Structure" (cit. n. 8); *idem*, "On Some Points of the Physical Geography of North America in Connection with Its Geological Structure," *Geological Society of London Quarterly Journal*, 1851, *7:*212–15; *idem*, Arctic Searching Expedition, 2 vols. (1851; reprint, New York: Greenwood Press, 1969), vol. 2, app. I–V. See also Eduard Desor, "On the Parallelism of Mountain Chains in America," *American Journal of Science*, 1851, *12:*118–21; Zeller, "Classical Codes" (cit. n. 9), pp. 22–4. Richardson entrusted his biological specimens to another philosophical naturalist, the quinarian taxonomist William Swainson: see Zeller, "Gullivers and Crusoes" (cit. n. 3), pp. 210–11. On the Romantic background see Trevor H. Levere, "Elements in the Structure of Victorian Science or Cannon Revisited," in *The Light of Nature*, eds. J. D. North and J. J. Roche (Dordrecht: Martinus Nijhoff, 1985), pp. 433–49.

stances rushing from the north."[11] Bigsby shared Richardson's fascination with "rolling" rocks ("erratics") that scattered with lighter debris ("drift") for considerable distances from their sources. As G. L. Davies suggests, while Hutton had inspired considerations of ice's geological impact, Jameson's Wernerianism too noted Alpine geologists' glacial insights. Citing such European work, Richardson noted rolling rocks during both expeditions, as did John Ross on his third arctic voyage (1829–1833). Bigsby incorporated this and Edward Hitchcock's American data, mapping erratics' North American dispersion for the GSL in 1851 (Map 2).[12]

Diluvial assumptions informed H. W. Bayfield, R. N. (the Great Lakes' hydrographic surveyor), R. H. Bonnycastle and F. H. Baddeley (Royal Engineers stationed at Quebec), and others, with varying emphases. In 1829, Bayfield posited a "great convulsion" flowing southwestward from Hudson Bay. Bonnycastle agreed that Canadian clues might well explain transported boulders everywhere. Towards Labrador, Baddeley found not hints of the ancient deluges he sought, but deeply scoured rocks.[13] In Newfoundland, W. E. Cormack, another Jameson student, explored the island in 1822 but did not recognize even stronger ice effects. In Nova Scotia, C. T. Jackson and Francis Alger from New England concluded in 1828–1829 that an "overwhelming torrent" had "crossed this province from north east to south west." Jackson had once travelled with William McClure, a Scottish-American geologist who had visited Jameson and sought out Werner, but McClure questioned the ex-

---

[11] John J. Bigsby, "Notes on the Geography and Geology of Lake Huron," *Geol. Soc. London Transac.*, 1824, ser. 2, *1*:204–5, *2*:175–209; *idem*, "Notes on the Geography and Geology of Lake Superior," *Quarterly Journal of Science*, 1825, *18*:1–34. G. B. Morey, "Early Geologic Studies in the Lake Superior Region: The Contributions of H. R. Schoolcraft, J. J. Bigsby, and H. W. Bayfield," *Earth Sci. Hist.*, 1989, *8*:36–42; Zeller, "Gullivers and Crusoes" (cit. n. 3), pp. 207–8. It was likely Bigsby who sent William Buckland some of Lake Huron's fossiliferous specimens in 1821; see Torrens, "Transmission of Ideas" (cit. n. 3), p. 112.

[12] Richardson, "Geognostical Observations" (cit. n. 8), pp. 497, 505, 537; *idem*, "Topographical and Geological Notices" (cit. n. 8) pp. li–lii. See also Charles König, "Rock Specimens," in supplement to the appendix of *Captain Parry's Voyage* (London: John Murray, 1824), p. ccli; Sir John Ross, "Geology," in Sir John Ross, *Appendix to the Narrative of a Second Voyage* (1835; reprint, New York: Greenwood Press, 1969), p. cvi. For alpine glacial theory: Horace Bénédict de Saussure, *Voyages dans les Alpes*, 4 vols. (Neuchâtel-Geneva, 1779–96); Johann Gottfried Ebel, *Über den Bau der Erde in Alpengebirge*, 2 vols. (Zurich, 1808); Christian Leopold von Buch, *Versuch einer mineralischen Beschreibung von Landeck* (Breslau, 1797) and *Reise nach Norwegen und Lappland* (Berlin, 1810); and Louis Agassiz, *Studies on Glaciers*, trans. and ed. Albert V. Carozzi (1840; reprint, New York: Hafner Publishing Company, 1967), chap. 1; Richardson, *Arctic Searching Expedition* (cit. n. 10), p. 176; John J. Bigsby, "On the Erratics of Canada," *Geol. Soc. London Quart. Jour.*, 1851, *7*:215–52; James Robb, "Notice of Observations on Drift and Striae in New Brunswick," *American Association for the Advancement of Science Proceedings*, 1850, *4*:349–51; and Louis Agassiz, *Lake Superior* (1850; reprint, New York: Arno Press, 1970), p. 397; Davies, *Earth in Decay* (cit. n. 8), pp. 266–71; Sweet, "Instructions to Collectors" (cit. n. 7), pp. 405–6; Warkentin, *Western Interior* (cit. n. 3), pp. 125–6; Robert Leggett, "Early Canadian Record of Glacial Erratics," *Geoscience Canada*, 1983, *10*:133–4.

[13] Henry Wolsey Bayfield, "Outlines of the Geology of Lake Superior," *Literary and Historical Society of Quebec Transactions*, 1829, *2*:42–3; Richard Henry Bonnycastle, "Desultory Observations on a Few of the Rocks and Minerals of Upper Canada," *Lit. Hist. Soc. Quebec Trans.*, 1829, *1*:62; Frederick Henry Baddeley, "Geology of a Portion of the Labrador Coast," *Lit. Hist. Soc. Quebec Trans.*, 1829, *1*:71; *idem*, "On the Geognosy of a Part of the Saguenay Country," *Lit. Hist. Soc. Quebec Trans.*, *1*:164. Recent geomorphological theory in this region has returned to diluvial explanations involving subglacial lake water: see John Shaw, "A Qualitative View of Sub-Ice-Sheet Landscape Evolution," *Progress in Physical Geography*, 1996, *18*:159–84 and "A Meltwater Model for Laurentide Subglacial Landscapes," in *Geomorphology sans frontières*, eds. S. B. McCann and D. C. Ford, International Association of Geomorphologists, Publication No. 6 (Chichester and New York: John Wiley and Sons, 1996), pp. 181–239; the author thanks Ian Brookes for these references.

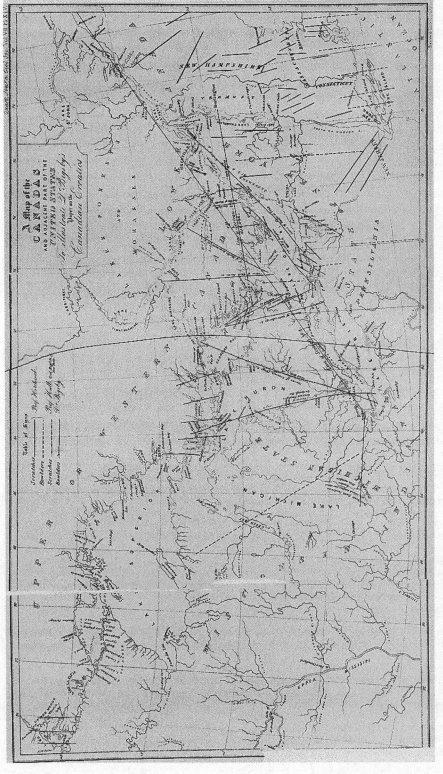

Map 2. *"A Map of the Canadas and Adjacent Part of the United States, to Illustrate Dr. Bigsby's Paper on the Canadian Erratics," Geological Society of London Quarterly Journal, 1851, 7:215–52. A remarkable collation mapping patterns of "Scratches" and "Boulders" in the United States (from Edward Hitchcock and James Hall) and Canada.*

planatory powers of the Neptunian geological framework. Following McClure's example, Jackson and Alger in Nova Scotia supplanted Neptunist predilections with renewed Vulcanist explanations.[14]

More than merely extracting information for the metropolis, the Wernerian community established colonial roots. Scottish immigration transported an "all-pervasive need to modernize the backward Scottish economy," motivating colonists towards similar goals in their new homeland. More powerfully than the more isolated Cormack, who was born in St. John's of Scottish parents, Bigsby animated his editor David Chisholme, of Montreal's influential Scottish business community, to demand colonial surveys as British subjects' rightful due.[15] Chisholme welcomed the Canada Land Company, chartered in 1826, to settle a million acres east of Lake Huron, and its warden of the woods and forests, an educated Scot with colonial surveying experience. William "Tiger" Dunlop ascribed the Huron Tract's origins to the recent "subsiding of an enormous mass of waters," but hesitated to tackle its pervasive granitic, unstratified Primitive formations.[16] In York (Toronto), Upper Canada's capital, Dunlop echoed Lower Canadian lobbies in Quebec and Montreal in their call for colonial mineralogical surveys.[17] Growing local interests hoped thereby to transform British North America into substantially more than Neptune's playground.

LAURENTIAN LAND

While Roderick Impey Murchison's (1792–1871) biographers have illumined his "science of empire," the "King of Siluria" was not "the only geologist of his day with the breadth of vision to think in terms of entire continents."[18] From beyond the metropolis, Murchison appears less Titanic and more the tip of an imperial iceberg, whose hidden foundations of colonial geologists not only buoyed his enterprise but navigated with similar success. Among them, Canada's W. E. Logan (1798–1875) evinced territorial appetites no less voracious than Murchison's.

[14] Charles T. Jackson and Francis Alger, "A Description of the Mineralogy and Geology of a Part of Nova Scotia," *American Journal of Science and Art*, 1828, *14:*305–30 and 1829, *15:*138, 204. Ian A. Brookes, "Ice Marks in Newfoundland: A History of Ideas," *Géographie Physique et Quaternaire*, 1982, *36:*140–2. George Edmond Gifford, Jr., "Jackson, Charles Thomas," in Gillispie, *Dictionary of Scientific Biography* (cit. n. 6), vol. 7, pp. 44; Newcomb, "Ideas of Werner and Hutton" (cit. n. 6), pp. 98–9. Jackson went on to refine his geological understanding at the Sorbonne, and with Léonce Élie de Beaumont at France's Royal School of Mines.

[15] William E. Cormack, "Account of a Journey Across the Island of Newfoundland," *Edinburgh Phil. Jour.*, 1823–24, *10:*156–62; John Jeremiah Bigsby, "On the Utility and Design of the Science of Geology," *Canadian Review and Literary and Historical Journal*, 1824–25, *1:*377–95; David Chisholme, editorial, *Canadian Review and Magazine*, 1826, *4:*249–51, 319–21, 332–3. Porter, *Making of Geology* (cit. n. 6), p. 153; Brookes, "Ice Marks" (cit. n. 14), p. 141; Zeller, *Inventing Canada* (cit. n. 2), pp. 16–9; Zeller, "Gullivers and Crusoes" (cit. n. 3), p. 210.

[16] William Dunlop, "Report of the Warden of the Forests" (1827 and 1828), reprinted in *Report on Canadian Archives*, comp. D. Brymner (Ottawa: Public Archives of Canada, 1899), pp. 9, 13–16; idem, "A Paper on Peat Mosses," *Canadian Literary Magazine*, 1833, *2:*98–100. Zeller, *Inventing Canada* (cit. n. 2), pp. 19–22.

[17] Zeller, *Inventing Canada* (cit. n. 2), pp. 21–39.

[18] Robert A. Stafford, *Scientist of Empire: Sir Roderick Murchison, Scientific Exploration and Victorian Imperialism* (Cambridge: Cambridge Univ. Press, 1989), esp. p. 188; James A. Secord, "King of Siluria: Roderick Murchison and the Imperial Theme in Nineteenth-Century British Geology," *Victorian Studies*, 1981–82, *28:*413–42; Martin J. S. Rudwick, *The Great Devonian Controversy* (Chicago: Univ. of Chicago Press, 1985); James A. Secord, *Controversy in Victorian Geology: The Cambrian-Silurian Dispute* (Princeton: Princeton Univ. Press. 1986).

Even as Murchison's army reworked Werner's Transitional into Silurian, Devonian, and Cambrian systems, Charles Lyell's (1797–1875) *Principles of Geology* (3 vols., 1830–1833) targeted Werner as the new geological era's straw man. However controversial his uniformitarianism, Lyell's accessible, heuristic synthesis made a worldwide appeal for "actualistic" observable forces and fragmentary evidence. His transatlantic social connections opened doors, captivating colonial geologists who admitted coveting "not only the encouraging notice of (their country's) inhabitants, but that of the whole civilized world."[19] Lyell offered a pipeline to the GSL, presenting H. W. Bayfield's evidence of erratics' transportation by ice in 1835. While recognizing erratics as northern phenomena in 1836, Lyell's resistance to catastrophic explanations hardened during his North American tour in 1841, delaying for decades the general acceptance of Louis Agassiz's (1807–1873) ice-age theory.[20]

One interesting colonial publication challenged Lyellian assumptions. Titus Smith, Jr., a Loyalist naturalist commissioned in 1801 to assess Nova Scotia's resources, published his geological results in 1836. Derived at least partly from his Sandemanian faith, Smith's precocious environmental sensibility saw nature's kingdoms interconnected ecologically. A natural philosopher, he explained minerals as inelastic fluids whose "internal motions" formed metamorphic compounds. Smith gained broad respect for his erudition, even though his wholesale rejection of industrialization as unnatural was diametrically opposed to contemporary views.[21]

Lyell's reformulations instead enhanced broad confidence in geology's ability to

[19] Alexander Ospovat, "The Distortion of Werner in Lyell's *Principles of Geology*," *British Journal for the History of Science*, 1976, *9:*190–8; Martin J. S. Rudwick, "The Strategy of Lyell's *Principles of Geology*," *Isis*, 1970, *61:*5–33; Philip Lawrence, "Charles Lyell Versus the Theory of Central Heat: A Reappraisal of Lyell's Place in the History of Geology," *Journal of the History of Biology*, 1978, *11:*125–8; Martin J. S. Rudwick, "Historical Analogies in the Geological Work of Charles Lyell," *Janus*, 1977, *64:*102–4; Leonard G. Wilson, "Geology on the Eve of Charles Lyell's First Visit to America, 1841," *Proceedings of the American Philosophical Society*, 1980, *124:*168–202. Frederick Henry Baddeley, "An Essay on the Localities of Metallic Minerals in the Canadas," *Lit. Hist. Soc. Quebec Trans.*, 1831, *2:*333. For Wernerian-Huttonian waverings in British North America: Frederick Henry Baddeley, "Mineralogical Examination of the Sulphate of Strontian, from Kingston," *Amer. J. Sci. Art*, 1830, *18:*104–7; idem, "Additional Notes on the Geognosy of Saint Paul's Bay," *Lit. Hist. Soc. Quebec Trans.*, 1831, *2:*76–94; H. D. Sewell, "A Few Notes Upon the Dark Days of Canada," *Lit. Hist. Soc. Quebec Trans.*, 1831, *2:*242, reprinted in *Edinburgh New Phil. J.*, 1833, *14:*229; idem, "Notes Upon the Country in the Vicinity of Quebec," *Lit. Hist. Soc. Quebec Trans.*, 1837, *3:*308; Richard Henry Bonnycastle, "On the Transition Rocks of the Cataraqui," *Amer. J. Sci. Art*, 1830, *18:*85–104; 1831, *20:*74–93; 1833, *24:*80–104; 1836, *30:*233–48.

[20] Henry Wolsey Bayfield, "Notes on the Geology of the North Coast of the River and Gulf of St. Lawrence," *Geol. Soc. London Proc.*, 1833–34, *2:*4–5; idem, "A Notice on the Transportation of Rocks by Ice," *Geol. Soc. London Proc.*, 1835–36, *2:*223. Charles Lyell, "Remarks on Some Fossil and Recent Shells, Collected by Capt. Bayfield, R. N., in Canada," *Geol. Soc. London Proc.*, 1839, *3:*119–20; idem, "Address to the Geological Society," *Geol. Soc. London Proc.*, 1837, n.s., *2:*381–4; Louis Agassiz, "Discourse of Neuchâtel" (1837) reprinted in Agassiz, *Studies on Glaciers* (cit. n. 12), pp. lxiii–lxviii; Robert H. Silliman, "Agassiz vs. Lyell: Authority in the Assessment of the Diluvium-Drift Problem by North American Geologists," *Earth Sci. Hist.*, 1994, *13:*180–6. William Edmond Logan, "On the Packing of Ice in the River St. Lawrence," *Geol. Soc. London Proc.*, 1842, *3* (part 2):767; William Hopkins, "Anniversary Address of the President," *Geol. Soc. London Quart. J.*, 1852, *8:*xxiv–lvii.

[21] William Smith, "Some Account of the Life of Titus Smith," *Nova Scotia Institute of Natural Science Transactions*, 1866, *1:*149–52; Terence M. Punch, "Smith, Titus," in *Dictionary of Canadian Biography*, gen. ed. Francess G. Halpenny (Toronto: Univ. of Toronto Press, 1988), vol. 7, pp. 814–16; A. Hill Clark, "Titus Smith, Junior, and the Geography of Nova Scotia in 1801 and 1802," *Annals of the Association of American Geographers*, 1954, *44:*291–314; Titus Smith, "A Lecture on the Mineralogy (and the Geology) of Nova Scotia," *Magazine of Natural History*, 1836, *9:*578–83; idem, "Conclusions on the Results on the Vegetation of Nova Scotia," *Magazine of Natural History*,

locate and develop industrial resources. This power cut both ways in Britain's first colonial surveys. Abraham Gesner's (1797–1864) prediction of too much coal in New Brunswick (1838–1842) and J. B. Jukes's (1811–1869) vision of too little in Newfoundland (1839–1840) cost both men their public funding. In the Canadas, political factors and armed rebellions delayed public geological surveys, distilling these valuable lessons in institutional survival.[22] In 1841, the newly United Province of Canada funded a survey just before its governor general, Lord Sydenham, suddenly died. The choice for survey director, William Logan, in 1842 stemmed less from imperial preferences than from the Montreal business community's plans for one of its own. Transatlantic campaigns persuaded Sydenham's successor, Sir Charles Bagot, to favor the Montreal-born, Edinburgh-instructed, self-styled and experienced "practical coal miner of education" Logan, who was respected by both British miners and geologists despite local concerns that he lacked palaeontological training.[23]

Logan undertook the Geological Survey of Canada (GSC) with ambitions, like those of his colleagues in Nova Scotia and New Brunswick, stoked by Charles Lyell. As J. W. Dawson's (1820–1899) mentor, Lyell opened geological avenues when mining monopolies precluded geological surveys of Dawson's native Nova Scotia. Introducing Dawson to Logan, Lyell praised their colonial homelands as "destined to surprise us yet" with valuable data, adding Logan's widely lauded explanation of coal's *in situ* origins to his *Principles of Geology*.[24] Logan envisaged the GSC and its British counterpart as "mutually serviceable": British palaeontology would com-

1835, *8*:661. Cf. Smith's chemico-physical explanation to R. H. Bonnycastle, "Transition Rocks of the Cataracqui," *Amer. J. Sci. Art*, 1836, *30*:246–7.

[22] Abraham Gesner, *Remarks on the Geology and Mineralogy of Nova Scotia* (Halifax, 1836); *idem*, First [to Fourth] Report on the Geology of New Brunswick (Saint John, 1839–42). James W. Ross to John William Dawson, 1 May 1846, John William Dawson Papers, McGill Univ. Archives (hereafter cited as Dawson Papers). James F. W. Johnston, *Report on the Agricultural Capabilities of the Province of New Brunswick*, 2nd ed. (Fredericton, N.B., 1850), p. 14; geological section by Andrew Robb, professor of natural history, University of New Brunswick. John Beete Jukes, "Report on the Geology of Newfoundland," reprinted in *Edinburgh New Phil. J.*, 1840, *29*:103–11; *idem, General Report of the Geological Survey of Newfoundland* (London, 1843), p. 152. Moses H. Perley, "Observations on the Geological and Physical Characteristics of Newfoundland," *Canadian Naturalist and Geologist*, 1862, *7*:art. 31. "Report of the Committee on Geological Surveys," 1 Mar. 1836, Upper Canada, Legislative Assembly, *Journals*, 1836, vol. 3, app. 126. F. Bond Head, *Rough Notes Taken During Some Rapid Journeys Across the Pampas* (London, 1826), p. v; *idem, Descriptive Essays*, 2 vols. (London, 1857), vol. 1, pp. 18–19, 21. Rudwick, "Historical Analogies" (cit. n. 19), pp. 102–4; Allison Mitcham, *Prophet of the Wilderness* (Hantsport, N.S.: Lancelot Press, 1995); Alex McEwen, ed., *In Search of the Highlands: Mapping the Canada-Maine Boundary, 1839* (Fredericton, N.B.: Acadiensis Press, 1988), p. 107; Zeller, "Gullivers and Crusoes" (cit. n. 3), pp. 214–15; Zeller, *Inventing Canada*, (cit. n. 2), chap. 1.

[23] Contrast Stafford, *Scientist of Empire* (cit. n. 18), pp. 65–8; for details of campaigns for a Geological Survey of Canada and Logan's appointment, Zeller, *Inventing Canada* (cit. n. 2), chaps. 1–2; on Logan's mining background, Hugh Torrens, "How, When and Where Did William E. Logan Learn his Geology in Britain 1831–1841?" forthcoming in *Geoscience Canada*. William E. Logan to James Logan, 14 Aug. 1841, Sir William E. Logan Papers, McGill Univ. Archives (hereafter cited as W. E. Logan Papers).

[24] The standard GSC history is Morris Zaslow, *Reading the Rocks* (Ottawa: Macmillan, 1975); see also Robert H. Dott, Jr., "Lyell in America—His Lectures, Field Work, and Mutual Influences, 1841–1853," *Earth Sci. Hist.*, 1996, *15*:101–40; Lyell tempered Gesner's exaggerations: Henry De la Beche, "Anniversary Address of the President," *Geol. Soc. London Quart. J.*, 1849, *5*:liv. Charles Lyell to J. W. Dawson, 2 Feb. and 2 May 1843, 30 May 1854; Dawson to Lyell, 5 Sept. 1845, 17 Feb. 1849; R. Brown to Dawson, 18 Nov., 9 Dec. 1845, 10 Mar. 1846; S. Cunard to Dawson, 28 Nov., 8 Dec. 1845, 19 Feb. 1846; Dawson to Hon. G. R. Young, n.d. 1846, 20 Apr. 1848, 17 Jan. 1849, Dawson Papers. W. E. Logan to James Logan, 16 Aug. 1841, W. E. Logan Papers.

pensate for his deficiencies, gaining "unity in Nomenclature" as Canada became "the measure of a correct geological comparison" between Europe and America. Remarkably, Logan saw himself in Canada "only learning my lesson as it were in preparation" to survey Britain's "Great Country" to the northwest, still geologically unknown.[25]

Grand dreams became permanent themes. Frustrated early on by erroneous GSL classifications of Canadian specimens, Logan felt "mortified" years later at finding his specimen boxes, still unopened, in the British Survey's basements. "I wish," he confided to Dawson, "I had trusted more to myself." Since consulting New York's Geological Survey meant using American nomenclature after all, Logan vowed to let evidence alone challenge Murchison's stratigraphical declensions. He also chafed at North American penchants for annual reports: "Avoiding politics as I would poison," he planned to offer "facts & no theory" while "enabl[ing] others to anticipate any results . . . beneficial to me."[26]

The crucial pitfall was coal, that desideratum of Victorian progress, still undiscovered in Canada. Logan feared the worst, having recognized in the geological strata near Montreal "a coal field, with the coal left out." His first report deflected his dreaded indictment of the province as geologically below the carboniferous, emphasizing instead metallic ores near the Upper Great Lakes. The GSC thus secured funding for five more years, with a chemist, Thomas Sterry Hunt (1826–1892) (and soon a palaeontologist, Elkanah Billings [1820–1876]) joining Logan's one assistant, Alexander Murray (1810–1884).[27]

With skeptics fraudulently adding coal to their purported "mines" to embarrass him, Logan needed to consolidate science's authority in the public eye. Here Murchison's aggressive stratigraphical campaign threatened to subsume his accomplishments, annexing Logan's Nova Scotian (and Jukes's Newfoundland) strata to Per-

[25] W. E. Logan to Sir Henry De la Beche, 19 Oct., 3 Dec. 1841; 24 Apr. 1843; 12 May 1845, Sir Henry De la Beche Papers, National Museum of Wales, Cardiff (hereafter cited as De la Beche Papers). William Edmond Logan, "On the Character of the Beds of Clay Lying Immediately Below the Coal Seams of South Wales," *Geol. Soc. London Proc.*, 1840, *3*:275–7; *idem*, "On the Coal-Fields of Pennsylvania and Nova Scotia," *Geol. Soc. London Proc.*, 1842, *3* (part 2):707–12. Charles Lyell, "On the Upright Fossil Trees Found at Different Levels in the Coal Strata of Cumberland, Nova Scotia," *Geol. Soc. London Proc.*, 1843, *4* (part 1):176–8; *idem*, "On the Coal-Formation of Nova Scotia," *Geol. Soc. London, Proc.*, 1843, *4* (part 1):184–90, reprinted in *Amer. J. Sci. Art*, 1843, *45*:356–9. John William Dawson, "On the Lower Carboniferous Rocks, or Gypsiferous Formations of Nova Scotia," *Geol. Soc. London Proc.*, 1843–44, *4* (part 2):272–81; Richard Brown, "On the Geology of Cape Breton," *Geol. Soc. London Proc.*, 1843–44, *4* (part 2):269–81, 424–30; Charles Lyell, *Travels in North America*, 2 vols. (New York, 1845); Zeller, *Inventing Canada* (cit. n. 2), chap. 2.

[26] Logan to Dawson, 4 June 1843, 10 Jan. 1853, Dawson Papers; Logan to De la Beche, 20 Apr., 11 Nov. 1844; 12 May, 27 Dec. 1845, De la Beche Papers; Zeller, *Inventing Canada* (cit. n. 2), pp. 62, 77.

[27] Baddeley, "Geognosy of a Part of the Saguenay Country" (cit. n. 13), pp. 163–4; *idem*, "Geognosy of St. Paul's Bay" (cit. n. 19), pp. 91–93; *idem*, "A Geological Sketch of the Most South-Eastern Portion of Lower Canada," *Lit. Hist. Soc. Quebec Trans.*, 1837, *3*:280–1. Logan to De la Beche, 5 Oct. 1840, 31 May 1843, 12 May 1845, De la Beche Papers. Geological Survey of Canada, "Preliminary Report," 6 Dec. 1842, and "Report of Progress for the Year 1843," in Canada, Province, House of Assembly *Journal*, 1844–45, app. W. John T. Brondgeest, "On Mines and Mining, Especially Referring to the Mining Districts on the Great Lakes of North America," *Simmonds's Colonial Magazine and Foreign Miscellany*, 1848, *13*:402–12; Zeller, *Inventing Canada* (cit. n. 2), chap. 3; M. J. Copeland, "Elkanah Billings (1820–1876) and Joseph F. Whiteaves (1835–1909): The First Two Palaeontologists of the GSC," *Earth Sci. Hist.*, 1993, *12*:107–10; R. W. Boyle, "Geochemistry in the Geological Survey of Canada 1842–1952," *Earth Sci. Hist.*, 1993, *12*:129–41.

mian formations that Murchison had named in Russia. In the same breath, Murchison relegated Logan's coal theory to a case study of his own preferred neo-Neptunist explanation. His successor as GSL president, Leonard Horner, eased this latter issue in 1846, attributing to Lyell, his son-in-law, a middle position: while Murchison clung to marine depositions, and Logan propounded coal's *in situ* formation on land, Lyell postulated an *in situ* submergence, inviting further research.[28]

Logan spent the 1850s solidifying GSC foundations, earning a sterling international reputation at the London and Paris Exhibitions in 1851 and 1855, respectively. The first Canadian to attain such rarefied heights in scientific circles abroad, he garnered a knighthood, induction into the Légion d'Honneur, and the GSL's Wollaston Medal.[29] He also demarcated his own distinctive Laurentian strata in the Canadian Shield. Murchison, now directing Britain's Survey, recognized "the very able manner in which he ha[d] . . . separated the grand series of fundamental sedimentary unfossiliferous rocks" from his own Silurian. Logan transcribed these Precambrian formations, once a Wernerian puzzle construed by F. H. Bonnycastle as a "Lacustrian Chain," into Lyellian terms as metamorphic. Dating them to "God knows when, prior to the creation [of] Siluria," he now illuminated "the most ancient [rocks] yet known on the continent."[30]

Further colonial developments stabilized the GSC. University renewal, with several scientific appointments, included J. W. Dawson's to McGill College, Montreal

---

[28] William Edmond Logan, "Report of Progress for the Year 1843," in Canada, Province, House of Assembly *Journal*, 1844–45, app. W. "Literary Soirée of the Natural History Society," *Montreal Gazette*, 18 April 1853; Robert Lachlan to Logan, 27 Dec. 1858, W. E. Logan Papers; Roderick Impey Murchison, "Presidential Address," *Geol. Soc. London Proc.*, 1843, 4:121–7, reprinted in *Edinburgh New Phil. J.*, 1843, 35:115–21. John William Dawson, "On the Newer Coal Formations of the Eastern Part of Nova Scotia," *Geol. Soc. London Proc.*, 1844–45, 4:504–12; *idem*, "Notice on Some Fossils Found in the Coal Formations of Nova Scotia," *Geol. Soc. London Quart. J.*, 1846, 2:132–7. Richard Brown, "On a Group of Erect Fossil Trees in the Sydney Coal-Field of Cape Breton," *Geol. Soc. London Quart. J.*, 1846, 2:96. C. J. F. Bunbury, "Notes on the Fossil Plants Communicated by Mr. Dawson from Nova Scotia," *Geol. Soc. London Quart. J.*, 1846, 2:136–9; *idem*, "On Fossil Plants from the Coal Formations of Cape Breton," *Geol. Soc. London Quart. J.*, 1847, 3:423–37. Leonard Horner, "Anniversary Address of the President," *Geol. Soc. London Quart. J.*, 1846, 2:170–81; Richard Brown, "Description of an Upright Lepidodendron with Stigmaria Roots," *Geol. Soc. London Quart. J.*, 1848, 2:46–51. John William Dawson, "Notice of the Occurrence of Upright Calamites near Pictou, Nova Scotia," *Geol. Soc. London Quart. J.*, 1851, 7:194–6; *idem*, "Notice of the Discovery of a Reptilian Skull in the Coal of Pictou," *Geol. Soc. London Quart. J.*, 1855, 11:8–9; *idem*, "On a Modern Submerged Forest at Fort Lawrence, Nova Scotia," *Geol. Soc. London Quart. J.*, 1855, 11:119–22. When Dawson later tried to split the difference between Lyell and Logan, he was taken on by Logan's chemist: cf. Thomas Sterry Hunt, "On Prof. J. W. Dawson's Papers on the Coal," *American Journal of Science*, 1861, 31:290–2; John William Dawson, "On the Conditions of the Deposition of Coal," *Geol. Soc. London Quart. J.*, 1866, 22:95–169.

[29] William Edmond Logan, "On the Occurrence of a Track and Foot-Prints of an Animal in the Potsdam Sandstone of Lower Canada," *Geol. Soc. London Quart. J.*, 1851, 7:247–52; Charles Lyell, "Anniversary Address of the President," *Geol. Soc. London Quart. J.*, 1851, 7:lxxv–lxxvi; William Edmond Logan, "On the Foot-Prints Occurring in the Potsdam Sandstone of Canada," *Geol. Soc. London Quart. J.*, 1852, 8:199–225; William Hopkins, "Anniversary Address of the President," *Geol. Soc. London Quart. J.*, 1852, 8:lxxx. Eagan, "Canadian Geological Survey" (cit. n. 3), pp. 103–4; Zeller, *Inventing Canada* (cit. n. 2), chap. 4.

[30] Memorial of Canadian Institute [1856], W. E. Logan Papers; "Reply Made by Sir Roderick Murchison . . . on the Grounds on which the Council had Awarded the Wollaston Medal to Sir William Logan," 1856, W. E. Logan Papers. Bonnycastle, "Transition Rocks of the Cataraqui" (cit. n. 19), pp. 234–5; Logan to De la Beche, 27 Dec. 1845, De la Beche Papers; William Edmond Logan, "Sketch of the Geology of Canada," in Canada, Province, House of Assembly *Journal*, 1856, app. 46; *idem*, "Report of Progress for the Years 1853–56," *ibid.*, 1857, app. 52.

in 1855 as principal and Logan as professor of geology and palaeontology. Murchison had rejected Dawson, a "mere colonist," for Edinburgh's natural history chair, despite Lyell's support. The author of *Acadian Geology* (1855), Dawson gave the Canadas an influential scientific spokesman, administrator, educator, and palaeobotanist, supplementing Logan's stratigraphical strengths. Bequeathing maritime colonial fieldwork to younger Lyellians, Dawson later admitted he had never enjoyed surveying anyway.[31]

The scientific newcomers publicly championed the GSC, both in print and before a Canadian government Select Committee in 1854. Welcoming also James Hall's (1811–1898) favorable testimony in assuring the GSC permanent status, Logan reciprocated when Hall's New York Survey came under review by the state legislature soon thereafter. Dawson and Logan hosted the American Association for the Advancement of Science's (AAAS) Montreal meeting in 1857, with Hall presenting his geosynclinal theory and Logan his Laurentian analysis. Murchison, granting the Laurentian's contemporaneity with Scotland's Fundamental Gneiss, nevertheless asked to rename it "Lewisian after the Scottish stratum." Surviving this imperial challenge, Logan's nomenclature swelled Canadians' pride in their Laurentian land.[32]

METAPHOR AND MYTH

Continental ties bound transversely, too. Logan's Laurentian system lent physical backbone and daring vision to a colonial world undergoing profound reconceptualization. With Britain's adoption of free trade from 1846, and its negotiation of North American reciprocal trade from 1854 to 1866; and with coalbeds resting east and west in British territory and south in the United States, where did the Canadas' industrial future lie?

[31] John William Dawson, *Acadian Geology* (Edinburgh, 1855); reviewed anonymously in *Edinburgh New Phil. J.*, 1855, n.s., *2*:380–92. Clipping, *Scottish Press*, July 1855, Dawson Papers. Dawson would suffer more such slights from his British colleagues: see Susan Sheets-Pyenson, *John William Dawson: Faith, Hope, and Science* (Montreal and Kingston: McGill-Queen's University Press, 1996). Randall F. Miller and Diane N. Buhay, "The Steinhammer Club: Geology and a Foundation for a Natural History Society of New Brunswick," *Geoscience Canada*, 1988, *15*:221–6; William R. Brice, "Charles Frederick Hartt (1840–1878): The Early Years," *Earth Sci. Hist.*, 1994, *13*:160–7; J. W. Dawson to Charles Lyell, 10 Oct. 1868, Sir Charles Lyell Papers, University of Edinburgh Library (hereafter cited as Lyell Papers). Suzanne Zeller, "'Merchants of Light': The Culture of Science in Daniel Wilson's Ontario, 1853–1892," in *Thinking With Both Hands: Sir Daniel Wilson in the Old World and the New*, ed. Elizabeth Hulse (Toronto: Univ. of Toronto Press, 1999), pp. 115–38.

[32] Edward John Chapman, "A Popular Exposition of the Minerals and Geology of Canada," *Canadian Journal of Industry, Science, and Art*, 1860, n.s., *5*:1–9. John William Dawson, "Geological Survey of Canada: Report of Progress for 1857," *Can. Natur. and Geol.*, 1859, *4*:62–9; "Geological Survey of Canada: Report of Progress from Its Commencement to 1863," *Can. Natur. and Geol.*, 1864, *1*:65–7. Comte de Rottermund, "Second Report of the Explorations of Lakes Superior and Huron," in Canada, Province, House of Assembly *Journal*, 1857, app. 5. Edward John Chapman to Logan, 13 Feb., 4 Mar., 10 Mar., 9 Apr. 1855; 22 Mar. 1856, W. E. Logan Papers. Robert H. Dott, Jr., "The Geosyncline—First Major Geological Concept 'Made in America,'" in *200 Years of Geology in America*, ed. Cecil J. Schneer (Hanover, N.H.: Univ. Press of New England, 1979), pp. 239–64; Kathleen Mark, "From Geosynclinal to Geosyncline," *Earth Sci. Hist.* 1992, *11*:68–9; Beryl M. Hamilton, "British Geologists' Changing Perceptions of Precambrian Time in the Nineteenth Century," *Earth Sci. Hist.*, 1989, *8*:141–9; Eagan, "Canadian Geological Survey" (cit. n. 3), p. 104; Zeller, *Inventing Canada* (cit. n. 2), pp. 94–6.

In the east, Logan's need to map the province of Canada topographically as well as geologically required longitudinal coordinates calculated from those established nearby. Moreover, he could "scarcely" exclude the neighboring colonies of Nova Scotia and New Brunswick, "both of which have important relations to Canada." For his inclusive Paris Exposition map (1855) he appealed to J. W. Dawson, who was just completing his map of Acadian geology, and James Hall at Albany. Adopting British Survey colors, Logan coordinated contiguous American strata with British representations as far as possible.[33] Nor did his Murchisonian expansions stop there, as gold discovered in Nova Scotia and New Brunswick in 1858 revived hopes for colonial surveys. Fearing the "bad hands" of certain local geologists, Dawson preferred GSC staff under Logan's supervision, with "the whole brought into one great work at the close." Yet both colonial governments demurred for the moment, despite (and perhaps because of) these enthusiastic Canadian solicitations.[34]

Farther west, the HBC's charter in Rupert's Land—the Laurentian rump and beyond—would expire in 1869. British and Canadian commercial and political interests, especially in Toronto, hoped to stymie American Manifest Destiny in these vast lands. With rumors of coal's abundance in the Saskatchewan territory persisting since the earliest explorations, Logan welcomed confirmation of its existence from even the HBC's defensive governor, George Simpson. Canadian expansionists disseminated evidence collated for the GSL in 1855 by A. K. Isbister (1822–1883), a fur trader's Métis son and an Edinburgh graduate, of "a vast coalfield, skirting the base of the Rocky Mountains for a very great extent, and continued probably far into the Arctic Sea" (Map 3). With the British government's assignment in 1857 of both a Select Committee and John Palliser's expedition to assess the territory's potential, the Canadas followed suit on both counts in order to revise negative impressions that shrouded in secrecy the HBC's monopoly.[35]

Concomitant geological analyses reflected their respective patrons' agendas. Dr. James Hector, a University of Edinburgh graduate selected by Murchison, offered Palliser detached scientific analyses of Tertiary and Cretaceous formations, identifying potential coal supplies for Britain's modern Northwest Passage, a transcontinental railway. H. Y. Hind (1823–1908), professor of chemistry at Trinity College, Toronto, won Logan's nomination to Canada's G. Gladman-S. J. Dawson expeditions, largely because GSC staff members were unavailable. A well-known ex-

[33] Logan, *Report of Progress 1853–56* (cit. n. 30). William Brydone Jack to Logan, 15 Oct. 1856, 24 Feb., 25 Mar., 5 May 1857; Logan to Dawson, 17 Feb. 1855; Dawson to Logan, 6 Mar. 1855; Robert Barlow to Logan, 1862, W. E. Logan Papers.

[34] John William Dawson, "On the Recent Discoveries of Gold in Nova Scotia," *Can. Natur. and Geol.*, 1861, *6:*417. Dawson to Logan, 18 Mar. 1858, W. E. Logan Papers. Dawson to Lyell, 10 Oct. 1861, Lyell Papers. Henry How to George W. Hill, 10 Sept. 1858; Hill to Prov. Secretary, 22 Sept. 1858, Public Archives of Nova Scotia, Provincial Secretary's Papers, RG7 vol. 40 (hereafter cited as N.S. Prov. Sec. Papers). How to Prov. Secretary, 12 Nov. 1861, N.S. Prov. Sec. Papers, RG1 Vol. 462. David Honeyman to Prov. Secretary, 19 Apr. 1864, N.S. Prov. Sec. Papers, RG7 vol. 50. Nova Scotia, House of Assembly, *Debates and Proceedings*, 18 Apr. 1864, p. 229.

[35] Richardson, "Topographical and Geological Notices" (cit. n. 8), pp. 68–9; Logan to De la Beche, 12 May 1845, De la Beche Papers; Logan, "Sketch of the Geology of Canada" (cit. n. 30); Alexander Kennedy Isbister, "On the Geology of the Hudson's Bay Territories," *Geol. Soc. London Quart. J.*, 1855, *11:*513; see also "Canadian Supply of Coal," *Toronto Globe*, 1 Feb. 1855; Alfred R. Roche to Logan, 15 May 1857, W. E. Logan Papers; *Report*, Select Committee on the Hudson's Bay Company, House of Commons, Parliament, Great Britain (London, 1857), esp. ques. 2898, 2935, 3083–3113; Canada, Province, House of Assembly *Journal*, 1858, app. 3.

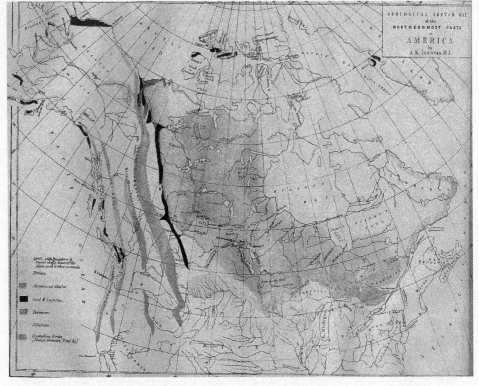

Map 3. *"Geological Sketch Map of the Northernmost Parts of America, collation to illustrate A. K. Isbister, 'On the Geology of the Hudson's Bay Territories,'"* Geological Society of London Quarterly Journal, *1855,* 11:497–520. *Estimated coal and lignite formations prominently marked.*

pansionist publicist, Hind waxed more enthusiastic than scientific about more limited terrain, exciting public notice as Palliser's sober reports could hardly do.[36]

Hind also became New Brunswick's consulting geologist in 1864, referring formations to Logan that he deemed "repetitious of rocks which occur in Canada." By contrast, his local colleagues, L. W. Bailey, the University of New Brunswick's

[36] James Hector, "Notes on the Geology of Captain Palliser's Expedition in British North America," *Edinburgh New Phil. J.*, 1860, *12:*225–8; *idem*, "On the Geology of the Country between Lake Superior and the Pacific Ocean," *Geol. Soc. London Quart. J.*, 1861, *17:*388–445, reprinted in *Can. Natur. and Geol.*, 1861, *6:*330; *idem*, "Physical Features of the Central Part of British North America," *Can. Natur. and Geol.*, 1861, n.s., *14:*212–40; *idem*, "On the Physical Features of the Central Part of British North America, and on Its Capabilities for Settlement," *Report*, British Association for the Advancement of Science, 1861, pp. 195–6 and *Edinburgh New Phil. J.*, 1861, *4:*160. Irene M. Spry, "Palliser, John," in *Dictionary of Canadian Biography* (cit. n. 21), vol. 11 (1982), pp. 661–4; William O. Kupsch, "Hector's Geological Explorations of the Canadian Prairie West (1857–1859)," *Proceedings*, Geological Association of Canada, 1971, *23:*31–41. Henry Youle Hind, editorial in *Can. J. Indust., Sci. and Art*, 1852, *1:*75–7; *idem, Narrative of the Canadian Red River Exploring Expedition*, reprint ed., 2 vols. (Edmonton: Hurtig, 1971), vol. 2, pp. 234, 354, 357, 370; *idem*, "Canadian Expeditions to the North-West Territory," *Can. J. Indust., Sci. and Art*, 1860, *5:*550. Anon. review of Hind, *Narrative*, in *Can. J. Indust., Sci. and Art*, 1861, n.s. *6:*68; Zeller, *Inventing Canada* (cit. n. 2), chap. 5; William L. Morton, *Henry Youle Hind* (Toronto: Univ. of Toronto Press, 1980), pp. 26, 78; Zeller, "Gullivers and Crusoes" (cit. n. 3), pp. 232–5; Zeller, "Classical Codes" (cit. n. 9), pp. 26–30.

professor of geology and natural history; G. F. Matthew, and C. F. Hartt struggled to correct earlier maps by unraveling the colony's unusually contorted strata.[37] In 1864, Logan also seconded Alexander Murray to crack Newfoundland's "harsh and forbidding" geological codes. "Newfoundland," Murray soon agreed, was "formed of the chippings of the world." Advancing his Laurentian empire, Logan kept Murray working there even after Newfoundland rejected Confederation. Indeed, as Robert Bell (1841–1917), Logan's younger GSC colleague, astutely recognized, Logan's "great map" (Map 4) in 1865 appeared "at the right nick of time . . . as the old detached maps of the province did not tend to instruct the provincialists in one section on the geography of the others." Extending inquisitive tentacles ever outward before Confederation warranted it politically, Logan's GSC normalized perceptions of transcontinental dominion as a natural development.[38]

In other ways, too, Victorian science fanned Canadian expansionist embers, through both Lorin Blodget's isothermal prognostications in 1857 of a temperate Northwest—partly based on John Richardson's Humboldtian inquiries—and the timely "discovery" of *Eozoon canadense*. GSC staff in 1858 reported curious markings in the "Laurentian limestones" near Ottawa, where there had seemed "little hope of finding anything very new or striking." Logan hesitated to declare them organic remains: fossils were not his forte, and his palaeontologist, Elkanah Billings, rejected the notion outright. By 1861, however, Dawson was steering the fossil bandwagon, interpreting the imprints as "the beginning of life" on earth. Seeing in the imprint a "higher Foraminifer," he soon dubbed it the "dawn animal of Canada"— priceless creationist support for his own ongoing anti-Darwinian campaign. Logan now found reasons to agree: his chemist, Sterry Hunt, had long held the theoretical possibility of finding evidence of life in the Laurentian system; an embarrassing public dispute between Billings and James Hall over strata that Logan had dubbed the "Quebec Group" called into question Billings's judgment; and, moreover, Logan respected Dawson. Accepting palaeontologists' affirmations over crystallographers' denials, Logan proclaimed his British presentations "altogether a great success," with Lyell pronouncing *Eozoon* "one of the most remarkable steps made in our time" and Murchison adopting it as Siluria's baseline.[39]

[37] Hind to Logan, 31 May, 10 July, 30 July 1864; 19 Dec. 1865, W. E. Logan Papers. Henry Youle Hind, "Observations on Supposed Glacial Drift in the Labrador Peninsula," *Can. Natur. and Geol.*, 1864, *1*:325–7; [John William Dawson?], Review of Loring Woart Bailey, Henry Youle Hind *et al.*, *Geology of New Brunswick, Can. Natur. and Geol.*, 1865, *2*:204–7. See also n. 32 above.

[38] Robert Bell to Logan, 27 Nov. 1864; Logan to A. Daubie, 17 Sept. 1866; Alexander Murray to Logan, 23 Apr. 1864, 4 June 1866; Logan to Murray, 16 July 1866; H. W. Hayles to Logan, 19 Apr. 1865; Edward Carter to Logan, 21 May 1866; Edward Morris to Logan, 16 Oct. 1866; Logan to Edward Morris, 5 Nov. 1866, W. E. Logan Papers. Isaac Chipman to Dawson, 15 Oct. 1846; Logan to Dawson, 28 Jan. 1867, 1 Dec. 1869, Dawson Papers. Murray quoted in Thomas C. Weston, *Reminiscences Among the Rocks* (Toronto, 1899), p. 93.

[39] Dawson to Lyell, 20 Mar. 1858, 28 Jan. 1859, 26 Apr. 1860, 13 Nov. 1860, 15 May 1861, 1 Mar. 1864, 28 Apr. 1864, Lyell Papers. Logan to Dawson, 24 Sept., 13 Oct. 1864; Charles Moore to Logan, 25 Sept. 1864; M. Salter to Logan, 8 Oct. 1864; James Hall to Logan, 6 Aug., 6 Dec. 1863, W. E. Logan Papers. Logan to Bell, 15 Dec. 1864, Robert Bell Papers, McGill Univ. Archives, Montreal. William Edmond Logan, "On the Occurrence of Organic Remains in the Laurentian Rocks of Canada," *Geol. Soc. London Quart. J.*, 1865, *21*:45–50. Lyell to Dawson, 18 Feb. 1865; Hall to Dawson, 5 May 1862, 7 Nov. 1863, Dawson Papers. Logan to Hall, 8 Dec. 1861, 14 Dec. 1863, 19 Oct., 21 Nov. 1866, 11 Jan. 1867, fol. 824, James Hall Papers, New York State Library, Albany (hereafter cited as Hall Papers). Hall to Logan, 11 Oct. 1864, n.d. 1864, fol. 197, Hall Papers. Elkanah Billings, "On the Date of the Report on the Geology of Wisconsin," *Can. Natur. and Geol.*, 1862, 7:art. 15; *idem*, "Remarks Upon Professor Hall's Recent Publication," *Can. Natur. and Geol.*, 1862, 7:art. 36.

Although *Eozoon*'s scientific supporters eventually dwindled to Dawson alone, its ideological cachet lingered. While the pioneer-naturalist Catharine Parr Traill lamented of Canada in 1836 that "there is no hoary ancient grandeur in these woods, no recollections of former deeds connected with the country," Romantic Canadian nationalists in 1865 revered the Laurentian features "which stretch from the sterile coast of Labrador to the fertile regions of the far West":

> Thus do the stones cry out, and in their sublime majesty preach strange sermons . . . [in] a land whose antiquities have been explored by [Logan's] genius, illustrated by his pencil, described by his pen, and reproduced in a manner [that has] awakened admiration in the old world as well as in the new, and has taught [both] to think of the attractions of a land on whose surface are some of the newest settlements of the human family.[40]

Logan's magnum opus, the *Geology of Canada* (1863), fulfilled his original mandate, upholding science's disturbing verdict that the province, surrounded by coal, was bereft of it. The GSC marked time while colonists debated their futures, political deadlock in Canada having necessitated constitutional renewal. Amid uncertainty, however, the GSC enjoyed higher ground, Logan having cultivated both sides so successfully that his few remaining critics were dismissed as fossils by their own colleagues.[41] Double rewards in 1867 clinched the point: besides the Royal Society of London's Gold Medal, Logan earned Murchison's personal dedication of the revised *Siluria*. Murchison appreciated Logan's application of his Silurian system to "the vast regions of British North America," as well as his research on the Laurentian "foundation-stones of all Palaeozoic deposits in the crust of the Globe, wherever their foundations are known." Rather than termination, the GSC saw Confederation that year extend it to all the other provinces.[42]

Under an expansionist new dominion government, Logan's staff traced Laurentian boundaries—and potential transcontinental railway routes—westward into what remained HBC territory. Now seventy years old, Logan shrank from longstanding ambitions, anointing as his successor A. R. C. Selwyn (1824–1902), who was trained in stratigraphy by the British Survey and was founding director of the Geological Survey of Victoria, Australia. Unacquainted with even the settled provinces' geology, Selwyn inherited a rapidly distending enterprise as Canada swallowed Rupert's Land (1870); British Columbia (1871) and Prince Edward Island (1873) stretched Confederation from sea to sea. Interest in Newfoundland's resources persisted, with British fears of impending "coal exhaustion" galvanizing further searches while petroleum and electricity were cultivated as alternative fuels.[43]

See also William E. Eagan, "'I would have sworn my life on your interpretation': James Hall, Sir William Logan and the 'Quebec Group,'" *Earth Sci. Hist.*, 1987, 6:47–60. Roderick I. Murchison, *Siluria*, 4th ed. (1839, 1854, London: 1867), pp. v, 12–13.

[40] C. Parr Traill, *The Backwoods of Canada*, reprint ed. (London, 1846), p. 154; Fennings Taylor (1865) quoted in Zeller, *Inventing Canada* (cit. n. 2), see pp. 103–5.

[41] Zaslow, *Reading the Rocks* (cit. n. 24), p. 83; Zeller, *Inventing Canada*, (cit. n. 2), pp. 101–3.

[42] Murchison, *Siluria* (cit. n. 39), pp. iii–iv. Logan to Bell, 13 Apr., 1 Oct. 1865; Logan to Murchison, 29 Nov. 1867, W. E. Logan Papers. Logan to Bell, 25 Jan. 1867, Geological Survey of Canada Director's Letterbooks, National Archives of Canada, Ottawa, p. 233 (hereafter cited as GSC Letterbooks).

[43] Logan to Bell, 14 May 1868; John Bell to Bell, 28 June 1869, Robert Bell Papers, National Archives of Canada, Ottawa. *Report of Progress from 1866 to 1869*, Geological Survey of Canada (Montreal, 1870), pp. 315, 363–4. Logan to Andrew Crombie Ramsay, 27 Nov. 1867, p. 287 and 2

Selwyn's considerable challenges in the transcontinental arena included oversee-ing distant surveyors; balancing expansive reconnaissance with detailed observa-tions; defending and extending the GSC's pioneering work on Archaean formations; accommodating a budding civil service; moving GSC headquarters from Montreal to the new capital city in Ottawa; and tackling new natural history responsibilities. Following Logan, he spearheaded coordinations of international geological nomen-clature and mapping schemes; and promoted, with J. W. Dawson, a renewed British imperialism, with Montreal hosting the British Association for the Advancement of Science (BAAS) in 1884.[44]

Selwyn shrewdly added highly qualified staff. J. W. Dawson's son, G. M. Dawson (1849–1901), fresh from London's Royal School of Mines (ironically as T. H. Hux-ley's prize student), observed, extrapolated, and theorized accurately over large ex-panses of complex terrain. Invaluable service with the International Boundary Com-mission prepared him for GSC fieldwork from 1875.[45] Beginning in British Columbia, Dawson explicated Cordilleran, Cretaceous, and Pleistocene formations, coordinating metamorphic effects with global volcanic activity. Like his elders pre-ferring iceberg drift to ice-age explanations, especially on the prairies, Dawson nev-ertheless admitted the spectacular impact of a great Cordilleran ice sheet in the Rocky Mountains. In Lyell's uniformitarian tradition of centers of dispersion, and inspired by the American geologist Warren Upham, Dawson's colleagues J. B. Tyr-rell, and A. P. Low, during the 1880s and 1890s, posited additional Keewatin, Labra-dorean, and Hudson Bay ice sheets emanating from the north. Dawson's original insights offered practical mining implications, including placer gold locations that extended public expectations to the Yukon.[46]

For all of geology's Victorian heyday, in appointing Selwyn, Logan had ignored his own precedent as a Canadian-born director. As a result, Selwyn's GSC suffered

Dec. 1868, p. 405; Logan to John Rose, 14 Feb. 1868, p. 305; 29 Feb. 1868, pp. 317–19; 30 Dec. 1868, p. 418; 29 Mar. 1869, p. 460; 16 Apr. 1869, pp. 474–5; Logan to William McDougall, 25 Aug. 1866, p. 152; 10 June 1869, p. 17; Logan to Albert Richard Cecil Selwyn, 4 Dec. 1868, p. 406; Logan to Bell, 16 June 1869, p. 22; Logan to W. Betts, 22 June 1869, p. 25; Selwyn to A. C. Ramsay, 23 Dec. 1870, p. 300, GSC Letterbooks. Charles Robb, "On the Petroleum Springs of Western Canada," *Can. J. Indust., Sci. and Art*, 1861, *6:*313; Thomas Sterry Hunt, "Notes on the History of Petroleum or Rock Oil," *Can. Natur. and Geol.*, 1861, *6:*art. 15; Sandford Fleming, "Note on the Present Condi-tion of the Oil Wells at Enniskillen," *Can. J. Indust., Sci. and Art*, 1863, *8:*249; William Stanley Jevons, *The Coal Question* (London, 1865); Robert Grant Haliburton, *The Coal Trade of the New Dominion* (Halifax, 1868); Richard Brown, *The Coal Fields and Coal Trade of the Island of Cape Breton* (London, 1871). William O. Kupsch, "GSC Exploratory Wells in the West, 1873–1875," *Earth Sci. Hist.*, 1993, *12:*160–79.

[44] Suzanne Zeller, "Selwyn, Alfred Richard Cecil," in *Dictionary of Canadian Biography*, gen. ed. Ramsay Cook (Toronto: Univ. of Toronto Press, 1994), vol. 13, pp. 934–5; David F. Branagan, "Alfred Selwyn—19th Century Trans-Atlantic Connections via Australia," *Earth Sci. Hist.*, 1990, *9:*143–57; David R. Oldroyd, "The Archaean Controversy in Britain: Part I," *Ann. Sci.*, 1991, *48:*407–52; Albert Richard Cecil Selwyn, "On the Origin and Evolution of Archaean Rocks," *P and T*, Royal Society of Canada, 1896, 2nd ser., *2:*lxxviii–xcix.

[45] Suzanne Zeller and Gale Avrith-Wakeam, "Dawson, George Mercer," in *Dictionary of Canadian Biography* (cit. n. 44), vol. 13, pp. 257–61.

[46] George Mercer Dawson, *On Some of the Larger Unexplored Regions of Canada* (Ottawa, 1890); idem, *The Progress and Trend of Scientific Investigation in Canada* (Ottawa, 1894). Warren Upham, *Glacial Lake Agassiz*, Monographs of the U.S. Geological Survey, U.S. Dept. of the Interior (Wash-ington, 1895), vol. 25; Zaslow, *Reading the Rocks* (cit. n. 24) pp. 159, 162, 164–5, 170, 193; J. A. Elson, "The Contributions of J. W. Dawson (Father) and G. M. Dawson (Son) to the Theory of Glaciation," *Geoscience Canada*, 1983, *10:*213–16; William Eagan, "The Multiple Glaciation De-bate: The Canadian Perspective," *Earth. Sci. Hist.*, 1986, *5:*144–51.

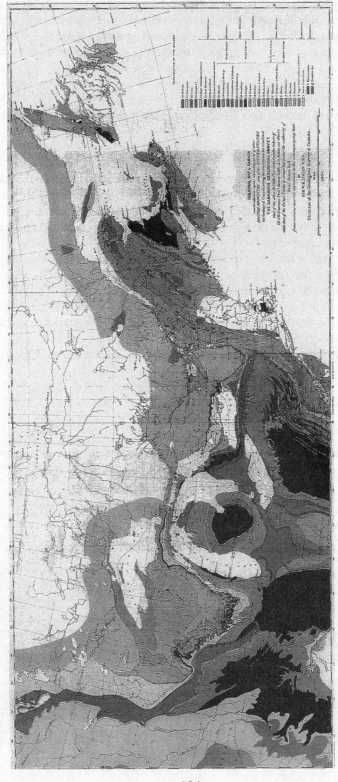

Map 4. Sir W. E. Logan, "Geological Map of Canada and Adjacent Regions Including Parts of the British Provinces and of the United States," in Atlas, *Geological Survey of Canada (1865), to accompany W. E. Logan, Geology of Canada (1863). Includes Laurentian and coal formations as they extend west, east, and south of the Province of Canada.*

from deep personal resentments that were left unresolved by a public inquiry in 1884. G. M. Dawson's succession as director in 1895, and Robert Bell's in 1901, dissolved these nativist tensions into familiar battles between theory and practice, in which Logan found no equal among his successors.[47]

CONCLUSION

With geology in its vanguard in Canada, Victorian scientific culture rested solidly upon empirical, utilitarian, and increasingly entrepreneurial foundations, promising to wrest widespread material prosperity from nature. Victorian scientists' Baconian identification of knowledge with power thus stood them in good stead. As producers of useful knowledge, they exercised effective control over its explication and dissemination. Geologists, in particular, helped to define both possibilities and limitations in colonial economic development, proclaiming their ability to locate, classify, and publicize a given territory's agricultural/industrial resources.[48]

A quintessential "scientist of empire," Roderick Murchison was by no means alone at the helm in this remarkable undertaking. Against substantial political, economic, and cultural odds, British North American geologists modeled their own empires, with important repercussions in each of these realms. Their efforts helped to modernize the expansive outlook that inspired their intellectual descendant, the political economist Harold Innis, in his famous claim that Canada developed "not in spite of geography, but because of it." Victorian geology, suggested powerful voices among its practitioners and supporters, was used to justify a wholesale reconstitution of British North America into a transcontinental Canadian dominion. Yet it did so on terms that saw the scientific independence prescribed by Basalla's model as a limitation more than as a goal.[49]

J. W. Dawson articulated most powerfully the bonds linking his culture, his politics, and his science. In 1858, in the *Canadian Naturalist and Geologist*, he explained that since "geological conditions . . . determine beforehand the resources and population" of any given land, it was geologists who could best inform settlement and development policy. The private, Anglo-Montrealer in Dawson who harbored doubts about Confederation nevertheless deferred to naturalists' experience of British America's "natural features as fixing its future destiny, and indicating its present interests," leading them to "regard its local subdivisions as arbitrary and artificial." Soon after Confederation, he marveled that "Nature" had indeed "already taken hold of the mind of Young Canada," and was "moulding it in its own image." The founding president of the Royal Society of Canada in 1882, Dawson nevertheless esteemed more highly his distinction in 1886 as the only person to have held

[47] Zaslow, *Reading the Rocks* (cit. n. 24), chap. 7. Dawson died suddenly in 1901, and was succeeded by Robert Bell.

[48] Richard Yeo, "An Idol of the Market-Place: Baconianism in Nineteenth-Century Britain," *History of Science*, 1985, *23*:251–98; A. D. Orange, "The Idols of the Theatre: The British Association and Its Early Critics," *Ann. Sci.*, 1975, *32*:277–94; Morris Berman, *Social Change and Scientific Organization* (Ithaca, N.Y.: Cornell Univ. Press, 1978); Susan Faye Cannon, *Science in Culture* (New York: Science History Publications, 1978); Richard Yeo, *Defining Science* (Cambridge: Cambridge Univ. Press, 1993); Donald Worster, *Nature's Economy*, 2nd ed. (Cambridge: Cambridge Univ. Press, 1994), pp. 30–1, 105–6.

[49] Harold Innis, *The Fur Trade in Canada* (Toronto: Univ. of Toronto Press, 1930); see also Suzanne Zeller, *Land of Promise, Promised Land: The Culture of Victorian Science in Canada* (Ottawa: Canadian Historical Association Historical Booklets, 1996), #56.

both AAAS and BAAS presidencies. Yet he went on in 1887 to promote the establishment of a more exclusive Imperial Geological Union.[50]

The geological myths and metaphors that undergirded a transcontinental Canadian discourse—between Britain in the Old World and the United States in the New—saw not contradiction but consistency in Dawson's varying positions. As Victorian geologists illuminated, unraveled, and transcribed the tangled stories in the strata they encountered, they presumed a natural coherence between local connections and distant geographical/historical links. The same held true in British North America. In the natural Laurentian heartland that unified the colonies from Newfoundland to Hudson Bay's western watershed—possibly extending to form the central axis of the Rocky Mountains, as G. M. Dawson theorized (and John Richardson would have loved); in the natural patterns of coal's patchwork dispersion; and in the continuing formal relationship with Britain even after Confederation, many Canadians envisioned their own natural ascendance, through transcontinental nationhood, to leadership in the empire itself. During the 1880s, the GSC's Robert Bell and John Macoun propelled new versions of these metaphors and myths ever north- and westward, in Canadian efforts to absorb Manitoba, Hudson Bay, and the territory beyond.[51]

The difference between this self-inclusive conception of empire—with science its faithful partner—and Basalla's presumed trajectory towards independence from the metropolis suggests a historical difference in the understanding of "natural" paths to progress. Whereas an Enlightenment preference for the universal and the categorical had sustained the American Revolution a century earlier, Victorian grapplings with variation in nature formed an important backdrop for Canadian Confederation. In 1846, well before Dawson's explication of the nature-culture interplay, the *British American Journal of Medical and Physical Science* appreciated that something special served "to lighten the toils of the practical geologist, and to form a sort of compensation for them, abstracting altogether from their value as a source of national wealth, or a professional occupation." Beneath the "hope of discoveries" and "the expectation of any direct utility" lay a bedrock of imagination—"that instinctive love of the miraculous, and the beautiful, that unconquerable delight which many persons have in beholding the pictures, and dwelling in the palaces, of nature." Imagination "inspired the enthusiasm which is prophetic, and, at the same time, productive of success," enkindling particular dreams generated and defined within equally particular historical contexts.[52]

More than a scientific hinterland between two metropolises, the Canada reshaped so effectively by Victorian geology presumed in its believers imagination sufficiently bold "to think, that as it is the latest developed portion of a new world—as it was the first, by millions of years, to nurse and cradle in her bosom the first spark of animal life in the eozoon—it may be the country where a last great, and fully

[50] John William Dawson, "Report of the GSC 1853–55," *Can. Natur. and Geol.*, 1858, *3*:34; [John William Dawson], "Pamphlets on British America," *Can. Natur. and Geol.*, 1858, *3*:392–3; Dawson to Lyell, 12 Dec. 1866, p. 909, Lyell Papers; MS speech (1868), "On the British American Mind," Dawson Papers; Sheets-Pyenson, *John William Dawson* (cit. n. 31), pp. 198–203.

[51] William Norris, "Canadian Nationality: A Present-Day Plea," in *Rose-Belford's Monthly and National Review*, 1880, *4*:118; see also Carl Berger, *The Sense of Power* (Toronto: Univ. of Toronto Press, 1970).

[52] "GSC: Report of Progress for the Year 1844," *British American Journal of Medical and Physical Science*, 1846, *2*:66; Zeller, "Classical Codes" (cit. n. 9).

developed humanity may find its fitting habitation and abode." Here not palm and pine, but coal and gold—and the tensions induced by their absence—redefined territories claimed and defended by geology's sultans of science, making Logan and Dawson lords of Laurentia just as surely as Murchison commanded his Siluria. Not only, as Roy MacLeod suggests, did imperial-colonial power structures in the history of science work both ways, they also bred new empires, replaying their cycles of dissemination and domination over and over again.[53]

Nor have Canadians yet abandoned their affinity for geological metaphors to describe their world. Reporting on an article from *Nature* recently, the country's national newspaper announced that "Canada as a geological entity probably began where Canada as a political entity is threatening to split apart." Indeed, geophysicists themselves ascribed political irony to the collision aeons ago of two nonconformable microcontinents—Opatica, a volcanic arc, and Abitibi, an ocean plateau—near James Bay, across the present Ontario-Quebec boundary. Masked superficially by subsequent reheatings of the earth's crust, the authors emphasized, the suture marking this violent ancient encounter never healed.[54]

[53] Norris, "Canadian Nationality" (cit. n. 52), [John William Dawson?], review of GSC, *Report of Progress 1858*, in *Amer. J. of Sci.*, 1861, *31*:122; see also Rudwick, "Historical Analogies" (cit. n. 19), pp. 89–107; Robert H. Dott, Jr., "The American Countercurrent—Eastward Flow of Geologists and Their Ideas in the Late Nineteenth Century," *Earth Sci. Hist.* 1990, *9*:158–62; MacLeod, "Moving Metropolis" (cit. n. 1), pp. 244–5.

[54] Stephen Strauss, "Quebec, Where It All Started—Maybe," *Toronto Globe and Mail*, 24 June 1995, pp. A1, A3; for the scientific report see A. J. Calvert, E. W. Sawyer, W. J. Davis, and J. N. Ludden, "Archaean Subduction Inferred from Seismic Images of a Mantle Suture in the Superior Province," *Nature*, 1995, *375*:670–4.

The Shaping of Latin American Museums of Natural History, 1850–1990

Maria Margaret Lopes and *Irina Podgorny***

ABSTRACT

This essay reflects upon the milieu and the character of Brazilian and Argentinean natural history museums during the second half of the nineteenth century. It argues that the museums were influenced not only by European and North American museums but by each other. Museum directors in the two countries knew each other and interacted. Some of the relationships between these museums were friendly and cooperative, but because they were in young, emerging nations, they also became deeply involved in the invention of nationality in their respective countries and interacted as rivals and competitors. Even through rivalry, however, they contributed to each other's development, as did rivalry among museums within each of the two countries. Later in the century they went well beyond the nationalist perspective, finding, through their research into paleontology and anthropology in their regions, a continental and uniquely South American scientific perspective, defined in reaction to North American and European views.

INTRODUCTION

IN RECENT YEARS, SEVERAL HISTORIANS HAVE STUDIED MUSEUMS of natural history and the ordering of nature in the eighteenth and nineteenth centuries.[1] Most of the history of South American museums has been written within spe-

---

* Instituto de Geociências, State University of Campinas, SP, Brazil.
** CONICET, Department of Archaeology, Museo de La Plata, Argentina.

Acknowledgments: Irina Podgorny acknowledges the support of an early career grant from Fundación Antorchas (Buenos Aires), and Maria Margaret Lopes acknowledges the support of FAPESP Brazil, and the Rockefeller Foundation Humanities Fellowship program, "Pro Scientia et Patria: Los Museos argentinos y la construcción de un patrimonio nacional," through the Museo Etnográfico de la Facultad de Filosofía y Letras de la Universidad de Buenos Aires. They are both especially grateful to the director of the Museo Etnográfico, José A. Pérez Gollán; Dr. Roy MacLeod, University of Sydney, Australia; and Dr. Silvia Figueirôa, University of Campinas, Brazil. This article is for Susan Sheets-Pyenson, *in memoriam*. (See David Allen, "Eloge: Susan Ruth Sheets-Pyenson, 9 September 1949–18 August 1998," *Isis*, 1999, *90*:168–9.)

[1] Sally G. Kohlstedt, "International Exchange and National Style: A View of Natural History Museums in the United States, 1850–1900," in *Scientific Colonialism: A Cross-Cultural Comparison*, eds. Nathan Reingold and Marc Rothenberg (Washington, D.C.: Smithsonian Institution, 1987), pp. 167–90; Mary P. Winsor, *Reading the Shape of Nature: Comparative Zoology at the Agassiz Museum* (Chicago: Univ. of Chicago Press, 1991); Claude Blanckaert, C. Cohen, P. Corsi *et al.*, eds., *Le Muséum au premier siècle de son histoire* (Paris: Muséum National d'Histoire Naturelle, 1997);

---

cific national traditions, however, and retains a hagiographic and parochial flavor. In 1988, Susan Sheets-Pyenson discussed two Argentinean museums, the Museum of La Plata and the National Museum of Buenos Aires, within the broader context of the expansion of colonial science.[2] The avenue of discussion that she opened up, however, has neither been followed nor contested. This silence cannot be explained by discontent with a revised framework. It suggests, rather, how highly a national scientific tradition may be valued by those who perceive themselves as its heirs. In this context, local institutions have continued to be seen as if they had emerged as isolated and independent phenomena.

In this essay, we do not attempt to write the history of Latin American museums generally or of Argentinean or Brazilian museums in particular.[3] Rather, we wish to reflect upon the milieu, the rhetorical character of scientific institutionalization and consolidation, from a comparative perspective. A Latin American point of view that aims at going further than the limits imposed by national boundaries quickly reveals that national histories have more in common than it is customary to assume. With regard to Argentinean and Brazilian museums of the second half of the nineteenth century, we want to show that the influence of contemporary European and North American institutions was mediated by South–South links among local naturalists.

THE LOST WORLD

As a consequence of the "enlightened" policies of Spain and Portugal at the turn of the eighteenth century, museums, cabinets, botanical gardens, and scientific societies arrived in Iberian America on the ships of metropolitan scientific expeditions. In return, the cabinets of America sent collections to the new or reorganized botanical gardens and museums of Madrid, Lisbon, and Coimbra. Within this framework came the first Cabinete de Historia Natural, in Havana, Cuba, the Casa Botánica de Bogotá in Colombia, the Casa de História Natural of Rio de Janeiro, as well as cabinets in Mexico and Guatemala.

The disruptive events that followed—scientific controversies, wars of independence, shortages of funding and staff—led to the dispersal of these early collections. Nevertheless, what remained was the basis on which museums were organized in the context of newly independent South American colonies during the first half of the nineteenth century.[4] Continuities with the scientific and cultural projects inherited from the colonizing powers did not mask the special character of these new institutions. The museums of natural history established in Buenos Aires (1812 / 1823), Rio de Janeiro (1818), Santiago de Chile (1822), Bogotá (1823), Mexico (1825), Lima (1826), and Montevideo (1837) were all framed in the process of building new nation states; national museums were found as former colonies became

Andreas Grote, ed., *Macrocosmo in Microcosmo: Die Welt in der Stube; zur Geschichte des Sammelns, 1450 bis 1800* (Opladen: Leske & Budrich, 1994).

[2] Susan Sheets-Pyenson, *Cathedrals of Science: The Development of Colonial Natural History Museums During the Late Nineteenth Century* (Montreal: McGill–Queen's Univ. Press, 1988).

[3] For an overview of Brazilian museums, see Maria Margaret Lopes, *O Brasil descobre a pesquisa científica: Os Museus e as ciências naturais no século XIX* (São Paulo: Hucitec, 1997).

[4] Cultural and academic exchanges between Spain and Portugal and their former colonies in America were reestablished at the end of the nineteenth century.

independent.[5] In the New World, museums were the loci of institutionalization of natural history. But as a standard measure by which to test the scientific culture of a country, they also became symbols of national identity.

The governments of the new political entities created in Brazil and in the Andean and River Plate regions had high hopes for the natural sciences and their museums. In particular, the Creole elites of the independent republics sought to overcome the economic, social, and cultural fragmentation resulting from the rupture of the colonial order by discovering and surveying new reserves of natural resources. One of the main roles of the natural history museum, especially in the very first years, was to collect and to display the mineralogical resources of these territories. Mineralogical collections were used for teaching in new university courses, but they were also used to create dazzling exhibitions of the country's mineral wealth. Interest in mineral research was at the core of the first museums of Latin American countries,[6] but in the second half of the nineteenth century other fields such as evolution took their place. This process can be understood partly as a consequence of the growing specialization of the natural sciences, but also in terms of the emergence of specific institutions for different scientific disciplines. In fact, by the second half of the nineteenth century the museums of Latin America had changed their priorities from mineralogical "El Dorado" exhibits to broader scientific purposes.

One common trait of natural history museums in this period—not limited to Latin America, as Sheets-Pyenson observed—tied the "builders" of science to their institutional and scientific settings.[7] On the other hand, alliances and conflicts among museum directors and scientific staff did not necessarily follow national lines. In fact, international scientific networks sometimes brought together local and "foreign" scientists, in defense of local institutional leaders and their museums.

Museums, as symbols of urban civilization, were also specific loci for displaying the histories of local nature and the histories of the extinct—or nearly extinct—indigenous inhabitants of the New World. In Latin America, as elsewhere, museums were also institutions where knowledge was produced, following the patterns of contemporary scientific practice. The nineteenth century witnessed many changes in this practice and new roles for museums in teaching, research, collecting, storage, and exhibitions. The mineralogical collections of the museums—which, by 1800, already included specialized collections in geology, anthropology, botany, zoology, and archaeology—were eventually replaced by paleontological collections illustrating the evolution of species. In this field, the Argentinean museums of La Plata and Buenos Aires became continental points of reference for research on the fossil remains of extinct mammals. Argentinean and Brazilian museum directors shared a common faith in science viewed as the warranty for progress, and they conceived of museums as centers for the nationalization of local nature. Their mission was assumed to have a civilizing aim, and native peoples were included in the naturaliza-

[5] Maria Margaret Lopes, "A construção de Museus Nacionais na América Latina Independente," *Anais Museu Histórico Nacional do Rio de Janeiro*, 1998, *32:*121–45.

[6] Silvia F. de M. Figueirôa, *As ciências geológicas no Brasil: Uma história social e institucional, 1875–1934* (São Paulo: Hucitec, 1997); Irina Podgorny, "Un Belga en la corte de Paraná," in *En los deltas de la memoria: Bélgica y Argentina en los siglos XIX y XX*, B. De Groof *et al.*, eds. (Louvain: Univ. of Louvain, Press, 1998), pp. 55–61.

[7] Irina Podgorny, *El Argentino despertar de las faunas y de las gentes prehistóricas* (Buenos Aires: Eudeba, in press).

tion of history. Adapted to cultural and scientific changes, Latin American museums became not only places for systematic research but also monuments. The Museo de La Plata, where glyptodonts became symbols of Argentina's glory, was the most conspicuous example.[8]

FRIENDLY EXCHANGES AND RIVALRIES

By the 1850s, it was received wisdom that every major European nation should possess—or already possessed—a national museum of natural history, aiming (or professing) to be a more or less complete epitome of the three kingdoms of nature: animals, plants, and minerals. Such was the extent and influence of the great Musée d'Histoire Naturelle in Paris and the Natural History Division of the British Museum in London. Debates took place about the extent to which a public museum of natural history should be supported by the state, on what scale, for what public, and with reference to what commercial and colonizing endeavors.[9]

The role of the state as patron was constantly invoked, along with the need for museums to be located in national capitals. At the core of all these debates arose the problem of storage and exhibition space, which was presented as a "natural" problem resulting from the richness of the country's natural resources and the vastness of its territories. In displaying the extent and variety of the Creative Power—and of the power of the state—exhibits of large animals became a trope in museum rhetoric. To have enough room for the mounted skeletons of a whale and a large extinct South American mammal was taken as the mark of a triumphant state.

During the 1860s and 1870s, in the South American lands that provided European museums with large fossil mammals, national institutions in Argentina, Brazil, Chile, and Uruguay were renewed by government funding. In 1862 Hermann Konrad Burmeister (1807–1892), a Prussian naturalist, was hired as director of the Museo Público de Buonos Aires, where he served until his death thirty years later. In 1868 his contemporary, Ladislau Netto (1838–1894), a Brazilian naturalist trained at the Musée d'Histoire Naturelle in Paris, was hired to lead the Museu Nacional do Rio de Janeiro, which he directed until 1892. Both headed fifty-year-old, federally supported institutions with considerable collections; nevertheless they consistently stressed the poverty of the museums they had to remake and were always seeking more funding.[10]

Netto left botanical studies to begin the classification of the Museu Nacional's anthropological and ethnographical collections. During his tenure he transformed the museum into a scientific institution of international standing. He was active not only in increasing state funding but also in promoting the double role of the museum as a place for both research and teaching. (Since Brazil had no university until the twentieth century—only isolated faculties of engineering, medicine, and law—the

---

[8] Irina Podgorny, "De razón a facultad: Ideas acerca de las funciones del Museo de La Plata en el período 1890–1918," *Runa*, 1995, 22:89–104.

[9] Richard Owen, *On the Extent and Aims of a National Museum of Natural History. Including the substance of a discourse on that subject, delivered at the Royal Institution of Great Britain, on the evening of Friday, April 26, 1861* (London: Saunders, Otley and Co., 1862).

[10] Hermann Burmeister, "Sumario sobre la fundación y los progresos del Museo Público de Buenos Aires," *Anales del Museo Público de Buenos Aires para dar a conocer los objetos de la História Natural nuevos o poco conocidos en este establecimiento.* Entrega primera (Buenos Aires: Bernheim & Bonneo, 1864), pp. 1–10. In the 1880s, the museum's name was changed to Museo Nacional.

Figure 1. *First building of Museu Nacional from 1818 to 1892. Archives Museu Nacional.*

teaching of science at the museum served a valuable function.) In 1876 he began the *Archivos do Museu Nacional*, in which scientific staff published research from the institution's different departments. In the 1880s his museum, as well as the Museo National de Santiago de Chile[11] and the Museo Público of Buenos Aires, were proud to exhibit skeletal whales in their halls. Netto wanted to have in Rio de Janeiro, the capital of the Brazilian Empire, a museum that was both metropolitan and universal. Side by side with Chinese porcelain from the Portuguese colonies in Africa and Asia, he displayed artifacts from Pompeii, Egyptian mummies, and Japanese herbaria—all essential parts of the metropolitan and universalistic regime that he imparted to his museum.

In order to stress the uniqueness of his Brazilian museum's natural, archaeological, ethnographical, and anthropological collections, Netto emphasized the museum's contribution to all of science. At a moment when it was important to enhance contacts with the most prestigious museums of the world, the Museu Nacional incorporated local findings and redefined them, presenting them as "national." Thus nationalized, they could compete with other territories equally exotic.

Netto's museum in Rio de Janiero was the national museum of Brazil. Because Argentina was not yet formally a nation, Burmeister's museum in Buenos Aires was still a provincial museum, sponsored by the province of Buenos Aires. It was associated with the local university, sharing rooms with it in the same building. Burmeister proposed a model of a museum with a local scientific character, focused on local zoology, with particular emphasis on paleontology. (The museum possessed rich paleontological collections.) His main goal was to organize collections and catalogues by describing new genera and species from the fossil remains in the museum's collection. To publicize his discoveries, he started the *Anales del Museo Publico de Buenos Aires* in 1864. Despite its name, the museum was seldom open to laymen. The institution was seen and used as the director's private cabinet, and Burmeister did not concern himself much with the public or with exhibits. The *Anales* became the main forum for presenting information about the museum's collections. (Netto's museum, by contrast, emphasized the public display of collections.) In Buenos Aires, teaching was not particularly emphasized, although Burmeister's staff did lecture in the university's medical and engineering faculties.

[11] Rudolph A. Philippi, "História del Museo Nacional de Chile," *Boletin del Museo Nacional*, 1908, *1*:3–30.

Despite their differences in theory and practice, Burmeister and Netto were friends and corresponded intensively. They also visited each other's countries. Burmeister, for example, went to Rio de Janeiro in 1886 and helped to mount a *Scelidotherium* skeleton that the Argentinean government was donating to Netto's museum.

The two directors were also rivals, however. Netto established close relations with a leading scientific society in Burmeister's city. In 1876 Netto was elected a corresponding member of the Sociedad Científica Argentina. In 1882, he visited Buenos Aires and gave a lecture on the theory of evolution to the young fellows of the society, who were enthusiastic Darwinists. The lecture was well received, and the society publicly eulogized Netto's museum, thereby sending an indirect message to Burmeister, the director of the museum in their own city. One of the younger Argentinean naturalists challenged the Argentinean Republic to take up seriously its "noble rivalry" with the admirable Brazilian museum, its more numerous personnel, and its scientific school—in short, to imitate Netto's museum.[12] The rhetoric of rivalry between the two South American institutions became a source of competition and support for both.

In 1880 Argentina became a unified nation, and Buenos Aires was made the national capital. (Formerly the city had simply been the capital of the province of Buenos Aires.) After federalization, some of the institutions in Buenos Aires were taken over by the new federal government, but others were still administered by the province. Despite the intrigues of the younger generation of naturalists in the Sociedad Científica, Burmeister's museum received federal sponsorship, and its name was changed from Museo Público de Buenos Aires to Museo Nacional de Buenos Aires.[13]

At this same time, a young member of the Sociedad Científica, Francisco Moreno—son of the *porteno* elite, explorer of Patagonia, and a former protégé of Burmeister—lobbied for the creation of a new museum-monument of natural history for Argentina as impressive as those of Paris and London. Partly because a new provincial capital was being established at La Plata, he was successful. In 1884, the Museo de La Plata was established in a magnificent building, a "Greek temple in the middle of the Pampas," and Francisco Moreno became its first director.[14]

In debating Darwinism, new models for museums, and careers for curators, the new generation employed a rhetoric of rupture with the past. Although they acknowledged their debt to their predecessors, they attacked their ideas, methods, research styles, and "foreignness." In so doing, the new generation followed international standards for professional scientific practices. These new practices involved the reclassification of collections, worldwide exchanges among museums and scientific institutions, and publication of modern-style research.

---

[12] Estanislau Zeballos, "El Museo Nacional de Rio de Janeiro," *Actas de la Sociedad Científica Argentina*, 1877, *3:*269–75. Some of the younger members of the staff of the Museu Nacional do Rio de Janeiro later headed provincial museums or other local scientific institutions. They included Barboza Rodrigues (1842–1909) and Emilio Goeldi (1859–1917), directors of Brazilian museums in the Amazon, as well as João Batista de Lacerda (1846–1915), Netto's successor in Rio de Janeiro, and Hermann von Ihering (1850–1930), in São Paulo.

[13] Their affiliations with the old museums and the new ones that were created in the 1880s and 1890s were as follows: Francisco Pascásio Moreno (1852–1919), at La Plata from 1887 to 1911; Florentino Ameghino (1853–1911), at Buenos Aires from 1902 to 1911; Carlos Berg (1843–1902) at Montevideo from 1890 to 1892 and at Buenos Aires from 1892 to 1902.

[14] Henry Ward, "Los Museos argentinos," *Revista del Museo de La Plata*, 1890, *1:*1–8, on p. 3.

Evolutionary theories had their monument in the new Museo de La Plata, which began with no collections but nevertheless aspired to become a center for national exploration and for the exhibition of national nature. In keeping with Darwinian principles, Moreno set out to gather collections and design exhibits that would illustrate the entire course of evolution in Argentina, covering everything from fossil remains in local sediments to contemporary industry and arts.[15] In 1890, the museum began the publication of two series—the *Anales* and the *Revista del Museo de La Plata*. The Museo de La Plata, as monument of natural history, consolidated the model of the public museum in Latin America. Moreno was criticized for using his elaborate exhibits to charm the public and attract the passing gaze of the province's politicians. The row of reconstructed skeletons of extinct glyptodonts and the whale skeletons hanging from the ceiling were described as a mercenary exploitation of science.[16]

Nevertheless, the 1890s witnessed his museum's success as a scientific icon. Its institutional journals and expeditions competed in the search for fossil and archaeological remains. Moreno was attacked but also envied by Burmeister in Buenos Aires: the Museo de La Plata had become a strong competitor for Argentinean resources. Neither the national museum of Argentina nor that of Brazil had enough state support or a proper building, and the contrast made them less attractive than La Plata.[17] Only by their publications could they ensure their status before an international audience that followed with interest debates between the Museo de La Plata and the most famous paleontologist in Argentina, Florentino Ameghino, on the classification of mammalian fossils. (In these years an independent scholar, he later became a museum director himself, heading the Museo Nacional de Buenos Aires from 1902 to 1911.) Ameghino's authority was viewed as greater than that of the state institutions themselves. Even the president of Argentina sent fossils to him for examination, instead of sending them to the museums that his ministries subsidized.[18]

In Brazil, the 1890s witnessed the founding of provincial museums. Hermann von Ihering, a German naturalist, formerly on the staff of the Museo Nacional in Rio de Janiero, established the Museu Paulista in São Paulo, with the help of Orville A. Derby (1851–1915), an American geologist who had started his career at the museum in Rio de Janiero. Derby had developed close ties with the wealthy elite of São Paulo province—then, because of coffee, the wealthiest province in Brazil. Through these connections Derby became the director of the Comissão Geográfica e Geológica (Geographical and Geological Commission) in the city of São Paulo.[19] From this position he was able to help Ihering become, in 1894, the director of a new museum there, the Museu Paulista. Its magnificent building, designed as a monument to Brazilian independence, had far more space than was needed to house the collections of the Comissão and the private collections with which it began. Like

[15] Irina Podgorny, "De razón a facultad" (cit. n. 8).
[16] Eduardo L. Holmberg, *El Joven coleccionista de Historia Natural en la República Argentina* (Buenos Aires: Ministerio de Instrucción Pública de la Nación, 1905).
[17] "We have no space for displaying new specimens. . . . Where to arrange the whales' skeletons stored in the attic and in the corridors of the Museum?" Carlos Berg, "Informes sobre el Museo Nacional," in *Obras completas y correspondencia científica*, ed. Florentino Ameghino (La Plata: Taller de Impresiones Oficiales, 1934), vol. 1, pt. 18, pp. 464–7.
[18] Irina Podgorny, *El Argentino despertar* (cit. n. 7).
[19] Silvia F. de M. Figueirôa, *Um Século de pesquisas em Geociências* (São Paulo: Instituto Geológico, 1985).

La Plata, this new provincial museum would also challenge the national museum in the capital.

Ihering's directorship of the Museu Paulista would last almost twenty years. Following principles laid down by George Brown Goode, the famous secretary of the Smithsonian Institution,[20] Ihering asserted that specialization in particular domains of knowledge was the unique solution to the "crisis" that large and complex museums everywhere were experiencing around 1900. Ihering limited his territory to the natural sciences of Brazil and South America. From this perspective he launched an attack against the universalistic museum in Rio de Janiero, which aspired to represent all of nature. Ihering opposed the uniqueness of his museum in São Paulo to everything that the older Brazilian museum stood for.[21] The national museum, he claimed, was not genuinely scientific. The board of that institution never forgave him for his criticism and ridiculed the Museu Paulista for its majestic but still rather empty building. In their view, Ihering's boasts were made merely to gratify his own ego and to please his wealthy *paulista* patrons.[22]

Despite these rivalries at home, Ihering reached out across national borders and made strong alliances with Ameghino in Argentina and also with his counterparts in Chile at the museums in Valparaiso and Santiago do Chile. He expanded the boundaries of the study of South American natural history at his museum in keeping with the times, encouraging empirical research into the evolution of South American mollusks (both fossil and living).[23]

BOUNDARIES AND SOUTH–SOUTH LINKS

Two directors quite literally helped to draw national boundaries in South America. In the mid-1980s, Moreno was selected to head the commission that would determine the border between Argentina and Chile. Emilio Goeldi, from the Museu Paranaense de Historia Natural e Etnografia, in the Amazon, headed the commission charged to determine the border between Brazil and French Guiana. The directors' normal work also involved putting the national stamp on new, unexplored territories in their countries. To expunge the label "unknown" or "unexplored" from the maps of Brazil and from Brazilian (and all of South American) nature was a goal that museum directors assigned themselves. Adding to Argentina and Brazil thousands of square kilometers of "unknown" land and nature included collecting the material culture of the indigenous peoples—and the peoples themselves—in La Plata, Buenos Aires, São Paulo, the Amazon, and Rio de Janeiro.

This process tied the building of science to the invention of national identity. The museums participated in this latter process in two ways—explicitly, by exploring the territories to be annexed, and implicitly, by giving value to the objects acquired

[20] George Brown Goode, "The Principles of Museum Administration," *Proceedings of the Sixth Annual General Meeting of the British Association of Museums* (Newcastle-upon-Tyne, July 1895), pp. 69–148.

[21] In the same spirit, Ihering advocated a South American mollusks museum. Maria Margaret Lopes, "Viajando pelo mundo dos museus: Diferentes olhares no processo de institucionalização das ciências naturais nos museus brasileiros," *Imaginário*, 1996, 3:59–78.

[22] João B. de Lacerda, "Ao sr. dr. Von Ihering, director do Museo Paulista," *Archivos do Museu Nacional do Rio de Janeiro*, 1895, 5:ix–xix.

[23] Hermann von Ihering, "Les Mollusques fossiles du tertiaire et du crétacé supérieur de l'Argentine," *Anales del Museo Nacional de Buenos Aires*, 1907, 8:1–68.

for collections and by giving legitimacy to extermination policies. The catalogue of native peoples' skeletons, skulls, and material culture as "exotic" or "unique" became—after their scientific baptism—part of the precious treasure of Brazilian and Argentinean natural history. In this framework, the museums collected the archaeological and physical remains of primitive cultures and—before they vanished—their habits and languages. The more extensive the comparative collection of skulls and languages, the more quickly could the question of the origins of humankind be solved.

Another idea, shared from the River Plate to the Amazon, was that America held the key to the past as well as the future of human life. Ameghino claimed that South America, specifically Argentinean territory, was the birthplace of humankind. This idea was taken up by part of the Argentinean cultural elite, and the Argentinean "nationality of humankind" was defended with the same ardor as was the controversial claim that Ameghino himself had been born in the Pampas rather than in Europe.[24] Ihering asserted that the extermination of "savage Indians" from São Paulo was an inevitable consequence of progress and civilization. João Batista da Lacerda, director of the Museu Nacional do Rio de Janeiro at the turn of the century, took the Brazilian Botocudos Indian nation as the standard of inferiority in the scale of the development of the human races, which also made them closer to the possible origins of humankind. This physician-physiologist and craniometric anthropologist proposed transforming the "Brazilian race" from black to white as the only path to civilization.[25]

Argentinean and Brazilian museums cooperated in the search for the origins of people, animals, and territories. In cases where the origins of humankind could not be linked to a national Argentinean identity, the Argentineans appealed to fossil and geological evidence. Both Burmeister, in the new edition of his *History of Creation*, and Ameghino, in his book *The Antiquity of the Peoples of La Plata*, claimed that from a geological perspective America was no younger than the "Old World," and that human beings who were contemporary with the large mammals that became extinct after the Deluge must have "existed simultaneously and before our times on both the Western and Eastern continents [the Americas and Europe]."[26]

Cooperation bore fruit in other ways. From his study of fossil remains of South American mollusks, for example, Ihering established paleontological links between the present-day continents of South America, Africa, and Australasia.[27] Because of their special friendship, Ihering and Ameghino exchanged mollusk and mammalian fossils, each trying to understand the geological formation of the Southern Hemisphere.[28] At the same time, they worked to emancipate South American geology from

[24] In fact, Ameghino probably was born in Moneglia, Italy, close to Genoa. He emigrated to Argentina with his parents as a child. After his death in 1911 a branch of the Catholic movement accused him of being Italian. See Irina Podgorny, "De la santidad laica del científico: Florentino Ameghino y el espectáculo de la ciencia en la Argentina moderna," *Entrepasados, Revista de Historia*, 1997, *13:*37–61.

[25] For a detailed account of this theme, see Thomas Skidmore, *Black into White: Race and Nationality in Brazilian Thought* (Oxford: Oxford Univ. Press, 1974).

[26] Hermann Burmeister, *Historia de la Creación: Exposición Científica de las fases que han presentado la tierra y sus habitantes en sus diferentes periodos de desarrollo*, 9th ed. (Madrid: Gaspar, n.d.), p. 310.

[27] Hermann von Ihering, *Archhelenis und Archinotis* (Leipzig, 1907).

[28] Maria Margaret Lopes and Silvia F. de M. Figueirôa, "Horizontal Interchanges in Geological Sciences," *Nineteenth International Symposium of the International Commission on the History of*

the preconceived theories of North American paleontologists concerning Patagonian fossils.[29] Their work is an example of unparalleled scientific cooperation in the continent.[30] Although removed from the exchange because of his work on the boundary commission, Moreno also appealed, as if it were a boundary problem, to science with regard to questions about Southern Hemisphere geology and paleontology.[31]

Here it is important to underline some aspects of the relationship between Brazil and Argentina with metropolitan centers. The idea of South–South geological links initially appeared in opposition to the ideas of northern institutions, especially in North America; Europe was not a "neutral" authority in these controversies, which were of essential importance in deciding which European institutions and which countries should dominate the field.

CONCLUSION

Within Latin America there was no doubt that, at the end of the century, the ideal natural history museums were those located south of the River Plate.[32] North of the river, from Montevideo[33] to the Amazon, across the Andes in Santiago,[34] and in Valparaiso, the Argentinean museums of Buenos Aires and La Plata fused into a prevailing model of the museum. In this model, whales of the southern seas were to be seen suspended from the ceiling, and large, extinct South American mammals marched down the halls. These museums were the most envied, because they constituted exemplars. On the other side of the river, Carlos Berg and Ameghino, Burmeis-

---

Geological Sciences [INHIGEO]: Useful and Curious Geological Enquiries beyond the World (Sydney: INHIGEO, 1994), pp. 1–6.

[29] Concerning his controversy with the North Americans from the Princeton expeditions to Patagonia about the Patagonian collections, for example, Ihering wrote to Ameghino, "Between us, I believe that our position regarding the American gentlemen is, scientifically, the same that unfortunately prevails in politics. I would expect impartial and proficient help only if European geologists, who agree with your point of view, would proceed to examine those samples again." The manuscript, Carta 1546, São Paulo, 30/08/1902, is reprinted in Florentino Ameghino, *Obras completas y correspondencia científica*, ed. A. J. Torcelli (La Plata: Taller de Impresiones oficiales, 1937), p. 79.

[30] Another example of the close scientific and personal relationship among Latin American museum directors was that sustained by Carlos Berg, a Russian-born naturalist, and his colleagues from the Museo Nacional de Montevideo, in Uruguay, which he also directed between 1890 and 1892.

[31] "Interesting problems, which can only be solved by a systematic examination of the Argentine country by an experienced geologist. In the course of my paper on Patagonia, read before the Royal Geographical Society (May 29), I proposed that this Society, the Royal Society, and the British Museum, with other scientific institutions, should proceed to carry out these necessary investigations. . . . If these expeditions be made, how many changes may be produced in actual and general ideas on the age of South American fossiliferous strata, on the disappearance of the lost southern lands, and on the affinities of extinct faunas so distant in time and space as those of South America and Australia!" Francisco Moreno, "Note on the Discovery of Miolania and of Glossotherium (Neomylodon) in Patagonia," *Nature*, 1899, *1566*, 60:397–8.

[32] José A. Pérez Gollán, "Mr. Ward en Buenos Aires: Los Museos y el proyecto de nación a fines del siglo XIX," *Ciencia Hoy*, 1995, *28*:52–8.

[33] In his vast correspondence with Carlos Berg, José Arechavaleta (1838–1912), a Spanish botanist who directed the Museo Nacional de Montevideo from 1893 until 1912 and made it one of the most distinguished Latin American museums by the turn of the century, expressed his admiration for the collections and library of the Museo Público de Buenos Aires, and also for details of small exhibitions presented there.

[34] The Argentinean museums were also mentioned by Eduardo Moore, director of Chile's Museo Nacional de Santiago, in his "Report of Activities for the years 1910–1911," *Boletin del Museo Nacional*, 1912, *3*:1–14. He said that the Santiago Museum could attain the level of similar institutions in foreign countries, "especially in the Argentine Republic," only if it could increase the salaries of staff members so that they could become full-time researchers.

ter's successors at the Argentinean Museo Nacional, spent their days stressing, amidst innumerable problems, the need to store the big bones and to construct a new building to rival the sumptuous palace at La Plata.

What were envied and disputed were not merely the accidents of collections and buildings, but also the principles of museum design and investigation that reached beyond national limits. The social roles that museums would play in the new century turned upon their being conceived not as local, circumscribed to specific regions, but as incorporating continental dimensions. They "musealized" natural environments that political frontiers between countries did not divide; they shared scientific interests that united South America; and they identified a common basis of intellectual culture in the South that could finally be recognized in the North.

The affinities and rivalries among Brazilian and Argentinean museums did not reduce the value that naturalists working in Latin America attached to their North American and European contacts. Indeed, as we have noted, they took great pride in establishing and keeping those relationships.[35] However, in thinking about the "mondialization" of science, besides the relationship between metropolis and colony we must also consider the dynamics of Latin American scientific integration. The case of museums is but one aspect of such integration. More evidence is furnished by the Latin American Scientific Congresses that were held in Buenos Aires in 1898, Montevideo in 1901, Rio de Janeiro in 1905, and Santiago in December 1908 and January 1909, in which directors of museums participated. These congresses were considered the first attempts at building a scientific community within the regional context of Latin America.[36]

Our purpose has been to advance beyond the discussion of acclimatization, reception, and translation of science around the world. Further investigation of the "mondialization" of science focusing on relationships among countries that do not belong to the North Atlantic axis is needed. Within this framework, it is possible to understand how the directors of Brazilian and Argentinean museums, far from permanently assuming a colonial discourse, could instead dispose their institutions to serve as symbols of a new national identity, using science "as a nationalist enterprise."[37] Despite the specific circumstances of each museum, of each country, the praise of the unique, the proper, the peculiar that characterizes the whole, the essence of these new museums was the definition of species from type, a basic principle of taxonomy. This perspective united Brazilian and Argentinean museums. At the turn of the century, museums in London, Paris, and Washington were, no doubt, centers of reference, but ones perceived from a local perspective, stimulated by the carapaces of glyptodonts, and by the hanging bones of whales from the remote southern seas.

[35] Maria Margaret Lopes, "Brazilian Museums of Natural History and International Exchanges in the Transition to the Twentieth Century," in *Science and Empires: Historical Studies about Scientific Development and European Expansion,* eds., Patrick Petitjean, Catherine Jami, and Annie Marie Moulin (Dordrecht: Kluwer, 1992), 193–200.

[36] Francisco R. Sagasti and A. Pavez, "Ciencia y tecnología en América Latina a principios del siglo XX: Primer Congreso Científico Panamericano," *Quipu,* May–Aug. 1989, *6,* 2:189–216.

[37] Roy MacLeod, "Reading the Discourse of Colonial Science," in *Les Sciences coloniales: Figures et institutions,* ed. Patrick Petitjean (Paris: Organization pour la Recherche Scientifique des Territoires d'Outre-Mer, 1996,) pp. 87–96, on p. 95.

Colonial Encounters and the Forging of New Knowledge and National Identities: Great Britain and India, 1760–1850

*Kapil Raj**

ABSTRACT

In opposition both to the dominant vision of colonial science as an hegemonic European enterprise whose universalization can be conceived of in purely diffusionist terms, and to the more recent perception of it as a simple reordering of indigenous knowledge within the European canon, this essay seeks to show the complex reciprocity involved in the making of science within the colonial context. Based on the example of India during the first century of British colonial conquest, it examines the specificities of intercultural encounter in the subcontinent, the formalized institutions that were engendered, and the kinds of knowledge practices that emerged in the case of the geographical survey of India. The essay suggests that the knowledge created in this context is not just local in character, but participates wholly in the emergence of universal science, as well as of other institutions of modernity.

INTRODUCTION

OVER THE LAST DECADE, PROMINENT IMPERIAL HISTORIANS have called into question the concept of simple diffusion of the fundamental values of modernity from Britain to its colonies—values such as democracy, justice, and the welfare state. They have argued that modernity and its institutions are not a simple emanation from a well-defined center, but are the result of adaptations and accommodations of British institutions confronted with the social, political, and economic organization of the countries Britain came to dominate, including Ireland, Scotland, and India. They thus imply that Britain, its modern institutions, and its empire were co-constituted.[1]

However, modern knowledge and its making did not figure among the domains these authors studied. This lacuna has now been partially filled by Christopher Bayly

---

* Centre de Recherche en Histoire des Sciences et des Techniques (CNRS UMR 2139), Cité des Sciences et de l'Industrie, 30, avenue Corentin Cariou, 75930 Paris Cedex 19, France.

[1] I refer here to historians like David Washbrook, Burton Stein, David Cannadine, and, most notably, Christopher Bayly. See his *Imperial Meridian: The British Empire and the World, 1780–1830* (London: Longman, 1989).

---

with his recent *Empire and Information*.[2] In a move away from the coproductivist perspective of his earlier works, Bayly here surveys the complex indigenous information-gathering networks of precolonial India—ranging from gossipmongers in the bazaars, marriage makers, and midwives to astronomers, physicians, and philosophers—and the historical contingencies that led to their partial, though informal, inclusion in the surveillance systems set up by the British following their rise to power in the latter half of the eighteenth century. "The colonial information order," Bayly states, "was erected on the foundations of its Indian precursors . . . reclassified and built into hierarchies which reflected the world view of the Britons."[3] However, riven by mutual suspicion, distortion, and violence between the British officials and their indigenous informants, the new colonial state's intelligence systems were fragile. The whole enterprise resulted in a monumental failure when the British were caught almost completely unawares by the popular rebellions and mutinies of 1857, which almost cost them their South Asian empire.

Bayly, however, does not deal with the workings of other, more successful and resilient institutions devoted to knowledge making and dissemination on which the colonial information order equally depended. Indeed, the late eighteenth century saw the rise, both in Britain and in its colonies, of a number of field sciences that at once fed on and reinforced the colonial order, such as geographical surveying, agriculture, botany, forestry, and anthropology.[4]

To be sure, Bayly does discuss debates *about* science and the status of scientific knowledge among learned Indians and British in the nineteenth century, but this is a second-order discussion, one step removed from the making of new knowledge.[5] Moreover, Bayly's approach is inadequate for studying the development of these sciences during this period. For a start, Britons themselves were in the process of forging a national identity; to speak of a single British world view at the time, into which indigenous knowledge was incorporated, is anachronistic.[6] In addition, the field sciences developed in a much tighter, more formal, and stratified institutional context than the informal networks of intelligence-gathering at the heart of Bayly's new work. Indeed, the successful functioning of these institutions presupposed the imposition of a certain authority by the state, a degree of control that was beyond the means (and ambitions) of individuals and their informal networks.[7] And, although colonial institutions grew out of preexisting administrations of indigenous regimes and inherited much of their workforces, they were transformed by the new situation through mechanisms of accommodation and negotiation, producing novel forms of knowledge that were not simply linear offshoots of past practices and traditions. Study of colonial institutions thus calls for an approach that, by bringing negotiation

---

[2] Christopher Alan Bayly, *Empire and Information: Intelligence Gathering and Social Communication in India, 1780–1870* (Cambridge: Cambridge Univ. Press, 1997).

[3] *Ibid.*, p. 179.

[4] Simon J. Schaffer, "Field Trials, the State of Nature and British Colonial Predicament," paper presented at the "Science and Empire" seminar, Centre de Recherche en Histoire des Sciences et des Techniques, Cité des Sciences et de l'Industrie, La Villette, Paris, 11 June 1999.

[5] Bayly, *Empire and Information* (cit. n. 2), chaps. 7 and 8.

[6] See Linda Colley, *Britons: Forging the Nation 1707–1837* (New Haven and London: Yale Univ. Press, 1992). See also Roy Porter and Mikuláš Teich, eds., *The Scientific Revolution in National Context* (Cambridge: Cambridge Univ. Press, 1981).

[7] See Svante Lindqvist, "Labs in the Woods: The Quantification of Technology During the Swedish Enlightenment," in *The Quantifying Spirit in the Eighteenth Century*, eds. Tore Frängsmyr, John L. Heilbron, and Robin E. Rider (Berkeley: Univ. of California Press, 1990), pp. 291–315.

and coproduction back to center stage, accounts for the complex character of knowledge making and circulation during this period—an approach closer to Bayly's earlier works as well as to recent work in the history and sociology of science.

From the mid-eighteenth to the mid-nineteenth centuries, the Indian subcontinent played an important and active role in this coemergence inasmuch as it became a space for multiple cultural encounters in the context of empire: encounters between different groups from the British Isles, and between them and different sectors of the subcontinent's own population. Corresponding to these two different types of encounters, the first part of this essay, situated mainly in Calcutta, presents the specificities of scientific practices and the circumstances that led to their institutionalization. The second focuses on one institution—the Survey of India—to show its hybrid nature and the coproduction of geographical knowledge that emerged.[8] I shall conclude by addressing questions related to the nature and scope of the knowledge produced in the co-construction of modernity across the globalized space of empire.

THE EAST INDIA COMPANY, CALCUTTA, AND THE COLLEGE

Direct contact between England and India dates from the establishment of the English East India Company in 1600. Coming to participate in the lucrative spice and luxury commodity trade, the English initially represented no more than a few hundred civilians and a couple thousand troops. Even at the apogee of empire in the twentieth century, the British presence in India never exceeded a few tens of thousands of civilians, a number at all times too small not to rely heavily upon autochthonous intermediaries for most administrative and technical tasks.[9] In fact, from their arrival in the subcontinent, a collaboration was established between the British and segments of the region's population: *banians* (bankers) and *munshis* (interpreter-secretaries), and skilled workmen like weavers, jewelers, carpenters, shipbuilders, and sailors. In the face of inter-European rivalries in the second half of the eighteenth century, especially vis-à-vis the French, this collaboration extended to the establishment of an army that included indigenous troops, artificers, and gunsmiths.

The conquest of Bengal in 1757 put the British firmly on the road to territorial and political power. However, consciousness of this new role was slow in coming for, in the years that followed, East India Company officials devoted their attention to unbridled personal profiteering through looting and extorting exhorbitant taxes from the local peasantry. But after ten million lives had been lost in the space of three years—victims of a famine that was a direct consequence of the ruthless policies of the company's servants—parliament in Britain pressured the company to establish more orderly and permanent forms of exploitation and government.

So it was that Warren Hastings, governor-general of Bengal from 1772 to 1785, received orders from London to take over the whole civil administration of the

---

[8] I have chosen this institution partly to provide a counterpoint to Matthew Henry Edney, *Mapping an Empire: The Geographical Construction of British India, 1765–1843* (Chicago and London: Univ. of Chicago Press, 1997), the argument of which is largely founded on Bayly's central thesis in *Empire and Information* (cit. n. 2).

[9] One estimate, made for the Madras presidency during the first half of the nineteenth century, put the proportion of Britons to South Asians directly serving the company's civil administration at 1 to 180. See Robert E. Frykenberg, *Guntur District 1788–1848: A History of Local Influence and Central Authority in South India* (Oxford: Clarendon Press, 1965), p. 7.

province. As the emerging state of Great Britain held that civil justice, public order, transport, and communications depended upon taxation, Hastings took the orders to mean the entire management of the province's revenues.[10] The collaboration between Britons and South Asians thus broadened to include tax collection and running a civil government. And although the British set up a variety of new intermediary relationships, their interlocutors remained in large part the indigenous "under civil servants"—land-revenue officials, minor judges, and police officials inherited from the Mughal and other princely administrations.

To Hastings's mind, successful administration required drawing up a kind of Domesday Book of the company's territories. "Every accumulation of knowledge," he wrote, "and especially such as is obtained by social communication with people over whom we exercise a dominion founded on the right of conquest, is useful to the state. . . . "[11] In addition to taxation and law, this knowledge was to include natural history and antiquities, local customs, diet and general living conditions—in short, all that would, by the end of the century, go under the name of "statistics." Giving the highest priority to a knowledge of languages, Hastings devised handsome monetary incentives for those officials willing to study Indian languages and culture. This policy constituted the first step in the transformation of the European study of exotic peoples from an individual avocation to a massive and institutionalized activity, reflecting how vital a concern it was for the emerging rulers of the subcontinent.

However, not all of the East India Company's agents had the wherewithal to take up Hastings's offer. The vast majority of recruits to the company arrived in India between the ages of fourteen and eighteen.[12] The only prerequisite for recruitment was knowledge of "the rule of three and merchants accounts."[13] Few of the English had been to university. Engrossed in fortune making in this "fine country for a gentleman to improve a small fortune in," most had little curiosity about the subcontinent's inhabitants nor, indeed, the culture to acquire learning.[14]

Of the minority of Englishmen who did have a penchant for intellectual pursuits, most were, in the fashion of the "great school" and Oxonian, High-Church elite to which they generally belonged, obsessed with classical thought and scripture. Indeed, their education was dominated by the study of Greek and Latin, and by the "grand tour" of Italy and Greece.[15] Their attitudes towards politics and government,

---

[10] See John Brewer, *The Sinews of Power: War, Money and the English State 1688–1783* (London: Unwin Hyman, 1989), and Georg Robert Gleig, ed., *Memoirs of the Life of Right Honourable Warren Hastings*, 3 vols. (London: Richard Bentley, 1841), vol. 1, p. 214.

[11] Warren Hastings to Nathaniel Smith, chairman of the East India Company, 4 Oct. 1784, reprinted in Charles Wilkins, trans., *Bhagavad Gita* (London, 1785), preface, p. 13.

[12] India Office Records, Court Minutes 1784–85: B/100, p. 216, British Library, London (hereafter cited as India Office Records).

[13] Writers' petitions and educational testimonials for recruitment into the East India Company quoted in Anthony Farrington, *The Records of the East India College Haileybury & Other Institutions* (London: Her Majesty's Stationery Office, 1976), p. 4.

[14] James Rennell to the Rev. Gilbert Burrington, 7 Nov. 1763, India Office Records, MSS Eur/ D1073.

[15] For a description of the daily life of the British in India during this period, see Percival Spear, *The Nabobs* (Oxford: Oxford Univ. Press, 1963). For eighteenth-century classical education in Britain, see Robert Maxwell Ogilvie, *Latin and Greek: A History of the Influence of the Classics on English Life from 1600 to 1918* (London: Routledge and Kegan Paul, 1964), pp. 46 ff. See also John Harold Plumb, *Men and Places* (London: Cresset Press, 1963), and John Lawson and Harold Silver, *A Social History of Education in England* (London: Methuen, 1973).

conduct, manners, and style testify to the influence of the classical disciplines. Not surprisingly, in their eyes, Sanskrit was to Indian vernaculars what Greek and Latin were to contemporary European languages. Like their virtuoso contemporaries in Europe who invested in recovering the works of ancient Greece and Rome, they concentrated, in their exploration of Indian learning, on ancient Sanskrit literary, philosophical, and scientific works. Naturally, they sought as informants and privileged interlocutors their Indian counterparts—those of the Brahminical upper castes who had mastered Sanskrit, and *maulavis* (Muslim law-officers) and *munshis* adept in Arabic and Persian—a reliance that reinforced their classical inclinations. Their understanding of the contemporary society that they were supposed to govern was shaped by a scrutiny of classical texts and of texts specially commissioned to be written, in Sanskrit, by their Brahminical collaborators.

The *Vivādabhangārnava* (Treatise on putting an end to litigation) by Jagannātha Tarkapanchānana, a venerable professor of law and legal expert to the Calcutta Supreme Court, is a typical example of the new type of collaboration. A hybrid text constructed jointly by the British jurist Sir William Jones and Jagannātha's team of court-pundits (native legal assistants), it went well beyond the scope of preexisting Hindu law manuals.[16] Because of its comprehensiveness and the immense authority Jagannātha commanded among his colleagues, pupils, and the Bengali public, this work was widely used by Indian court assistants and British judges alike. It was also to serve as a model for a spate of supplementary compositions by pundits in the employ of British-administered courts. These works, in turn, formed a corpus prescribed in the syllabus of the Sanskrit College in Banaras, established in 1794 to breed a new generation of pundits "to assist the European judges in the due, regular, and uniform administration of [the] genuine letter and spirit [of Hindu law] to the body of the people."[17]

If the classicists were mainly trained in the English public schools and at Oxford and Cambridge, men of science, law, and medicine were mostly trained either in the dissenting academies or in the Scottish and Dutch universities.[18] Indeed, in Scotland's more egalitarian Presbyterian tradition, many more men went to university than in England, and at a much earlier age. Moreover, Scottish education, both at school and university, was much broader than in England, covering (besides Latin and Greek) history, navigation, geography, mensuration, and natural and moral philosophy.[19] However, Scotland itself did not have the capacity to absorb its qualified workforce, which consequently emigrated to England and beyond.[20] A large number

---

[16] For a list of expressly commissioned works, see J. Duncan M. Derrett, "Sanskrit Legal Treatises Compiled at the Instance of the British," *Zeitschrift für vergleichende Rechtswissenschaft*, 1961, *63*:72–117. For an evaluation of the nature of the *Vivādabhangārnava*, see *idem, Religion, Law and the State in India* (London: Faber & Faber, 1968), pp. 247–8.

[17] Jonathan Duncan, Resident Benares, to the Earl of Cornwallis, Governor-General, 1 Jan. 1792, in *Selections from Educational Records*, ed. Henry Sharp, 2 vols. (Calcutta: Superintendent Government Printing, 1920), vol. 1, p. 11.

[18] See Nicholas Hans, *New Trends in Education in the Eighteenth Century* (London: Routledge and Kegan Paul, 1951), pp. 32–6.

[19] See Paul Wood, "The Scientific Revolution in Scotland," in Porter and Teich, *The Scientific Revolution*, (cit. n. 6), pp. 263–87.

[20] Francis Buchanan is a good illustration of this. See Marika Vicziani, "Imperialism, Botany and Statistics in Early Nineteenth-Century India: The Surveys of Francis Buchanan (1762–1829)," *Modern Asian Studies*, 1986, *20*, 4:625–60.

Figure 1. *"An European Gentlemen with his Moonshee, or Native Professor of Languages" from Charles Doyley,* The European in India *(London: Edward Orme, 1813), frontispiece. (British Library, India Office Collections.)*

were absorbed into Britain's ever-expanding colonial services, to occupy senior technical positions as engineers, soldiers, veterinarians, diplomats, doctors, and naturalists. Indeed, it was predominantly the Scots who manned the highly successful operational, scientific, and technological aspects of British activity in India.

Thus, Scottish medical officers of the company were the first to systematically make meteorological recordings in India, while another Scot, Robert Kyd, set up the Botanic Gardens in Calcutta that William Roxburgh consolidated, assisted by William Hunter, James Anderson, and Francis Buchanan. John McCleland headed the first committee for the exploration of mineral resources. Buchanan and Colin Mackenzie were among the pioneers of large-scale topographical surveys in the subcontinent, and David Ross was called to teach natural and experimental philosophy when the Hindu College was set up in Calcutta in 1817. As army medics, Scots came to learn the vernaculars of the subcontinent. Many mastered Persian and Arabic, the court languages of Mughal India, and with the help of Persian *munshis* compiled bilingual dictionaries and translated texts. Diplomatic missions were thus often entrusted to them: Alexander Hannay to the Mughal court; George Bogle, Alexander Hamilton, and Samuel Turner to Tibet.

However, linguistic ability was not the prerogative solely of the High-Church Englishmen or opportunistic Scots. Another group—Baptist missionaries—was also busy discovering the languages of the subcontinent's inhabitants. Persecuted like other nonconformists in England during this period, a few sought refuge in India, establishing themselves at the Danish colony of Serampore near Calcutta.[21] Under William Carey, a Baptist fugitive turned indigo factory owner and small-time tradesman, they tried to introduce the Bible to the crafts-oriented lower castes through an understanding of their languages and ways of life.[22] With their populist notions, they mastered a large number of the subcontinent's vernaculars and gained deep insights into its cultures. They, too, were to write grammars of Indian languages and collect folk tales and other lore, the better to understand the people they set out to proselytize. In 1800, the Serampore missionaries founded a printing press where, with their indigenous interlocutors, they cast fonts of many Indian vernaculars. This press was the first and most important in its time for books in living oriental languages.

Developments at the turn of the nineteenth century were to bring these different British groups and their respective indigenous collaborators together in a formalized institutional context. In an effort to stem the spread among its employees of the "erroneous principles of the same dangerous tendency [as the doctrines of the French Revolution]," which "had reached the minds of some individuals in the civil and military service of the Company in India," and instead "to fix and establish sound and correct principles of religion and government in their minds at an early period of life," the East India Company set up a college at Fort William, Calcutta in 1800.[23] Newly arrived, covenanted officers of the East India Company were to spend three years in residence at this "University of the East", "removed from the danger of profusion, extravagance and excess."[24] They studied Hindu, Islamic, and English law; civil jurisprudence, political economy, general history, world geography, and mathematics. The rigorous curriculum also included natural history, botany, chemistry, astronomy, Latin, Greek, Sanskrit, Arabic, Persian, and modern European languages in addition to the culture and the six major languages of their South-Asian subjects (Hindustani, Bengali, Telugu, Marathi, Tamil, and Kannada).[25] Many of these subjects and languages had never before been taught in Britain or in Europe at any level. The college, which cost £250,000 in its first three years alone, soon grew to a size comparable to its models—contemporary Oxford and Cambridge.[26]

In order to teach these various subjects, staff members were recruited from among

[21] See Michael Watts, *The Dissenters from the Reformation to the French Revolution* (Oxford: Oxford Univ. Press, 1978).

[22] William Carey, *An Enquiry into the Obligations of Christians to Use the Means for the Conversion of the Heathens*, new facsimile ed. (London: Carey Kingsgate Ltd., 1961), p. 74.

[23] Quotations from Wellesley's "Minutes on the Foundation of a College at Fort William, 10 July 1800," reprinted in *The Despatches, Minutes and Correspondence, of the Marquess Wellesley, K. G., During his Administration in India*, ed. Robert Montgomery Martin, 5 vols. (London: W. H. Allen & Co., 1836), vol. 2, p. 346.

[24] *Ibid.*

[25] "Regulation for the Foundation of a College at Fort William . . . Passed by the Governor-General in Council, on the 10th July, 1800 . . . " in *ibid.*, vol. 2, p. 359 ff.

[26] Farrington, *Haileybury Records* (cit. n. 13), p. 6. See also David Kopf, *British Orientalism and the Bengal Renaissance: The Dynamics of Indian Modernization 1773–1835* (Calcutta: Firma KLM,

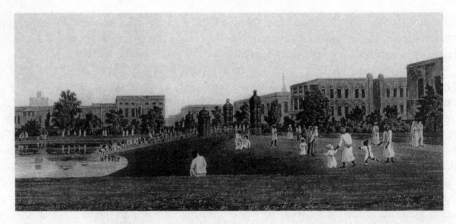

Figure 2. *The Public Exchange and Coffee-House Building, behind Tank Square, where the College of Fort William was first housed before being shifted (around 1806) to the Writers' Buildings (below), on the opposite side of the square. From James Baillie Fraser,* Views of Calcutta and its Environs *(London, 1824–1826), plates 22 and 6. (British Library, India Office Collections.)*

the British in Calcutta. Persian was entrusted to Neil Edmonstone, Arabic to John Baillie, Hindustani to John Gilchrist (Scotsmen all), Sanskrit to Henry Thomas Colebrooke, and the five remaining Indian vernaculars to the Baptist William Carey (thereby giving the Baptists a fig leaf of respectability in exchange for their knowledge about indigenous cultures that was inaccessible to official Indo-British culture). Natural and experimental philosophy were taught by a Scotsman, James Dinwiddie. A number of Indians, both Hindus and Muslims, were recruited to assist the Euro-

1969), and Kapil Raj, "L'Orientalisme en Inde au tournant du XIXe siède: La Réponse britannique à la Révolution Française," *Annales Historiques de la Révolution Française*, 2000, *320:*89–99.

pean staff members and very often taught in their place.[27] It was this institution that provided the first sustained, professional contact between the different "confessional nations" of the British in India.

In addition to teaching, the college organized and sponsored expeditions in company-controlled territories in order to discover and catalogue manuscripts for its library. By 1805, its Indo-British staff had succeeded in encoding a number of spoken languages of the subcontinent into grammatical forms, and translating them into English. The College of Fort William was the first of a series of institutions in which these different knowledge traditions and their corresponding skills were brought together, standardized, and rendered teachable. When in 1806 the company's Court of Directors set up the East India College at Hertford Castle (later transferred to Haileybury) in England as a preparatory school for new recruits before they left for India, some of the staff members of Fort William and their hybrid networks were transferred to the metropolis to teach alongside the mathematician Bewick Bridge, FRS and fellow of Peterhouse, Cambridge; the jurist Edward Christian, Downing Professor of Law at Cambridge, and the political economist Thomas Malthus.[28]

This focus on Calcutta reveals some of the contingencies that shaped the multicultural encounters, and the integration of the resulting networks into the newly emerging academic system of metropolitan Britain. A brief look at one of the principal, and undoubtedly the most prestigious, colonial scientific institutions—the Survey of India, also established in the latter half of the eighteenth century—throws similar light on the nature of intercultural collaboration and the kind of knowledge that resulted.

INDIA AS A SOURCE OF NEW SCIENTIFIC PRACTICES AND KNOWLEDGE MAKING

In the course of the seventeenth and eighteenth centuries the British, like other Europeans trading with the East, charted the seas and coasts between western Europe and Asia. However, they had little knowledge of the geography of mainland South Asia. For this, they relied principally upon information culled from travelers and missionaries, and on the occasional map, like that of the French armchair mapmaker Jean-Baptiste Bourguignon d'Anville, who in 1752 had published a *Carte de l'Inde* based on European travelers' accounts. Territorial acquisition changed needs and, in the wake of the conquest of Bengal, surveys of the new possessions were ordered to defend frontiers, to ascertain the extent and revenue potential of cultivated lands, and to ensure the safety and regularity of communications.

Like other British colonial institutions, the Survey of India had to rely upon indigenous staff members and their skills. Not only were the British too few to undertake surveys, but those few had little or no experience in countrywide terrestrial surveying. In the 1760s, when survey work was first undertaken in India, there was no unified, detailed map of the British Isles—with the notable exception of a map of Scotland, made by Scotsmen in the aftermath of the 1745 uprising—although there was no dearth of coastal, harbor, and fortification maps made for the Board of Ord-

---

[27] See Bernard S. Cohn, "The Recruitment and Training of British Civil Servants in India, 1600–1860," in *idem, An Anthropologist Amongst the Historians and Other Essays* (New Delhi: Oxford Univ. Press, 1989), pp. 500–53, especially pp. 529 ff.

[28] "Staff of the East India College, Haileybury," in Farrington, *Haileybury Records* (cit. n. 13), pp. 104–6.

nance, and estate, route, and county maps in the civilian domain. The latter were based on measurements made by estate and county surveyors whose skills and instruments, besides being unavailable in India, were inadequate for the purposes of extensive surveying. Indeed, the Ordnance Survey of Great Britain and Ireland was founded only in 1791, and it was not until 1801 that the first ordnance map appeared.[29]

Like the British, the Indians possessed no detailed maps of the whole of the subcontinent, although there is evidence of maps for the northwestern, central, and western parts dating to the sixteenth and seventeenth centuries. These do not show a uniform scale or orientation and their exact use is not known.[30] At any rate, the need for maps in precolonial India seems to have been obviated by gazetteers and manuals, used for administration and revenue collection, that provided systematic descriptions, in tabular form, of provinces and their subdivisions, noting their general location and territorial extent. The most famous of these was the *Ain-i Akbari* (Institutions of Akbar) made for the Mughal Emperor Akbar (1542–1605; reigned from 1556), who had his roads measured with great care. The Jesuit Antonio Monserrate (1536–1600), who spent many years in Akbar's court, describes the measurement of the latter's march to Kabul in Afghanistan in 1581:

> [Akbar] orders the road to be measured, to find the distance marched each day. The measurers, using ten-foot rods, follow the king, measuring from the palace. By this one operation he learns both the extent of his dominions, and the distances from place to place, in case he has to send embassies or orders, or meet some emergency. A distance of 200 times the ten-foot rod, called the *coroo* in Persian, or *cos* in the Indian language, equal to two miles, is the measure for calculating distances.[31]

There was also a well-established tradition of land measurement and surveying for the purposes of establishing property rights and fiscal dues. An eighteenth-century Sanskrit manuscript on land measurement from peninsular India, translated for the Moravian missionary, naturalist, and surveyor for the East India Company Benjamin Heyne (1770–1819), describes a method based on the use of corporeal and other techniques:

> The fundamental measure is that of an Inch which is determined in three different ways.

> *First*, By placing three rice corns in a line length ways—the place they occupy is called an Inch.

---

[29] Almost all of the 184 eighteenth-century British surveyors in India picked up surveying and mapmaking techniques on the job. See Reginald Henry Phillimore's authoritative study, *Historical Records of the Survey of India*, 5 vols. (Dehra Dun, India: Survey of India, 1945–1968), vol. 1, "Biographical Notes," pp. 307–400. For a history of the survey of Scotland and the techniques employed, see Raleigh Ashlin Skelton, "The Military Survey of Scotland 1747–1755," *The Scottish Geographical Magazine*, 1967, *83*:1–15. For the early history of surveying in England, see Charles Close, *The Early Years of the Ordnance Survey* (Newton Abbot, U.K.: David and Charles Reprints, 1969) and W. A. Seymour, ed., *A History of the Ordnance Survey* (Folkestone, U.K.: William Dawson, 1980).

[30] See Joseph E. Schwartzberg, "South Asian Cartography," in *The History of Cartography*, eds. John Brian Harley and David Woodward, vol. 2, book 1: *Cartography in the Traditional Islamic and South Asian Societies* (Chicago and London: Univ. of Chicago Press, 1992), pp. 400 ff.

[31] Antonio Monserrate, "Mongolicae Legationis Commentarius," trans. and cited by Phillimore, *Historical Records* (cit. n. 29), vol. 1, p. 10.

Secondly, By measuring the circumference of the second joint of the thumb, half of the length of which is an Inch.

Thirdly, By measuring the second joint of the middle finger, the half of which is called an Inch.

12 of these Inches are One Jana (literally translated as paw)—32 Janas are One Ghada (or Bamboo)—4 Ghadas (or One Square Bamboo) is One Kunta.

These measures . . . are universally understood.[32]

The mapping of India started by mobilizing available resources. The French savant-traveler Abraham-Hyacinth Anquetil-Duperron (1731–1805) has left the following amusing account of an early (1758) European military route survey:

I have traveled in the interior of India alone, with others and with the army. The commanding officer spends the better part of his day sleeping in his palanquin. At dinner he asks his *Dubash* [interpreter] . . . what distance they have traveled and which places they have passed. The latter in turn asks the porters or else replies himself, for reply one must; and the distances and place names are inscribed on the itinerary, on the map . . . (which, by the way, I found perfectly well made).[33]

James Rennell (1742–1830) can be considered the first Englishman to have brought these disparate traditions together on the same map. Rennell started his career as an ensign on a British naval vessel off the coast of Brittany during the Seven Years' War (1756–1763). There he learned the art of coastal and harbor surveying, a skill he was to use to great advantage during his thirteen years in the service of the East India Company as surveyor-general of Bengal. Indeed, since the Ganges-Brahmaputra delta forms a large part of this territory, Rennell used its navigable distributaries in the same way as one would a sea coast, tracing an outline of the whole delta.[34]

For the terrestrial cartography, Rennell sent both Indian and British soldiers on long route marches. From their accounts, as well as those of other Asian and European travelers and missionaries, he began compiling his map of the whole subcontinent. Foremost among the contributors were Ghulam Mohammad for peninsular India, Mirza Mughal Beg for northwestern India, and Sadanand for Gujarat.[35] His European informants consisted mainly of the Jesuit Fathers Antonio Monserrate and Joseph Tieffenthaler (1710–1785), and Frenchmen in India like Claude Martin (1735–1800)—who themselves relied on "native" surveyors.[36] And, of course, Rennell extensively used the tables of the *Ain-i Akbari*. Interestingly, he acknowledged

[32] "The 'Chetrie Ganietam'—A Sanskrit Work on Land Measurement Translated by Benjamin Heyne," Memoirs of the Survey of India (1773–1866), Memoir 3, f. 2, National Archives of India, New Delhi.

[33] Translated by the present author from the original French published in Jean Bernoulli, ed., *Description historique et géographique de l'Inde*, 3 vols. (Berlin: Pierre Bourdeux, 1786–1788), vol. 2, pp. 466–7.

[34] See Rennell's correspondence with his guardian, the Rev. Gilbert Burrington, India Office Records, MSS Eur/D1073; and his manuscript maps held at the Royal Geographical Society Archives, London.

[35] James Rennell, *Memoir of a Map of Hindoostan, or the Mogul's Empire*, 1st ed. (London: for the author, 1783), pp. vi, 66 n., 69; and *idem, A Bengal Atlas: Containing Maps of the Theatre of War and Commerce on that Side of Hindoostan* (London, 1781), p. x.

[36] Bernoulli, *Description historique* (cit. n. 33), vol. 1, p. ix.

Figure 3. *Frontispiece to James Rennell,* Memoir of a Map of Hindoostan *(London: for the author, 1783). (British Library, India Office Collections.)*

all his sources in the introduction to the memoir that accompanied his map of India, published in 1783. In the frontispiece to the first edition, one can see an open acknowledgement of the cooperation between Indian and British elites: a Brahman giving sacred manuscripts (*Shastras*) to Britannia while other Brahmans, each loaded with manuscripts, patiently await their turn. This map was much more dense with information than any so far made of Britain or of its overseas territories, and would serve as a model of detail and accuracy for the future mapping of Britain itself.

In recognition of his achievements, Rennell was awarded the Copley Medal of the Royal Society in 1791. On the occasion, Sir Joseph Banks, the society's president, proclaimed:

> Would I could say that England proud as she is of being esteemed by surrounding nations the Queen of Scientific improvement, could boast of a general Map as well executed as the Majors [Rennell's] delineation of Bengal and Baher [*sic*], a tract of Country considerably larger in extent than the whole of Great Britain and Ireland . . . the accuracy of his particular surveys stands unrivalled by the most laboured County Maps this nation has hitherto been able to produce.[37]

And if Banks's plea was answered that very year with the founding of the Ordnance Survey, the surveying techniques and instruments used in Britain were very different from those developed in India. While triangulation was adopted as the sole technique of extensive surveying in Britain, it was Rennell's composite method of data collection that was extended by his successors in India. Thus, Thomas Call (surveyor-general of Bengal, 1777–1788) employed at least forty Indians to collect information for his "Grand Atlas of India."[38] And when triangulation was introduced to the subcontinent, it was just one—albeit important—technique, used alongside others such as pacing and reckoning distance as a function of time (with the day's march as the common unit). The task of translating and arranging reports into maps was not a simple one, as a whole gamut of special procedures and protocols had to be constructed. Charles Reynolds (surveyor-general of Bombay, 1796–1807), who organized a series of survey teams composed exclusively of South Asians to crisscross the subcontinent, wrote to superiors who were anxious at the size of his budget, "The[ir] surveys cannot be rendered to use if they are taken down and translated by any other than a person conversant with the business."[39]

In the following decades, new methods for the reproduction of maps were developed for use in the Survey of India. For instance, the first-ever use of lithography in map printing was in Calcutta in 1823.[40] The adaptation, maintenance, and repair of instruments often involved modifications of their structure and protocols for use, and hence recalibration. For instance, the English perambulators were found to be "flimsy, bad in principle, and incapable of working except on a smooth road or bowling green; across country they go to pieces in a mile or two."[41] In the 1780s a Captain John Pringle of the Madras Infantry designed an instrument that was more resilient and better suited to the stature and gait of Indian *lascars* (footmen). By the mid-nineteenth century, the instrument, having undergone continuous modification, was still in use, but was very different in looks and operation from its English cousin.

[37] *Royal Society Journal*, 1790–1793, *Book 34*: 389–90.

[38] India Office Records, Bengal Public Consultations, 6 Oct. 1783 and 29 Nov. 1784: P/2/63 and P/3/7, respectively. A number of maps from similar surveys are held in the British Library, London, Add. MSS 13907 (a, b, c, d, e).

[39] India Office Records, Bombay Military Consultations, 13 Jan. 1807, cited by Phillimore, *Historical Records* (cit. n. 29), vol. 1, p. 288.

[40] See Andrew S. Cook, "The Beginning of Lithographic Map Printing in Calcutta," in *India: A Pageant of Prints*, eds. Pauline Rohatgi and Pheroza Godrej (Bombay: Marg Publishers, 1989), pp. 125–34.

[41] Ralph Smyth and Henry Landour Thuillier, comps., *A Manual of Surveying for India, Detailing the Mode of Operations on the Revenue Surveys in Bengal and the North-Western Provinces* (Calcutta: W. Thacker and Co., 1851), pp. 360–1.

Figure 4. *Some survey perambulators of the Survey of India: The Madras Pattern Perambulator introduced in 1780 (above) and the Everest Pattern Perambulator devised between 1832 and 1836 (below). Adapted by the author from Ralph Smyth and Henry Landour Thuillier, comps.,* A Manual of Surveying for India, Detailing the Mode of Operations on the Revenue Surveys in Bengal and the North-Western Provinces *(Calcutta: W. Thacker and Co., 1851).*

Novel survey methods had at times to be forged for terrains and circumstances that precluded the use of standard techniques—the mapping of central Asia in the 1860s using the rigorously calibrated pace of Indian surveyors is a good example.[42] Indeed, so distinct were the practices of the Survey of India that when, in 1851, the Thomason Engineering College was established in Roorkee (northern India) to train surveyors, an entirely new manual had to be written, for "scarcely any of the English works on Geodesy extant, touch on, or afford any practical insight into, the system of Survey, as carried on and as peculiarly applicable to this country."[43] More than half the book was written by Radhanath Sikhdar, the survey's chief computer.

CONCLUSION

Beyond pointing to the inadequacy of the diffusionist, confrontationist, and "reordering of traditional knowledge" approaches to the spread of Western science, the case of India stresses the complexity and reciprocity involved in the construction of modern science, even in the asymmetrical colonial situation. It also points to the active role that the heterogeneous knowledge networks developed in India played in the forging of a British identity and research-and-teaching tradition in the early nineteenth century, a role that needs to be worked out in much greater detail than the scope of this essay allows.

However, the perspective adumbrated here implies an apparent paradox. As pointed out above, the material practices of the Survey of India differed in crucial ways from those of the Ordnance Survey of Great Britain and Ireland. Yet the knowledge produced out of these differing practices and protocols could be rendered commensurable and placed on the same map. How does the switch from these local practices to universal science operate?

In order to answer this, one has to distinguish between two aspects of science: its material and social practices on the one hand, and the knowledge to which they give rise on the other. While the former are always local—mapping procedures and protocols, for instance, have varied, and indeed continue to vary, between countries institutions, and even individual laboratories—knowledge is *made* universal through a series of mediations, measurement and its calibration being among them. It is through calibration that instruments or techniques developed in a given place can be compared with those developed in another place, that the results obtained through one set can be compared with those obtained through another set—and thus that contemporary science can claim to be universal.[44] In the case of the Indo-British surveys of India, it was precisely this operation of calibration that allowed data collected through such varied methods as Rennell's translation of tabular data and Reynold's assistants to be legitimately incorporated into a single map of the subcon-

---

[42] See Kapil Raj, "La construction de l'empire de la géographie: L'Odysée des arpenteurs de Sa Très Gracieuse Majesté, la reine Victoria, en Asie centrale," *Annales Histoire, Sciences Sociales,* 1997, *5:*1153–80; and *idem,* "When Humans Become Instruments: The Indo-British Exploration of Tibet and Central Asia in the Mid-19th Century," in *Instruments, Travel and Science: Itineraries of Precision in the Natural Sciences, 18th–20th Centuries,* eds. Marie-Noëlle Bourguet, Christian Licoppe, and Hans Otto Sibum (Chur, Switzerland: Harwood Academic Publishers, forthcoming).

[43] Smyth and Thuillier, *Manual of Surveying* (cit. n. 41), p. iii.

[44] See Harry M. Collins, *Changing Order: Replication and Induction in Scientific Practice* (London: Sage, 1985), pp. 100–6.

tinent, which in turn could be commensurated with other local, regional, and national maps into one of the world.

However, a word of caution needs to be immediately spelled out: if Indians and Britons mobilized and transformed their specialized practices for the common resolution of problems, this does not mean, as some historians have recently argued, that they participated equally in a dialogic process in an idyllic "commonwealth of letters."[45] On the contrary, the kinds of knowledge discussed in this essay could only be constructed and sustained within a strong framework of formalized institutions with their imperatives of teamwork and a stratified division of labor. Hence, while the British and Indians collaborated in the making of new knowledge, their respective specialized practices were distinguishable and perfectly well hierarchized in the formal structures of teaching establishments, like the College of Fort William, and of science-dependent administrative organizations, like the Survey of India, creating a commonly made knowledge while creating different identities.

[45] See, for instance, Eugene Irschick, *Dialogue and History: Constructing South India, 1795–1895* (Berkeley, Los Angeles, London: Univ. of California Press, 1994) and Thomas R. Trautmann, "Hullabaloo about Telugu," *South Asia Research*, 1999, *19*, 1:53–70.

Acclimatizing the World: A History of the Paradigmatic Colonial Science

*Michael A. Osborne**

ABSTRACT

This paper examines the institutions, personages, and theories that informed accli-matization activities in nineteenth-century France, England, and the two colonies of Algeria and Australia. Treating acclimatization as a scientific concept and activ-ity, the essay begins with the conditions of its emergence in Enlightenment France. Subsequent sections trace the growth of the acclimatization movement and its translation to the British context, and consider reasons for its decline in the last third of the nineteenth century. Efforts are made to show why many perceived acclimatization to be *the* paradigmatic colonial science with applications as di-verse as agriculture, settlement schemes, field sports, and human health. Emphasis falls on the French and British cultural spheres, as these were the dual epicenters of both modern colonialism and organized acclimatization activity.

THE ZOOLOGICAL, BOTANICAL, AND MEDICAL CONCEPT OF ACCLI-matization was an important issue for nineteenth-century European science, particularly in the colonies of Africa and Australasia. Acclimatization is intimately entwined with the rise of modern imperialism and with the marginalization and alteration of indigenous ecosystems and peoples. In both the French and British Empires, acclimatization discourses influenced politics, settlement schemes, and regulations for the transport, hygiene, and length of duty of European armies in the colonies. Physicians and anthropologists pondered the ability of Europeans to sur-vive in exotic environments, while colonial functionaries, landowners, zookeepers, and naturalists formed acclimatization societies to promote the rational exchange of aesthetically pleasing and "useful" flora and fauna. Unintended plant, animal, and disease introductions accompanied European colonization, and they were a bane to later farmers.[1] In the nineteenth century, however, acclimatization generally denoted an intended and "scientifically" mediated transplantation of organisms.

Beyond its importance in questions of human health, acclimatization also func-tioned at other levels of Europe's colonial enterprise. By the 1830s, its constellation

* Department of History, University of California, Santa Barbara, California 93106-9410.

I wish to thank Christophe Bonneuil, Anita Guerrini, Roy MacLeod, James E. McClellan III, and Kapil Raj for critical readings of this paper. Funding was provided by the Humanities Division of the College of Letters and Science, University of California, Santa Barbara; the Centre National de la Recherche Scientifique, Paris; and the Centre Alexandre Koyré, Paris.

[1] Alfred W. Crosby, *Ecological Imperialism: The Biological Expansion of Europe, 900–1900* (Cambridge: Cambridge Univ. Press, 1986) treats mainly the unintentional introduction of biota.

of heterogeneous practices seemed to embody the utilitarian and manly ethos that permeated colonial science. Moreover, acclimatization touched on such major colonial settlement issues as animal health, labor management practices, modes of development, and the cultural needs of Europeans far from home. In 1860, the French colonial botanist Auguste Hardy, who directed the *jardin d'essai* at Algiers, summarized his views on the subject by declaring that "the whole of colonization is a vast deed of acclimatization."[2] More recently, an historian of anthropology has typified Darwin's era as "an age obsessed with the problem of acclimatization," and claimed that "[t]he utilitarian objectives of colonialism made acclimatization the fundamental scientific question it raised."[3]

By 1900, more than fifty acclimatization societies had formed around the globe, and most were in the European colonies.[4] They were driven by many motives— from scientific curiosity about the mutability of species, to nostalgia for the game birds and sport fishes of Europe. On the peripheries, and in Europe, these societies promoted colonization and functioned as a vast network for the exchange of ideas, techniques, and organisms. In metropolitan Paris and London, and even in the French provinces at Grenoble, it was common to find acclimatization societies linked with menageries, natural history museums, and agricultural and botanical societies. Kew Gardens and the Paris Muséum National d'Histoire Naturelle were hubs in the wheels of international scientific exchange and colonial agriculture.[5] Attempting to Europeanize the tropics and simultaneously render Europe more exotic and cosmopolitan, acclimatization organizations espoused a practical approach to science, one promising economic prosperity, improved diets and health, and aesthetic enjoyment.[6] This essay examines the social and scientific functions of acclimatization and acclimatization societies in Paris and London, and in two areas of colonial activity, French Algeria and the British dominions of Australasia.

FRENCH AND BRITISH THEORIES OF ACCLIMATIZATION

Today, the term "acclimatization" is used principally in respiratory physiology and exotic plant and animal management. In the nineteenth century, however, the word

---

[2] Auguste Hardy, "Importance de l'Algérie comme station d'acclimatation," reprinted from *L'Algérie agricole, commerciale, industrielle* (Paris, 1860), p. 7.

[3] Henrika Kuklick, "Islands in the Pacific: Darwinian Biogeography and British Anthropology," *American Ethnologist*, 1996, *23*, 3:611–38; quotation on p. 628.

[4] Christopher Lever, *They Dined on Eland: The Story of the Acclimatisation Societies* (London: Quiller Press, 1992), pp. 193–4. The appendix, "Chronological List of Principal Acclimatisation Societies," lists fifty-four groups founded from 1854 to the 1930s. Thirty-seven are in New Zealand, Australia, and Tasmania. Others, recorded mainly within the French cultural sphere, are in *Liste générale des membres de la Société Impériale Zoologique d'Acclimatation au 16 mai 1862* (Paris: L. Martinet, n.d.), pp. 93–5, and *Liste générale des membres de la Société Nationale d'Acclimatation de France au 3 mai 1884* (Paris: Imprimerie spéciale du Jardin d'acclimatation, 1884), pp. 71–2.

[5] See the papers presented by Marie-Noëlle Bourguet and Christophe Bonneuil in "Dossier thématique: De l'Inventaire du monde à la mise en valeur du globe: Botanique et colonisation," *Revue Française d'Histoire d'Outre-mer*, 1999, *86*, 322–23.

[6] For the general context of Australian scientific societies, see Michael Hoare, "The Intercolonial Science Movement in Australia," *Records of the Australian Academy of Science*, 1976, *3*, 2:7–28; *idem*, "Science and Scientific Associations in Eastern Australia, 1820–1890" (Ph.D. diss., Australian National Univ., 1974). On utilitarian science and its intersections with acclimatization in the French context, see Michael A. Osborne, "Applied Natural History and Utilitarian Ideals: 'Jacobin Science' at the Muséum d' Histoire Naturelle, 1789–1870," in *Re-Creating Authority in Revolutionary France,*

embraced an astounding range of uses and meanings. Many noted physicians and naturalists, including Charles Darwin and Alfred Russel Wallace in England, and the French zoologist Isidore Geoffroy Saint-Hilaire, attempted to clarify the concept's scope and definition. In France and its colonies, where the term came to signify a rationally forced adaptation to new environments, acclimatization connoted biological changes at physiological and sometimes structural levels. In the British sphere, the term tended to signify a transfer of so-called exotic organisms from one location to another with a similar climate. (In the parlance of the day, an "exotic" organism was one that originated nearly anywhere other than the country or place under study.) But if clarity of definition and precision of use were goals, they were seldom achieved. Enthusiasts often used the term "acclimatization" interchangeably with "naturalization" or "domestication," and definitions varied with cultural, temporal, and geographical context.

The term first appeared in eighteenth-century France, where it was associated with the botany of exotic plants and with Louis-Jean-Marie Daubenton's efforts to introduce merino sheep into the country. Daubenton traveled to Spain to study sheep breeding, experimented on wool quality and breeding at numerous sites in France, and dreamed of acclimatizing the tapir, peccary, and zebra.[7] During the Enlightenment, but also in the revolutionary era, the idea resonated with calls for utilitarian science. The Abbé Féraud wrote in 1787 that "acclimate" was a new word attributed to Guillaume-Thomas Raynal, a critic of European methods of colonization. According to Féraud, it signified being "habituated to a climate."[8] By 1835, the verb *acclimater* had gained legitimacy by appearing in the dictionary of the Académie Française, which defined it as "to accustom to the temperature and influence of a new climate," and provided examples of usage. These included the experiences of Spanish sheep in northern Europe and the difficulties facing European settlers in the West Indies.[9]

Simultaneous with Enlightenment calls for practical science, the writings of Jean-Baptiste Lamarck and a revival of Neo-Hippocratic perspectives on health and disease provided theoretical backing for the acclimatization doctrine. The French had

---

1789–1900, eds. Bryant T. Ragan, Jr. and Elizabeth Williams (New Brunswick, N.J.: Rutgers Univ. Press, 1992), pp. 125–43; and Claude Blanckaert, "Les Animaux 'utiles' chez Isidore Geoffroy Saint-Hilaire: La Mission sociale de la zootechnie," *Revue de Synthèse*, 1992, *113*, 3–4:347–82.

[7] Histories of the term and changes in its definition and use may be found in Michael A. Osborne, *Nature, the Exotic, and the Science of French Colonialism* (Bloomington, Ind.: Indiana Univ. Press, 1994), pp. 62–72. On the related problem of domestication, see Jean-Pierre Digard, *L'Homme et les animaux domestiques: Anthropologie d'une passion* (Paris: Fayard, 1990), pp. 85–103. See also Lever, *They Dined on Eland* (cit. n. 4), pp. vii–ix; Warwick Anderson, "Climates of Opinion: Acclimatization in Nineteenth Century France and England," *Victorian Studies*, 1992, *35*:135–57, esp. pp. 137–8. On Daubenton, see Camille Limoges, "Daubenton, Louis-Jean-Marie," in *Dictionary of Scientific Biography*, ed. Charles Gillispie, vol. 15, pp. 111–14.

[8] Abbé Féraud, *Dictionnaire critique de la langue française*, 3 vols. (Marseille: Jean Mossy, 1787), vol. 1, p. 20. The term is absent, however, from Adolphe Hatzfeld and Arsène Darmesteter, with the assistance of Antoine Thomas, *Dictionnaire général de la langue française du commencement du XVIIe siècle jusqu'à nos jours*, 2 vols. (Paris: Librarie Ch. Delagrave, 1890–1893). See also Guillaume-Thomas Raynal, *L'Anti-colonialisme au 18ème siècle: Histoire philosophique et politique des établissements et du commerce des Européens dans les Deux-Indes*, introduction, choix de textes et notes par Gabriel Esquer (Paris: Presses Universitaires de France, 1951).

[9] *Dictionnaire de l'Académie Française*, 2 vols., 6th ed. (Brussels: J. P. Meline, 1835), vol. 1, pp. 17–18.

Figure 1. *Louis-Jean-Marie Daubenton and merino ram at the Jardin d'Acclimatation in Paris. The statue, executed by the sculptor Godin and dedicated on 13 November 1864, celebrates the Société d'Acclimatation's continuation of applying natural history to the problems of agriculture.*

long celebrated Hippocratic medicine, and the Hippocratic treatise "Airs, Waters, Places" informed French acclimatization theory, as well as their distinctive approaches to medicine and medical geography.[10] Among the major medical questions of the era was whether, and if so by what means, Europeans could adapt to life in the colonies. References to acclimatization and "seasoning" appeared with increasing

[10] Michael A. Osborne, "The Geographical Imperative in Nineteenth-Century French Medicine," in *Medical Geography in Historical Perspective,* ed. Nicolaas A. Rupke (*Medical History Supplement,* no. 20, in press).

frequency after 1830, when the French began grappling with the problems of settling Algeria, and British administrators pondered the future of their Australasian and Indian interests.[11] European ideas of civilization, medicine, and morality, evolving in tandem with imperialism, made acclimatization a problem of interest to geographers, physicians, naturalists, and amateurs.[12]

The acclimatization movement's godfather and major theorist was Isidore Geoffroy Saint-Hilaire, a naturalist at the Paris Muséum National d'Histoire Naturelle. This zoologist—and his more famous father, Etienne Geoffroy Saint-Hilaire—believed that animal life was constructed upon a single, unified plan. In contrast to Georges Cuvier, who tried to excise evolutionary thought from biology, the Geoffroy Saint-Hilaires promoted a variant of Lamarckian transformism and investigated human and animal teratology. Isidore Geoffroy Saint-Hilaire portrayed his transformist theory as a moderate alternative between the extremes of Cuvier and Lamarck. In his view, animals were endowed with vast adaptive potential, and could be forced to acclimate to a wide variety of environments.[13] This theory of limited variability of type provided the rationale for numerous experiments on the acclimatization of exotic animals. Faith in the malleability of animal and plant form and function typified the French approach to acclimatization, and helps explain why the French attempted to introduce everything from ostriches to yaks and llamas both in their own country and in its dependencies.

In midcentury France, acclimatization agreed comfortably with monogenist racial theories and seemed to provide an explanation for the diversity of humanity. Polygenists, however, like the military physician, medical geographer, and anthropologist Jean Ch. M. F. J. Boudin—and many other members of Paul Broca's Société de Anthropologie de Paris—denied that acclimatization could happen or doubted that its minor imprint on racial attributes could be passed on to future generations. Using vital statistics collected to monitor the success of European troops and settlers in Africa, Boudin argued that the various races of humanity maintained their health

[11] On medical statistics and the problem of "seasoning," see Philip D. Curtin, *Death by Migration: Europe's Encounter with the Tropical World in the Nineteenth Century* (Cambridge: Cambridge Univ. Press, 1989), pp. 44–7, 66, 109–111.

[12] There are several studies of the medical aspects of human acclimatization, especially in relationship to ethnicity and colonial activity. Among the more relevant and recent are David N. Livingstone, "Tropical Climate and Moral Hygiene: The Anatomy of a Victorian Debate," *British Journal for the History of Science*, 1999, *32*:93–100; *idem*, "The Moral Discourse of Climate: Historical Consideration on Race, Place, and Virtue," *Journal of Historical Geography*, 1991, *17*:413–34; *idem*, "Human Acclimatization: Perspectives on a Contested Field of Inquiry in Science, Medicine and Geography," *History of Science*, 1987, *25*:359–94; and Mark Harrison, "Tropical Medicine in Nineteenth-Century British India," *Brit. J. Hist. Sci.*, 1992, *25*:299–318. See also the special section on "Race and Acclimatization in Colonial Medicine," *Bulletin of the History of Medicine*, 1996, *70*, which includes: Warwick Anderson, "Disease, Race, and Empire," pp. 62–7; Mark Harrison, "'The Tender Frame of Man': Disease, Climate, and Racial Difference in India and the West Indies, 1760–1860," pp. 68–93; Warwick Anderson, "Immunities of Empire: Race, Disease, and the New Tropical Medicine, 1900–1920," pp. 94–118. For French perspectives, see Anne-Marie Moulin, "Expatriés français sous les tropiques: Cent ans d'histoire de la santé," *Bulletin de la Société de Pathologie Exotique*, 1997, *90*, 4:221–8; Michael A. Osborne, "European Visions: Science, the Tropics, and the War on Nature," in *Nature et environnement*, eds. Christophe Bonneuil and Y. Chatelin (Paris: Editions de l'Office de la Recherche Scientifique et Technique d'Outre-Mer, 1996), pp. 21–32; Richard Fogarty and Michael A. Osborne, "Constructions and Functions of Race in Late Nineteenth Century French Military Medicine," in *Race in France: A History*, eds. Sue Peabody and Tyler Stovall (forthcoming).

[13] Goulven Laurent argues that Geoffroy Saint-Hilaire was fully transformist. See Laurent, *Paléontologie et évolution en France de 1800 à 1860: Une Histoire des idées de Cuvier et Lamarck à Darwin* (Paris: Editions du Comité des Travaux Historiques et Scientifiques, 1987), pp. 467–89.

only by remaining within a very narrow geographical range, and that analogies between supposed animal and human acclimatization were based on faulty reasoning and therefore invalid.[14]

By the late nineteenth century, Europeans had reconsidered the ideal of the settlement colony and were pondering new ways to exploit, fund, and develop colonial resources with fewer European agents.[15] Simultaneously, newer biological concepts cast additional doubts on the heritability of acclimatization. Around 1900, for example, French colonial agricultural experiment stations, some of which tried to use Mendelism and other quantitative crop-selection methods, supplanted older botanical institutions and techniques based on the theory of acclimatization. In general, French *jardins d'essais*, like those in Algeria, adhered to acclimatization. Focusing mainly on the diffusion of gardening plants and of those fit for cottage industries, the practice was to experiment on a wide variety of plants within the confines of the institution and to present experimental results in qualitative fashion.[16]

Eventually, the acceptance and diffusion of a viable germ theory, the advance of parasitological studies of tropical diseases, and the insufficiency of Neo-Lamarckian ideas of transformism rendered acclimatization a rather outmoded theory. In France, the writings of the physician Louis-Adolphe Bertillon, and especially the 1884 publication of Alfred Jousset's *Traité de l'acclimatement et de l'acclimatation*, signaled a medicalization of the acclimatization discourse and a turning away from the Neo-Lamarckian and adaptationist thinking that had favored its extension.[17]

In Britain, zoologists adopted the term "acclimatization" and transformed its meaning to suit their own scientific and imperial agendas. With rare exceptions— such as Alfred Russel Wallace, who integrated explanations of acclimatization into a selectionist and evolutionary framework—British acclimatization theory tended to be antitransformist, antiprogressionist, and antievolutionist.[18] These theoretical issues, as well as the founding and history of the Acclimatisation Society of the United Kingdom (1860–1867), are treated below.

ENVIRONMENTAL AND POLITICAL ECOLOGIES

Acclimatization activity, like colonization itself, forced consideration of environmental issues, including the conservation and preservation of indigenous flora and

[14] Michael A. Osborne, "The Vagaries of Acclimatization Theory and Transformist Biology in Nineteenth Century France," in *Jean-Baptiste Lamarck, 1744–1829*, ed. Goulven Laurent (Paris: Editions du Comité des Travaux Historiques et Scientifiques, 1997), pp. 529–41.

[15] Susan Sheets-Pyenson, *Cathedrals of Science: The Development of Colonial Natural History Museums during the Late Nineteenth Century* (Montreal: McGill-Queens's Univ. Press, 1988).

[16] Christophe Bonneuil, "'Penetrating the Natives': Peanut Breeding, Peasants and the Colonial State in Senegal (1900–1950)," *Science, Technology, and Society*, 1999, *4*, 2:273–302; idem, *Des Savants pour l'empire: La Structuration des recherches scientifiques coloniales au temps de 'la mise en valeur des colonies françaises,' 1917–1945* (Paris: Editions de l'Office de la Recherche Scientifique et Technique d'Outre-Mer, 1991), pp. 42–7. See also Christophe Bonneuil and Mina Kleiche, *Du Jardin d'essais colonial à la station expérimentale, 1880–1930* (Paris: Centre de Coopération Internationale en Recherche Agronomique pour le Développement, 1993); Christophe Bonneuil, *Mettre en ordre et discipliner les tropiques: Les Sciences du végétal dans l'empire français, 1870– 1940* (Paris: Editions des Archives Contemporaines, forthcoming).

[17] See Osborne, *Nature, the Exotic* (cit. n. 7), pp. 90–7, esp. 95–6.

[18] Alfred R. Wallace, "Acclimatisation," *Encyclopedia Britannica*, 9th ed. (New York: Werner, 1898), vol. 1, pp. 84–90; idem, "Acclimatization," *Encyclopedia Britannica*, 11th ed. (Cambridge, 1910), vol. 1, pp. 114–19.

fauna.[19] Acclimatization was a part of colonial agriculture, whose practices have been identified as responsible for both massive environmental degradation and the birth of environmentalism.[20] Yet in the Maghreb, narratives of restoration and civilization, rather than illusions of a lost Eden, framed acclimatization.[21] In India, shortly before the First World War, Frank Finn, assistant director of the Indian Museum at Calcutta, wrote that habitat destruction and the unintended introduction of disease were considerably more damaging to indigenous species than acclimatized exotic organisms.[22] But fears of the coming extinction of "native" species and criticism of acclimatization were frequent from the 1870s onwards.[23] As a reviewer noted in *Nature*, "The English Acclimatisation Society fortunately came to an end, before it had time to do any harm here [in England]. . . . " But, he continued, as if to remember Frances Trevelyan Buckland's piscicultural prowess (discussed below), "its example has been mischievous in our dependencies."[24]

Colonial settlers employed exotic species to renovate the biota of their adopted countries. Thomas R. Dunlap has noted how acclimatization societies formed "part of the settler's continuing attempt to come to terms with their new lands, to find their place in the country and its place in them."[25] While it is probable that acclimatization societies introduced exotic animals that became pests and altered colonial ecosystems, the historical assessment of these events remains incomplete. Eric C. Rolls, for example, has concluded that Australian acclimatization societies can not be blamed for initiating the transfer of deer in 1806, or for the importation of more infamous biota, such as the fox, rabbit, and prickly pear, which came to plague settler agriculture.[26] A global history of acclimatization's environmental effects would be difficult to achieve, for each colony had a specific environment, and levels of integration with European cultural and economic orbits varied among locations. More-

[19] On environmentalism and acclimatization in the Russian context, see Douglas R. Weiner, "The Roots of 'Michurinism': Transformist Biology and Acclimatisation as Currents in the Russian Life Sciences," *Annals of Science*, 1985, *42*:243–60; *idem*, "The Historical Origins of Soviet Environmentalism," in *Environmental History: Critical Issues in Comparative Perspective*, ed. Kendall E. Bailes (Lanham, Md.: Univ. Press of America, 1985), pp. 379–411.

[20] Richard H. Grove, *Green Imperialism: Colonial Expansion, Tropical Island Edens and the Origins of Environmentalism, 1600–1860* (Cambridge: Cambridge Univ. Press, 1995).

[21] Michael A. Osborne, "La Renaissance d'Hippocrate: L'Hygiène et les expéditions scientifiques en Egypte, en Morée et en Algérie," in *L'Invention scientifique de la Méditerranée*, eds. Marie-Noëlle Bourguet, *et al.* (Paris: Editions de l'Ecole des Hautes Etudes en Sciences Sociales, 1998), pp. 185–204.

[22] Frank Finn, appendix to article on acclimatization, *Encyclopedia Britannica*, 11th ed. (Cambridge, 1910), vol. 1, pp. 119–21.

[23] Osborne, *Nature, the Exotic* (cit. n. 7), esp. pp. 145–71. On the conditions of colonial natural science in Australia, see Barry W. Butcher, "Darwin's Australian Correspondents: Deference and Collaboration in Colonial Science," in *Nature in its Greatest Extent: Western Science in the Pacific*, eds. Roy MacLeod and Philip F. Rehbock (Honolulu: Univ. of Hawaii Press, 1988), pp. 139–57; *idem*, "Darwinism and Australia, 1836–1914" (Ph.D. diss., Univ. of Melbourne, 1992).

[24] From excerpts of a review on the ornithology of New Zealand in Joyce M. Wellwood, comp. and ed., *Hawke's Bay Acclimatisation Society Centenary, 1868–1968* (Hastings, N.Z.: Cliff Press Printers, 1968), p. 24.

[25] Thomas R. Dunlap, "Remaking the Land: The Acclimatization Movement and Anglo Ideas of Nature," *Journal of World History*, 1997, *8*:303–19, quotation on p. 304. See also Ian Tyrrell, "Peripheral Visions: California-Australian Environmental Contacts, c. 1850s–1910," *J. World Hist.*, 1997, *8*:275–302.

[26] Eric C. Rolls, *They All Ran Wild: The Animals and Plants that Plague Australia*, rev. ed. (London: Angus and Robertson, 1984). Particularly relevant to acclimatization society activity are pp. 270–344.

over, it is likely that—except for Europeans themselves—indigenous species, rather that acclimatized exotic organisms, were much more disruptive to European-style agriculture in areas like the forests of eastern Australia.[27]

Even if questions remain about the impact of acclimatization on the environment, it is certain that its practice and its theoretical explanation advanced hand-in-hand with European colonialism. In Algeria, the French made sustained attempts to acclimatize tropical crops. The efforts found their rationale in mercantile economic theory, and in attempts to recover sources of cane sugar, exotic spices, and fruit. These products were what France had lost two decades earlier when the "Black Jacobins" of its Caribbean colony of St. Domingue (Haiti) claimed independence during the Revolution.[28] The later emergence of metropolitan and colonial acclimatization societies in the 1850s, 1860s, and 1870s coincided in the British Empire with what Roy MacLeod has termed "colonial science," when local scientific interests gained increasing importance in the colonies.[29] George Basalla, who vests "colonial science" with a somewhat different meaning, would also have colonial acclimatization societies emerging when colonial scientific activity was still largely dependent on metropolitan personnel, but European settlers had begun to found their own intellectual venues and scientific institutions.[30] In their criticisms of Basalla's account, both MacLeod and Ian Inkster are sensitive to the highly dynamic nature of colonial relations, and to the larger economic contexts and social matrices that shaped the genres of science taking root on Europe's periphery.[31] Whatever chronology one uses, the heyday of acclimatization was part and parcel of the founding and establishment of Europe's settler colonies.

Acclimatization was especially prominent during eras of economic protectionism, when tariffs favored new ventures such as llama culture and the cultivation of vanilla beans. The practice was common both in temperate climes, such as Britain's Australasian dominions, and in dependencies such as France's favored imperial outpost, Algeria. With llamas costing as much as 3,500 francs each, the projects could be expensive.[32] Draining private investment as well as colonial and metropolitan budgets, acclimatization also altered evolving colonial legal structures. In New Zealand, for example, laws protected acclimatized fish and game birds, and acclimatization societies assumed the role of issuing licenses for fishing. An 1895 amendment to the Animals Protection Act mandated that the minister of agriculture review and approve all importation of exotic fauna. So successful were the New Zealanders at acclimatization that, by the 1920s, one observer claimed, "The Game Animals of

[27] See, for example, Warwick Frost, "European Farming, Australian Pests: Agricultural Settlement and Environmental Disruption in Australia, 1800–1920," *Environment and History*, 1998, *4*:129–43.

[28] James E. McClellan III, *Colonialism and Science: Saint-Domingue in the Old Regime* (Baltimore, Md.: The Johns Hopkins Univ. Press, 1992).

[29] Roy MacLeod, "On Visiting the 'Moving Metropolis': Reflections on the Architecture of Imperial Science," in *Scientific Colonialism: A Cross-Cultural Comparison*, eds. Nathan Reingold and Marc Rothenberg (Washington, D.C.: Smithsonian Institution Press, 1987), pp. 217–49, esp. pp. 223–37.

[30] George Basalla, "The Spread of Western Science," *Science*, 1967, *67*:611–22.

[31] Ian Inkster, "Scientific Enterprise and the Colonial 'Model': Observations on Australian Experience in Historical Context," *Social Studies of Science*, 1985, *15*:677–704. On the concept of a "social matrix" of institutions and how they enable colonialism, see *idem*, "Prometheus Bound: Technology and Industrialization in Japan, China and India Prior to 1914—A Political Economy Approach," *Ann. Sci.*, 1988, *45*:399–426.

[32] Osborne, *Nature, the Exotic* (cit. no. 7), p. 26.

New Zealand consist wholly of species introduced from the continental areas of the Old and New Worlds."[33]

Frequently, political and financial support for acclimatization depended on a few key people. In France and Britain, the aristocracy opened its game parks to acclimatization societies and patronized the movement with influence and money. From the late 1860s, however, as detailed in the following sections, European acclimatization societies fell on hard times as their aristocratic patrons withdrew support, and tensions between amateurs and professionals began to emerge. In France, acclimatization first connoted association with Napoleon III's government and later, after 1871, with an outmoded royalism of uncertain parentage. By the century's end, links between metropolitan and colonial acclimatization societies had also weakened or disappeared altogether, as in the demise of London's acclimatization society. In the end, "[a] shortage of money and the failure of some animals to adapt saved the world from these societies' worst enthusiasms."[34]

Keeping in mind the intellectual, ecological, and political issues discussed above, the next sections examine the emergence and decline of acclimatization across two empires.

ACCLIMATIZATION AND COLONIALISM: THE CASE OF FRANCE AND ALGERIA

The first and largest of all acclimatization societies, the Société Zoologique d'Acclimatation, was founded in Paris in 1854. As noted above, the Paris group was tightly linked to Isidore Geoffroy Saint-Hilaire, and while he never visited the French colonies, his research on exotic species was important to all three zoos in the French capital.[35] A major goal of its members was to improve agriculture through the acclimatization and subsequent domestication of exotic animals. The origins of the Paris society were tied to colonial botany, and particularly to Algeria, where the French government operated some two dozen botanical gardens.[36] As in Britain, experience with colonial botanical gardens and plant exchanges informed acclimatization activity.[37]

Supporters of the new Société Zoologique d'Acclimatation included Baron Montgaudry, a nephew of Georges Buffon; a substantial cluster of Muséum naturalists; and landed and wealthy notables such as the Baron A. de Rothschild and Counts Eprémesnil, Séguier, and Sinety. By 1860, more than twenty-six hundred people— diplomats and heads of foreign states among them—had joined the society. Most importantly, the group secured the patronage of Napoleon III and received land for a large zoo on the western edge of Paris in the Bois de Boulogne. The zoo, or Jardin

[33] T. E. Donne, *The Game Animals of New Zealand: An Account of their Introduction, Acclimatization, and Development* (London: John Murray, 1924), p. v.; Wellwood, *Hawke's Bay* (cit. n. 24), pp. 229–47, lists a series of animal protection acts.

[34] Rolls, *They All Ran Wild* (cit. no. 26), p. 270.

[35] Michael A. Osborne, "Zoos in the Family: The Geoffroy Saint-Hilaire Clan and the Three Zoos of Paris," in *New Worlds, New Animals: From Menagerie to Zoological Park in the Nineteenth Century*, eds. Robert J. Hoague and William Deiss (Baltimore, Md.: The Johns Hopkins Univ. Press, 1996), pp. 33–42.

[36] Michael A. Osborne, "The System of Colonial Gardens and the Exploitation of French Algeria, 1830–1852," in *Proceedings of the Eighth Annual Meeting of the French Colonial Historical Society, 1982*, ed. E. P. Fitzgerald (Lanham, Md.: Univ. Press of America, 1985), pp. 160–8.

[37] Christophe Bonneuil, "Une Botanique planétaire," *Cahiers de Science et Vie*, April 1999, 50:48–57.

Figure 2. *When the Jardin d'Acclimatation opened in October 1860, Parisians found this artificial mountain populated with symbols of colonial conquest, including wild sheep from Algeria and Angora goats supplied by defeated Algerian patriot Abd-el-Kader.*

d'Acclimatation, opened in 1860 as a showcase of colonial flora, fauna, and peoples. The group obtained organisms through diplomatic and military channels, and by exchange with the Paris Muséum's menagerie and other European zoos. In this manner, the Jardin d'Acclimatation gave acclimatization a cultural presence not seen in other European capitals. Through ethnographic displays and lectures, school children and the polite public of Paris learned much of France's *outre-mer* and the peoples and resources contained within the colonial empire.

The work of the Paris society was conducted mainly by commissions and committees. Among the most active was a Permanent Commission on Algeria. This commission, composed of nineteen members in 1860, included metropolitan savants; military men such as General Eugène Daumas, director of Algerian affairs at the Ministry of War; and experts on tropical hygiene, agriculture, and botany. Commission members functioned as consultants to the French government on colonial agriculture and settlement. Like French geographical societies, the acclimatization society also provided a venue for the discussion of colonial topics. Admittedly, as in failed projects for yak and llama culture, acclimatization schemes tended toward the fantastic. Yet they held out the possibility that France would one day transform Algeria and make it French. The society examined crops such as bamboo and quinine; conducted experiments designed to stimulate a silk industry; and also tried to initiate other industries such as llama culture and resettlement schemes for exotic plants, animals, and sometimes peoples. Although the Algiers Jardin d'Essai functioned as the society's principal colonial "laboratory," other sites were offered. In the 1860s, for example, Archbishop Charles Lavigerie of Algiers offered the labor of the hundreds of orphans living on farm-schools in his archdiocese. The children were prepared, he wrote, to acclimatize and cultivate Egyptian cattle and other exotic ruminants.[38]

Acclimatization also played a social function in the lives of French colonists and colonial officials. Prince Jerôme Napoleon, honorary president of the Jardin d'Acclimatation as well as minister of Algeria and the colonies, joined with the minister of the navy to order the formation of colonial branches of the Paris society in Algeria, Cayenne, Réunion, Martinique, and Guadeloupe. State sponsorship had its benefits, and plants and animals flowing to and from the colonies usually traveled on government ships at no cost. The largest of the colonial branches was at Algiers, which in 1859 listed fifty-four members. The research program at the Algiers Jardin d'Essai, which hosted numerous experiments on useful organisms including ostriches, Chinese yams, bamboo, and cochineal beetles, was designed to complement and not compete with crops easily grown in France, such as grapes and wheat. Few of these ventures succeeded, but prior to about 1870, systematic exotic crop acclimatization seemed the ideal way to achieve a return on commercial investment. The Paris society and its colonial branches promoted colonization, even when its projects failed and its core ideology of biological transformism came under attack.

On the whole, the French tended to concentrate on acclimatization for agriculture, and sometimes on using familiar animals in new ways—such as eating horses at the end of their lives of labor. Hippophagia, and hippophagic banquets, were promoted by Isidore Geoffroy Saint-Hilaire and by acclimatizers in Paris, Nancy, Algiers, and throughout France. This practice was understandably less popular in Britain, where different ideas of "humane" behavior prevailed.[39] As the next section shows, not only hippophagia, but social goals and scientific ideas as well, distinguish the history of British and Australian acclimatization from that of France.

ACCLIMATIZATION AND COLONIALISM: THE CASE OF ENGLAND AND AUSTRALIA

In the British Empire, small clusters of businessmen, scientists, and publicists spurred acclimatization, although the aristocracy was not entirely absent from the movement. Since the eighteenth century, botanical acclimatizations had been conducted under the patronage of the East India Company, Joseph Banks, and other notables. When Victoria came to the throne in 1837, the empire possessed eight botanical gardens. At Victoria's death in 1901, there were more than one hundred, about fifty of which were in India and the Australasian colonies. These institutions—at Calcutta, Bangalore, and elsewhere—were brought under the nominal control of the Royal Botanic Gardens at Kew in the 1880s.[40] In contrast to France, British acclimatization had prospered on the periphery long before a metropolitan acclimatization society formed in London in 1860. Within a decade of settlement, game animals had been imported to Australasia. For example, Dr. John Harris imported

---

[38] [Charles] Lavigerie, "Essais d'acclimatation en Algérie," *Bulletin de la Société Impériale Zoologique d'Acclimatation*, 1869, 3rd. ser., 6:506–8.

[39] See Anita Guerrini and Michael A. Osborne, "Eating a Horse" (forthcoming).

[40] For the history of Kew and colonial botanical networks, see Lucile H. Brockway, *Science and Colonial Expansion: The Role of the British Royal Botanic Gardens* (New York: Academic Press, 1979). Estimates for British colonial gardens are drawn from Donal P. McCracken, *Gardens of Empire: Botanical Institutions of the Victorian British Empire* (London: Leicester Univ. Press, 1997), pp. 17, 19.

deer to Sydney in 1803. By the 1830s, partridge, hare, deer, monkeys, and other exotic animals had made the voyage to Hobart.[41] Acclimatization was also associated with Britain's natural history trade. The effort to procure exotic plants and animals for wealthy patrons or institutions, including the Zoological Society of London, brought collector-naturalists such as William Swainson and Alfred Russel Wallace to the far corners of the earth. Of all the introduced exotic organisms in Australasia, however, none would rival the Spanish merino sheep in terms of economic importance and success.

The connections between zoos and empire have long been explicit. Exotic animal collections constitute a record of the collectors' diplomatic activity and personal connections.[42] Members of the early Zoological Society of London benefited from the assistance of Sir Stamford Raffles. The chief architect of Great Britain's Far Eastern empire, as well as a fellow of the Royal Society, Raffles was instrumental in establishing a network of exotic animal collectors and colonial functionaries that sustained the London Zoo. Zoological Society fellows corresponded regularly with the Colonial Office, and with others who were well connected with the East India Company. For Englishmen at home and abroad, acclimatization often meant the importation of fish and birds for gentlemanly field sports. But the interests of zoologists and gentry could be quite different, and the Zoological Society of London failed to sustain a program of animal acclimatization intended to provide exotic game animals to the squirearchy.[43]

Around 1860 the subject of acclimatization was much in the news in Britain, and was aired at meetings of the British Association for the Advancement of Science. At least one luminary of British zoology, Richard Owen, advocated the importation of African ruminants whose flesh would improve the British diet, and a BAAS Committee on the Acclimatisation of Domestic Animals functioned briefly. But acclimatization, with its amateurish aspects and overtly utilitarian goals, gained only a tenuous place in the world of professional English zoology. Members of the BAAS committee did little, and one member, John Edward Gray, keeper of zoology at the British Museum, deprecated the sudden celebrity of acclimatization and cautioned against the importation of elands, alpacas, and llamas.[44]

The most prominent British acclimatizer was Francis Trevelyan Buckland, a surgeon-turned-writer and the son of the naturalist Reverend William Buckland. The Buckland family had a tradition of eating exotic animals, such as ostrich and crocodile, and young Frank had embraced his father's love of natural history as well as his eccentricities of diet and religion. Frank Buckland, and the majority of acclimatizers in Britain and the empire who recorded their rationales in print, approached acclimatization as merely a transfer of organisms between analogous climates. Natu-

[41] Rolls, *They All Ran Wild* (cit. n. 26), pp. 270–3.

[42] Michael A. Osborne, "The Role of Exotic Animals in the Scientific and Political Culture of Nineteenth Century France," *Colloques d'Histoire des Connaissances Zoologiques*, 1998, *9*:15–32.

[43] Adrian Desmond, "The Making of Institutional Zoology in London, 1822–1836," *Hist. Sci.*, 1985, *23*:153–85, 223–50. For the broader context, see David E. Allen, *The Naturalist in Britain: A Social History*, 2nd. ed. (Princeton, N.J.: Princeton Univ. Press, 1994). On animals and empire, see Harriet Ritvo, *The Animal Estate: The English and Other Creatures in the Victorian Age* (Cambridge, Mass.: Harvard Univ. Press, 1987), esp. pp. 205–88.

[44] On the BAAS activity, see Michael A. Osborne, "The Société Zoologique d'Acclimatation and the New French Empire: The Science and Political Economy of Economic Zoology During the Second Empire" (Ph.D. diss., Univ. of Wisconsin, 1987), pp. 343–47.

ral theology featured prominently in their ideas. Buckland's *Natural History of British Fishes* described the denizens of the "Great Fish Farm" of the North Sea and sought to show "the truth of the good old doctrines of the Bridgewater Treatises. . . . I steadfastly believe that the Great Creator . . . made all things perfect and 'Very Good' from the beginning. . . ."[45] Given that the Creation was good and perfect, man's only option was to slightly rearrange nature to suit his needs. Of the acclimatization of his favorite organism, Buckland wrote:

> By the acclimatisation of fish I mean that, not only is it possible to obtain from other countries fish not as yet known as British fish, but where as we have already in our waters some of the best fish in the world, that it would be desirable to improve their breed by transferring them from places where they are already found in abundance to other places having a similarity of soil and climate.[46]

W. Oldham Chambers, a fellow of the Linnaean Society, echoed these views. He argued that acclimatization was mainly a transfer of organisms, and one needed only "to select waters resembling as far as possible those from which the fish in the first instance were taken. . . ."[47] Buckland sought practical application of his ideas, and in 1864 placed salmon and trout ova on a ship bound for New Zealand and Australia. By 1880, brown trout, presumably all descended from the thousand or so ova he had sent, were established in Australasia. This accomplishment was surely, he modestly claimed, "the greatest feat of Pis[c]iculture of modern times. . . ."[48]

Inspired by the success of the Paris society, the Acclimatisation Society for the United Kingdom took form in the offices of the field-sports magazine *Field* in 1860. Buckland became the group's first secretary and later its naturalist-manager. Living mainly by his pen, he published in *Field*, founded a rival publication called *Land and Water*, and organized a Museum of Economic Fish Culture at South Kensington. He also brought out editions of his father's celebrated Bridgewater Treatise, Gilbert White's *Natural History of Selborne*, and several editions of his own extremely popular *Curiosities of Natural History*.[49] Like Buckland, the London acclimatization society focused mainly on introducing game birds and fish—animals of "practical and immediate utility to the country gentleman."[50]

The London acclimatization group funded Buckland's piscicultural investigations and began a fish breeding program that soon exhausted its treasury. Having contacts in France, where he had gone in 1849 to take surgical training at La Charité Hospital, Buckland established exchanges of fish ova and fry between the London society and the French government hatchery at Huningue, and made contact with members of

[45] Frank Buckland, *Natural History of British Fishes; Their Structure, Economic Uses, and Capture by Net and Rod* (London: Society for Promoting Christian Knowledge, [1880?]), p. x.

[46] *Ibid.*, section titled "Fish for Acclimatisation," pp. 344–72; quotation on p. 345.

[47] W. Oldham Chambers, *The Introduction and Acclimatisation of Foreign Fish* (London: William Clowes and Sons, Ltd., 1884), p. 3.

[48] Buckland, *Natural History* (cit. n. 45), p. 318.

[49] Biographical details may be had in George C. Bompas, *Life of Frank Buckland*, 2nd ed. (London: Smith, Elder, & Co., 1885); G. H. O. Burgess, *The Eccentric Ark: The Curious World of Frank Buckland* (New York: Horizon Press, 1967); "Buckland, Francis Trevelyan," *Dictionary of National Biography*, vol. III, pp. 204–5. See also Roy MacLeod, "Government and Resource Conservation: The Salmon Acts Administration, 1860–1886," *Journal of British Studies*, 1968, 7, 2:114–50.

[50] Grantly F. Berkeley in the *Dorset Country Chronicle*, quoted in Burgess, *The Eccentric Ark* (cit. n. 49), p. 105.

the Paris society.[51] After the London society's bankruptcy in 1867, Buckland became the government's inspector of salmon fisheries and continued to promote acclimatization of British salmon and trout in Australia and New Zealand.

On the periphery, the transfer of the Spanish merino sheep breed and the establishment of the wool industry in Australia predated the founding of Australasian acclimatization societies by some sixty years. The soldier, farmer, and politician John Macarthur (1766–1834) claimed credit for the establishment of the merino sheep in New South Wales. In fact, he was only one of a number of settlers who had obtained sheep from the Cape of Good Hope as early as 1793. Clashes with colonial governors and military superiors resulted in his return to England and subsequent resignation from the army. Promoting himself to British wool interests, he obtained eight merinos from King George III's herd and returned to New South Wales in 1805 with a letter ordering the governor to grant him four thousand hectares of land. Together with his wife Elisabeth and members of their extended family, Macarthur developed, publicized, and improved the Australian wool industry.[52] After 1820, infusions of merinos came from the French herd at Rambouillet, which had been developed by Daubenton, and herds in Saxony, England, and the United States. At midcentury, the pastoralists of New South Wales cared for more than twelve million sheep, and by 1891 the continent contained more than one hundred million sheep—the vast majority of them merinos.[53]

Macarthur had realized the conditions necessary for successful large-scale agriculture in early Australia. Although endowed with vast expanses of cheap and fertile land, Australia's population base and internal markets were small. Hence, commodities needed a large export market and had to maintain quality on long voyages. They also had to be economical in terms of volume and value ratios, and be produced with minimal labor costs. For several decades, only sheep grazing on native pastureland fulfilled these criteria. By the 1850s, however, as a series of gold rushes attracted thousands of prospectors who subsequently settled on the continent, the pastoral model began to falter. These new immigrants obtained grants of land far smaller than those of Macarthur's era, and could not survive by growing wool. It was within this context of economic transition and a search for new agricultural industries that Australia's businessmen and scientists formed acclimatization societies.[54]

Prominent in Australian acclimatization was Ferdinand von Mueller, director of the Melbourne Botanic Gardens and government botanist for Victoria from 1853 until his death in 1896. Von Mueller was well connected throughout the globe and exchanged seeds and plants with most of the world's scientific institutions. An advocate of transferring eucalyptus to places where it served as an antimalarial agent,

[51] Osborne, "The Société Zoologique d'Acclimatation" (cit. n. 44), chap. 6, "The Science and Activity of Acclimatization in the United Kingdom," pp. 320–89, esp. pp. 361–3.

[52] Biographical details from Rollo Gillespie, "Macarthur," *The Australian Encyclopedia*, 5th ed. (Terrey Hills, N.S.W.: Australian Geographic Society, 1988), vol. 5, pp. 1825–9; Margaret Steven, "Macarthur," in *Australian Dictionary of Biography*, ed. Douglas Pike (Melbourne: Melbourne Univ. Press, 1967), vol. 2, 1788–1850, I–Z, pp. 153–9.

[53] "Wool Industry," in *Australian Agriculture: The Complete Reference on Rural Industry/National Farmers Federation*, ed. Julian Cribb (Camberwell, Australia: Morescope, 1991), pp. 137–56. Figures from Anthony Barker, *When Was That: Chronology of Australia* (Sydney: John Furguson, 1988), pp. 125, 210.

[54] Bruce Davidson, "Developing Nature's Treasures: Agriculture and Mining in Australia," in *The Commonwealth of Science: ANZAAS and the Scientific Enterprise in Australasia, 1888–1988*, ed. Roy MacLeod (Melbourne: Oxford Univ. Press, 1988), pp. 273–91.

von Mueller also wrote a manual for plant acclimatization. In 1887, the Parisian society issued a revised edition of the manual as a book titled *Manuel de l'acclimateur ou choix de plantes recommandées pour l'agriculture, l'industrie et la médecine*. Von Mueller established contacts with the French consuls in Melbourne, such as the Comte de Castelnau, an honorary member of the Parisian society, who also held diplomatic posts in Brazil (Bahia) and Victoria; and with botanists such as Charles Naudin in France and Auguste Hardy in Algeria.[55] Academic scientists who joined von Mueller in promoting acclimatization included Frederick McCoy, government paleontologist for Victoria and the first professor of natural science at the University of Melbourne. McCoy, like von Mueller, Buckland, and Isidore Geoffroy Saint-Hilaire, rejected Darwinian evolution. However, like von Mueller and Buckland, McCoy also rejected French-style transformism and embraced natural theological explanations. The greatest achievements of acclimatization, he believed, had been merely to rearrange God's creation, that is, *"the bringing together in any one country the various useful or ornamental animals of other countries having the same or nearly the same climate and general conditions of surface"* (McCoy's emphasis).[56]

Von Mueller and McCoy were two of Australia's most famous naturalists. But no one worked harder or more successfully for acclimatization than Edward Wilson. Wilson, an owner of *The Argus* newspaper in Melbourne, played an important role in founding acclimatization societies in London, Victoria, New South Wales, Tasmania, and elsewhere. In Melbourne, Wilson used *The Argus* to lobby for introducing British songbirds and alpacas into the colony.[57] Wilson gained the ear of Thomas Embling, a member of the Victorian Parliament, who led a campaign with him to introduce alpacas and llamas. Wilson also secured government and private financing for an aviary for British songbirds. Still dreaming of a success on the scale of the Spanish merinos, in 1858 his efforts led to the importation of 276 alpacas, llamas, and alpaca-llama crossbreeds.

As Linden Gillbank has astutely noted, acclimatization societies could establish zoos as well as emerge from them. In Melbourne, members of the Acclimatisation Society of Victoria, which spun out of a faltering zoological society, would stimulate the foundation of Australia's first zoo, the Royal Melbourne Zoological Gardens. The Victorians organized their acclimatization society in 1861. Subsequently, the provincial government funded the group and granted it fifty acres of land with the proviso that it absorb a Zoological Gardens Management Committee. At center stage was the flock of alpacas and llamas, but success on the scale of the merinos could not be duplicated. The government withdrew support of the project in 1869 and

---

[55] A selection of von Mueller's correspondence has been published in R. W. Home and Sara Maroske, "Ferdinand von Mueller and the French Consuls," *Explorations: A Bulletin Devoted to the Study of Franco-Australian Links*, June 1995 (issued December 1997), *18*:3–50. For von Mueller's correspondence network, see Ferdinand von Mueller, *Regardfully Yours: Selected Correspondence of Ferdinand von Mueller*, ed. R. W. Home *et al.* (Bern; New York: Peter Lang, 1998–).

[56] Frederick McCoy, "Acclimatisation, its Nature and Applicability to Victoria," *Acclimatisation Society of Victoria, First Annual Report* (Melbourne, 1862), pp. 31–51, quotation on p. 36.

[57] Linden Rae Gillbank, "The Acclimatisation Society of Victoria," *The Victorian Historical Journal*, 1980, *51*:255–70, esp. pp. 259–62; *idem*, "The Origins of the Acclimatisation Society of Victoria: Practical Science in the Wake of the Gold Rush," *Historical Records of Australian Science*, 1986, *6*:359–74. For Wilson's activities, see also *idem*, "A Paradox of Purposes: Acclimatization Origins of the Melbourne Zoo," in *New Worlds, New Animals: From Menagerie to Zoological Park in the Nineteenth Century*, eds. R. J. Hoague and William A. Deiss (Baltimore, Md.: The Johns Hopkins Univ. Press, 1996), pp. 73–85, esp. pp. 74–82.

forced the acclimatizers to reconsider their goals. The Victorian society survived a few years longer than the London group, but in 1872 it was transformed into the Zoological and Acclimatisation Society. The name change signaled a change of emphasis, and brought with it an unintended legacy, the Royal Melbourne Zoological Gardens.[58]

CONCLUSION

Conceived in Paris, the organizational model of the acclimatization society—but not the theory of biological transformism that had engendered it—took root thoughout the globe.[59] Acclimatization societies emerged during a period when scientists in the French and British Empires collaborated across huge distances to develop the resources of their respective colonies. In so doing, the French and British described nature as being, respectively, predominately malleable or perfected but rearrangeable. The scientific activities of these groups reflected a Eurocentric—and often mainly French or British—vision of colonial agriculture, settlement, and development. Promoted as the incarnation of a cooperative and humanistic civilizing mission, acclimatization was also touted as a utilitarian activity that promised economic betterment and aesthetic enjoyment for Europeans. The colonized might benefit too, but only secondarily through such things as improved diets, which were themselves deemed necessary for labor control and colonial governance. In fact, the extension of export agriculture to French North Africa resulted in diminished diets and famine for the Algerian peoples.

In the colonies, the goals of the acclimatization movement anticipated, but fell short of, the universalistic scope for science proclaimed by the literature of late-nineteenth-century Europe. In French North Africa, contradictory leitmotifs infused acclimatization projects. The first was the crafting of an agricultural future that would be different from that of France and, in the case of Algeria, similar to that of the lost colony of St. Domingue. The second, which had the additional function of legitimating France's presence in the Maghreb, was that the French, as the rightful inheritors of Rome, would use their science and technology to restore the region to the fertility it had supposedly known under Roman rule. But Algeria, without vast expanses of natural pasture, never had exportable animal products to compare with the wool of Macarthur's merino sheep, the animal that so changed the early fortunes of Australia. Nor did Algeria have the mineral wealth and gold rushes that drew adventurers and settlers to Australia. What French North Africa had, of course, and Australia lacked, was a well-armed and tenacious resistance movement that bedeviled and circumvented European objectives for decades.

The rise of the acclimatization movement also signals a time when fascination with the exotic and a knowledge of colonial affairs had spread beyond Europe's scientific and administrative elites. The cultural semiotics of acclimatization were diverse. Even when projects failed, as most of them did, naturalists and amateurs

---

[58] Gillbank, "A Paradox of Purposes" (cit. n. 57).

[59] This theme is further developed in Michael A. Osborne, "A Collaborative Dimension of the European Empires: Australian and French Acclimatization Societies and Intercolonial Scientific Cooperation," in *International Science and National Scientific Identity: Australia between Britain and America*, eds. R. W. Home and Sally G. Kohlstedt (Dordrecht: Kluwer Academic Publishers, 1991), pp. 97–119.

gained greater knowledge of the care and physiology of exotic flora and fauna, and in some instances broadened their understanding of tropical hygiene. Moreover, the acclimatized exotic organism functioned as a symbol of Europe's power over nature and over far-off lands. Vested with visions of imperial superiority, acclimatized animals and plants provided material manifestations of science serving the interests of transplanted Europeans. Relying on exotic plants and animals, acclimatization schemes also tended to devalue indigenous methods of agriculture, and probably degraded colonial environments. By their very nature, acclimatization projects seemed to confirm that colonization was possible and that colonials were interested in science and had the abilities to conduct experiments. Thus, even in the face of considerable obstacles, acclimatization projects emboldened Europeans, enabled the continuation of colonial projects, and offered a reason to retain colonial possessions. Fortified by the hope of future success, France retained its colonies until public works projects, vaccination programs, tropical medicine, and newer methods of crop selection would render the colonies more profitable and habitable for Europeans.

The checkered history of acclimatization, like the persistence of epidemic disease, also signaled that much in nature was, in the end, beyond European control. The fact that so many acclimatization projects and societies failed served to mark the limits of European science. Of necessity, many of the social and cultural dimensions of acclimatization, such as the formation of scientific societies and the organization of exotic animal exchange, required a critical mass of settlers with disposable income and time. This circumstance, which peaceful Victoria had and colonial Algeria lacked, helps to explain why the Australasian settler colonies became successful epicenters of acclimatization activity. In this, the acclimatization societies occupied an important but ephemeral space within the evolving edifice of colonial science.

Part III: Science, Culture, and the
Colonial Project

Enlightenment in an Imperial Context: Local Science in the Late-Eighteenth-Century Hispanic World

*Antonio Lafuente**

ABSTRACT

This paper aims to assess the figuration of local and metropolitan scientific practices and theories in the eighteenth-century Hispanic Empire by focusing on two colonies: New Spain (Mexico) and New Granada (Colombia). In New Spain, Creole and metropolitan scientists negotiated the assimilation of old local wisdom with new European knowledge in their botanical studies of native plants. Through the openness of both groups of scientists to new ideas, the naturalization of standardized procedures, and the verbalization of old problems in new terminology, the globalization process of scientific practices was successfully integrated there at the local level. In New Granada, less favorably, the Royal Botanical Expedition (1783–1816) provoked disagreement between representatives of the viceroy and of the colony's Creole intelligentsia not only about plant classification systems, but about the proper relationship between scientific and political interests.

O VER THE LAST TWO DECADES, MUCH RESEARCH HAS BEEN dedicated to the study of the scientific and technological exchanges between different cultures and civilizations.[1] Within the framework of this literature, several models have been proposed.[2] In one of the earliest attempts to describe the expansion of Western science, George Basalla suggested that, in order for countries to reach a third phase of "independent science," they must pass through a phase of "colonial

* Centro de Estudios Históricos, Consejo Superior de Investigaciones Científicas (CSIC), Departamento de Historia de la Ciencia, Duque de Medinaceli 6, 28014 Madrid, Spain.

[1] Nathan Reingold and Marc Rothenberg, eds., *Scientific Colonialism: A Cross-Cultural Comparison* (Washington, D.C.: Smithsonian Institution Press, 1987); Patrick Petitjean, Catherine Jami, and Anne Marie Moulin, eds., *Science and Empires: Historical Studies about Scientific Development and European Expansion* (Dordrecht/Boston/London: Kluwer Academic Publishers, 1992); Antonio Lafuente, Alberto Elena, and María Luisa Ortega, eds., *Mundialización de la ciencia y cultura nacional* (Aranjuez: Doce Calles, 1993).

[2] Roy MacLeod, "On Visiting the 'Moving Metropolis': Reflections on the Architecture of Imperial Science," in Reingold and Rothenberg, *Scientific Colonialism* (cit. n. 1), pp. 217–49, David W. Chambers, "Period and Process in Colonial and National Science," in *ibid.*, pp. 297–321; Antonio Lafuente and María Luisa Ortega, "Modelos de mundialización de la ciencia," *Arbor*, 1992, *142*:93–117.

science."[3] In this phase, small groups of scientists carry out activities that follow programs laid down by the metropolitan centers.[4]

This model has been widely discussed and, in general terms, superseded. The reasons for this lie in its tacit identification of metropolitan science with science itself, thus implicitly accepting that science is universalized knowledge. Today, a clearer understanding of colonial science calls for a recognition of the unequal distribution of knowledge, both in space and time, as well as its concentration in a limited number of institutions and localization in a small number of countries.[5] Basalla also ignored the fact that exchange or communication of ideas may not simply be one-way but may take place between colonial centers without the intervention of the metropolis.[6] Besides, recent scholarship shows the importance of emphasizing local contexts, independent of their place within the structure of an empire.

In this essay, we take the colony as the focus of colonial history. From this point of view, the globalization of science not only involves the dissemination of scientific methods but also the transmission of ideas and scientific values. The process succeeds when the recipient has some means to reproduce these elements and become an independent center of scientific activity. This new center joins a network of scientific centers and their satellites.

In any process of globalization of science the receiver, far from being merely passive, selects fragments of the transmitter's broadcast and adapts them to its own circumstances. From the point of view of the transmitter, the reception is an incomplete and/or mediocre copy of what was broadcast. But seen from the point of view of the receiver, the phenomenon is much more complex: a preexisting cultural base has been enriched (and deformed) by something different and external. This means that a tradition must be "invented" in such a way that it can interface with a new element. Only through this interactive model of mutual renewal can novelty be accepted and—most of all—used to advantage. Thus the analysis of local acclimatiza-

[3] On independent science, see MacLeod, "On Visiting" (cit. n. 2), p. 224; Chambers, "Period and Process" (cit. n. 2), p. 310. George Basalla, "The Spread of Western Science," *Science*, 1967, *156*:611–22; George Basalla, "The Spread of Western Science Revisited," in Lafuente, Elena, and Ortega, *Mundialización de la ciencia* (cit. n. 1), pp. 599–603.

[4] Latour distinguished metropolitan centers from peripheral or colonial centers in that their institutions are more efficient and better equipped to concentrate, organize, distribute, and, in short, take full advantage of information received, thus to complete what he calls the "cycle of accumulation." See Bruno Latour, *Science in Action* (Cambridge, Mass.: Harvard Univ. Press, 1987), p. 219.

[5] Xavier Polanco, "Science in the Developing Countries: An Epistemological Approach on the Theory of Science in Context," *Quipu*, 1985, 2:303–18; Christopher Vanderpool, "Center and Periphery in Science: Conceptions of a Stratification of Nations and Its Consequences," in *Comparative Studies in Science and Society*, eds. Sal Restivo and Christopher Vanderpool (Columbus, Ohio: n.p., 1974), pp. 432–42.

[6] In recent years, other models for the globalization of science have been developed in order to solve the problem of the linearity of Basalla's model, and to give greater consideration to scientific and technical exchanges among the peripheral areas of the "science-world." Xavier Polanco has made progress in this direction. Polanco has stressed modern science's resemblance to a worldwide, multinational company, and its function in terms of regularizing borders. This science-world is organized hierarchically between a center and its peripheral and semiperipheral regions. Xavier Polanco, "Une science-monde: La Mondialisation de la science européenne et la création de traditions scientifiques locales," in *Naissance et développement de la science-monde*, ed. Xavier Polanco (Paris: Editions La Découverte/Conseil de l'Europe/UNESCO, 1990), pp. 10–52; Xavier Polanco, "World-Science: How is the History of World-Science to be Written?" in Petitjean, Jami, and Moulin, *Science and Empires* (cit. n. 1), pp. 225–42.

tion may be considered fundamental to the study of the mechanisms of international transmission of ideas and institutions.

We have selected two case studies, all taken from the eighteenth-century Hispanic world, a complex and rich colonial scenario in which modern science underwent grafts, mutations, resistance to innovation, and controversy. We make no secret of our wish to avoid current reductionist explanations. "Diffusionist" theories, the automatic identification of science with emancipation, or radical contrasts such as "Creoles versus metropolitans" or "ancient versus modern" obscure rather than shed light on the topic. They oversimplify processes that seem to us more dynamic, more obvious and, above all, more pluralistic. On the contrary, we feel it is necessary to highlight a series of phenomena that afford a less mechanical but more organic view, a less geometric but more historical view of the movements and metamorphoses experienced and described by science in the course of its chronological and spatial development.

In this connection it is not unreasonable to criticize the extrapolation of events that took place in the British Empire and their generalization to other colonial situations. Descriptions that may initially be considered appropriate for "settler societies" (British colonies in America, Australia and, for some, in Argentina and Chile) may be overly simplistic for societies with a high level of social complexity, like those in India, New Spain, Peru or even, in the opinion of Richard Jarrell, in Ireland and Canada.[7] Generally speaking, Basalla's model may be said to share the flaws of "modernization theory" that, in the mid-1960s, dominated North American approaches to the problems of developing countries.

From studying the movements and changes of science in Hispanic culture during the Enlightenment, one can draw a number of conclusions. First, modern science did not land on barren terrain but on separate centers—each with deeply rooted local scientific traditions—where it settled and regenerated.[8] Second, at least in the case of Spain's colonies in Latin America, modern science did not expand throughout the viceroyalties in spite of the metropolis or in defiance of it. Overuse of this argument has given rise to Basalla's assumption that modern science and political independence are the same thing. Third, neither Creoles (people born in the colonies to Spanish parents) nor metropolitans were as homogeneous as they are usually depicted.[9] Fourth, the criteria imposed by Western scientific culture to determine

[7] Louis Hartz, *The Founding of New Societies* (New York: Harcourt, Brace and World, 1964); Richard A. Jarrell, "Differential National Development and Science in the Nineteenth Century: The Problems of Quebec and Ireland," in Reingold and Rothenberg, *Scientific Colonialism* (cit. n. 1), pp. 323–50.

[8] A revised historiography can be found in Juan J. Saldaña, "Teatro científico americano: Geografía y cultura en la historiografía latinoamericana de la ciencia," in *Historia social de las ciencias en América Latina*, ed. Juan J. Saldaña (Mexico City: Porrúa Ediciones, 1996), pp. 7–41.

[9] At least towards the end of the century, many learned people in the Americas and in the Iberian Peninsula vacillated between their loyalty to a liberating scientific rationalism and loyalty to the king. They were forced to adopt an eclectic approach, since they had to choose between salvation or liberty—a difficult choice that rendered the Creoles a less-unified group than is usually admitted. There is a copious literature opposing traditional, uncritical historical interpretations based on the tension between Creoles and mainland Spaniards. Horst Pietschmann opposed this reductionist view in "Protoliberalismo, reformas borbónicas y revolución: La Nueva España en el último tercio del siglo XVIII," in *Interpretaciones del siglo XVIII mexicano: El Impacto de las reformas borbónicas*, ed. Josefina Zoraida Vásquez (Mexico City: Ed. Nueva Imagen, 1992), pp. 27–65. See also Horst Pietschmann, "Los Principios rectores de la organización estatal en las Indias," in *De los imperios a*

Figure 1. *Science, Monarchy, and Expeditions. From Augustín de Zuloaga*, Tratado instructivo y práctico de maniobras militares *(Cádiz, 1766).*

whether something was modern were the criteria of a victorious culture and, as such, tended to underestimate and ignore alternative ways of understanding and experiencing modernity.

NATURAL SCIENCES AND BOURBON IMPERIAL AMBITIONS

The introduction of new disciplines such as astronomy, mechanics, chemistry, or botany into the viceroyalties was inextricably linked to the political program pursued by Bourbon reformism during the second half of the eighteenth century. After the Seven Years' War (1756–1763), when England defeated both the French and the Spanish, Carlos III launched an offensive to regain control of colonial territories, because most Spanish reformist politicians thought of the Americas as being not only the monarchy's main problem but also its remedy.[10] The ultimate aim was

---

las naciones: Iberoamérica, eds. Antonio Annino, Luis Castro Leiva, and François-Xavier Guerra (Saragossa: IberCaja, 1994), pp. 75–103.

[10] There were few major politicians or intellectuals of the Spanish Enlightenment who, in order to deal with the problems of the metropolis, did not rack their brains to find imperialistic solutions. The two were inextricably linked, for the empire provided not only the opportunity to promote a national policy, but also the major threat to Spain's international stability, due to the rise of nationalist move-

to transform the old Universal Monarchy into a colonial empire. It was an attempt to update an obsolete structure, typical of the seventeenth century and successful then, but unable to resist the pressure that transatlantic commerce was putting on modern international relations. The longed-for regeneration to stop the crown's decline (the topic par excellence of seventeenth-century Spain) involved closer ties with the American territories and more extensive exploitation of their fiscal, political, and natural resources. Within this context of political and social reform, science was employed as a suitable instrument to fulfill imperial ambitions. Thus, the program of scientific expeditions to America promoted by the crown from the 1760s onwards was the clearest sign of the dynamic of Enlightenment Spanish science.[11] To create natural history collections and botanical gardens, to set up astronomical observatories and seminars on mining, to chart coasts and classify plants were all initiatives that, from the metropolitan perspective, constituted different parts of the same new policy.

However, effective implementation of this ambitious program entailed many tasks that significantly exceeded the program's original parameters. Those who thought that the American territories could be regained, and that their profitability could be increased a hundredfold, miscalculated the consequences. To begin with, the immediate outcome of this large-scale politicization of science was the politicization of scientists themselves. In a few decades, these new protagonists went from being mere savants, versed in the works of Newton or Linnaeus, to being agents of an international corporation serving a despotic monarchy. If at first scientists were content to play the part of warriors against the baroque, with their fight presented as a struggle between ancient and modern, they were soon to discover that such a bond with the new Bourbon dynasty limited their potential for institutional action. After becoming institutionalized as agents of the crown, they went on to ask for greater influence in government decisions. They soon began to consider themselves indispensable, and even dared to demand continuous funding for their activities. Since these approaches were not always accepted, they ceased to be unconditional allies and began to question the rationality of the crown's decisions, as well as the interest of its ministers in advancing the Enlightenment. The power struggle among the me-

---

ments in the colonies and the competition for trade control between European metropolises. See Pedro Pérez Herrero, "El México borbónico: ¿Un éxito fracasado?", in Vázquez, *Interpretaciones del siglo XVIII* (cit. n. 9), pp. 109–151. In the same book, see also Brian R. Hamnett, "Absolutismo ilustrado y crisis multidimensional en el período colonial tardío, 1760–1808," pp. 67–108. This inevitable linkage had its reasons: the metropolis was profoundly dependent on its colonies, and Britain, France, and Russia were disputing the Spanish hegemony in America and on its sea routes with renewed aggression. These considerations tend to shift the focus of research on the modern age of Spain away from explorations of Spain's influence on its empire, and towards study of the empire as the driving force of peninsular politics. This concept may well have influenced the work and intellectual career of J. H. Elliott: "I therefore felt that it might be more valuable to take a rather different approach, and look at the history of Spanish imperialism from the standpoint of its impact on the colonizing power, rather than on the colonized." See John H. Elliott, *Spain and its World 1500–1700* (New Haven: Yale Univ. Press, 1989), p. 3.

[11] Literature on the expeditions is so abundant that we will limit ourselves to citing a few books in which numerous references are to be found. For an overview, see Fermín del Pino, ed., *Ciencia y contexto histórico nacional en las expediciones ilustradas a América* (Madrid: CSIC, 1988); and the minutes of the two conferences held by the Ateneo of Madrid on "España y las Expediciones Científicas en América." See also Alejandro R. Diez Torre *et al., La ciencia española en Ultramar* (Aranjuez: Doce Calles, 1991), and Alejandro R. Diez Torre *et al., De la ciencia ilustrada a la ciencia romática* (Aranjuez: Doce Calles, 1995).

tropolis, the viceroy, and the scientists grew, with each lobby laying claim to the patriotic and utilitarian ideals of Enlightenment science, and with each waving the banner in its own way.[12]

Alongside these protagonists emerged new institutions, new languages, and new ways in which the scientists interfaced with the ruling elites. Over and above the international recognition science achieved in the peripheral territories was its ability to make statements about the colony's physical or social environment. What interests us is how the Creole scientific elite shaped and appropriated the political imagery, an ability that was evident everywhere, though with differences in emphasis and often with contradictory outcomes.[13] The introduction of Linnaeus's classification, Lavoisier's chemistry, or Newton's physics, nurtured with equal zeal in all the kingdoms of the monarchy, gave rise to a different set of institutions in each place, whose protagonists and set objectives differed.

Political language soon reflected publicly what had been widely rumored, and the idea of separate kingdoms under the same sovereign gave way to that of colonies with one metropolis. And naturally these changes had their correlation in colonial scientific institutions. It could not be otherwise, since the institutions had always been viewed as instruments of reform at the service of the state. Events only served to reinforce their statist nature; indeed, the changes that were introduced could be perceived as a process of *metropolitanization* of the scientific activities of the Spanish Enlightenment.[14]

In the following discussion, scientific practices in New Spain (Mexico) and New Granada (Colombia) are of particular interest. We shall confine our comments to the debates concerning the projects undertaken by José Mariano Mociño (1757–1819) and Luis José Montaña (1755–1820) in Mexico City, and by Francisco José Caldas or Francisco Antonio Zea (1766–1822) in Bogotá.[15]

NEW SPAIN: MOTHERLAND AND CULTURAL HERITAGE

Within the Spanish Empire, New Spain was economically the most profitable of the colonial dominions, and culturally the most dynamic. During the eighteenth century its prosperity, linked mainly to silver mining, increased considerably. Mexico City

[12] It was inevitable that this should happen, and on this point enlightened Spaniards and Americans were little different from their European contemporaries. Should we wish to draw further conclusions, it would be necessary to consider to what extent the design of policies for science affected the manners and measures of imperial policy. Our belief is that this process can be documented, and it will confirm that science policy served as both example and experiment for imperial policies. Antonio Lafuente and Leoncio López-Ocón, "Tradiciones científicas y expediciones ilustradas en la América hispana del siglo XVIII," in Saldaña, *Historia social de las ciencias* (cit. n. 8), pp. 247–81.
[13] Juan Pimentel has devoted two books to exploring the transposition of images from science to colonial policies. Using the case of the Malaspina expedition, he has established links between the two vernaculars and shown the impact of this relationship on the metropolis. Juan Pimentel and Manuel Lucena, *Los axiomas políticos de Alejandro Malaspina* (Aranjuez: Doce Calles, 1993). We particularly recommend Juan Pimentel, *La física de la Monarquía: Ciencia y política en el pensamiento colonial de Alejandro Malaspina (1754–1810)* (Aranjuez: Doce Calles, 1998).
[14] Antonio Lafuente, "Institucionalización metropolitana de la ciencia española en el siglo XVIII," in *Ciencia colonial en América*, eds. Antonio Lafuente and José Sala Catalá (Madrid: Alianza Editions, 1992), pp. 91–120.
[15] The biographies of these and other American scientists quoted in the text may be found in José M. López Piñero, Thomas F. Glick, Victor Navarro Brotóns, and Eugenio Portela, eds., *Diccionario histórico de la ciencia moderna en España*, 2 vols. (Barcelona: Ediciones Península, 1983).

was the second-largest urban settlement in America, and had the good fortune to be governed by viceroys who were strongly committed to the ideals of the Enlightenment. Patriotic and nationalist feeling took root among the Creole intelligentsia there and, although they complained unendingly of their isolation from Europe or of being ignored by the metropolis, they began to realize that they could achieve a culture that, while still modern, could also have roots in native traditions. Their efforts to claim a more important cultural role were replete with innovative proposals. For its part, Madrid imposed modernization on a scale that encompassed all sectors of the economy, in an attempt to strengthen political control and profitability. Mexico City was the jewel in the crown, a city in the midst of frenzied cultural activity that played host to a variety of scientific institutions, the most outstanding of which were the Real Colegio de Cirugía (Royal College of Surgery, established in 1768), the Real Jardín Botánico (Royal Botanical Garden, 1788) and the Real Seminario de Minería (Royal Mining Seminary, 1792). The metropolis was decisively involved with all of them, and they all had controversial beginnings, for local scientists wanted to play a greater role and were unwilling to be merely "extras." While fighting for greater institutional recognition, the scientists insisted on occupying the managerial posts, arguing not only that they knew as much science as the members of the scientific expeditions coming from Madrid but also that they were more familiar with the geographical, floral, and mineral peculiarities of the territory. These claims were backed by patriotic speeches, and they created an atmosphere favorable to intellectual initiatives that were sensitive to native cultural traditions. There are many examples, but perhaps none are more representative than those that assert the importance of native herbal lore.

The studies of Mexican plants conducted by Mariano Mociño and Luis Montaña in the Royal Native Hospital and the San Andrés Hospital, starting in 1801, are among the most innovative episodes of the new Creole science. They reveal a dual process: *negotiation* between Creole and metropolitan scientists regarding the conditions under which Linnaeus's botany was received; and the local *appropriation* of such ideas through the emergence of scientific procedures that stemmed from a European cultural tradition.[16] The wards of these hospitals became sites where the medical effectiveness of plants—prescribed by native tradition for centuries—was tested.[17] This required the exercise of some caution, given the vocal animosity towards these native practices, which were branded as primitive or superstitious. While respecting scientific rigor, however, a *scientific pragmatism* emerged that, apart from being endorsed by daily experience, was given legitimacy by contemporary studies by William Cullen and John Brown in Edinburgh.[18] In José Antonio

[16] José J. Izquierdo, *Montaña y los orígenes del movimiento social y científico de México* (Mexico: Ediciones Ciencia, 1955), pp. 200–5. See Donald B. Cooper, *Epidemic Disease in Mexico City 1761–1813: An Administrative, Social, and Medical Study* (Austin: Univ. of Texas Press, 1965).

[17] The Royal Native Hospital was founded in 1531 by the Franciscan Fray Pedro de Gante, leader of the first group of twelve Franciscans. It formed part of the cultural and evangelical complex that the first generation of Franciscans set up among the Indians as a forerunner to the creation of a utopian Mexican nation. In any case, it should be pointed out that in 1545 the Council of Trent established that all hospitals must answer to the Church, even if they were administered by laymen. See Josefina Muriel, *Hospitales de la Nueva España* (Mexico City: Editions Jus, 1959), pp. 80–5; Guenter B. Risse, "Medicine in New Spain," in *Medicine in the New World: New Spain, New France, and New England*, ed. Ronald L. Numbers (Knoxville: Univ. of Tennessee Press, 1987), pp. 12–63.

[18] In Edinburgh, great importance was given to the plant pharmacopoeia as well as to the clinical monitoring of the virtues of plants, following a process in which, according to Alzate, physicians

Alzate's (1737–1799) formulations of 1786, the method employed to test the validity of remedies was "to describe their natural history and pharmaceutical preparation," that is, to link their climatic conditions for growth and geographical origin with the mixture and dosage of plants employed in each case. The work program differed, according to Montaña, from "the circle of classification and nomenclature typical of cabinet physicists, overloaded with quotations from different authors and always contingent on a definite 'maybe.'" It was necessary to "ask again for the advice of the herbalist and the *ranchero* and rely on their information."[19]

During the nearly three years that the study lasted, three overlapping traditions converged. One of them, the medical tradition described in the writings of Cruz and Badiano, Sahagún, Hernández, Jiménez, and López emerged with the first contacts between Mexicans and settlers. The second, already mentioned, entailed the re-creation, in the periphery, of techniques stemming from Europe via Edinburgh; and the last one intended to explore the capacity of Linnaeus's binary model to make predictions.[20] It was, then, an effort to vindicate a medical heritage "dictated by pure tradition" (as Montaña fittingly puts it), on the one hand, and, simultaneously, to do the only thing that Alzate considered important: to see alleged healing properties with one's own eyes.[21]

Public discussion of the study began on the very same day that the Royal Botanical Expedition, led by the physician Martín Sessé, was announced. The expedition had been organized in Madrid without taking into account the colony's scientific merits and needs. Not only were the Creole scientists achievements ignored but their knowledge was scorned. The Creoles, meanwhile, regarded the metropolitan imposition of Linnaean botany as an arrogant gesture. We know that the pharmacist, and later professor of botany, Vicente Cervantes had in mind the new classificatory sys-

---

would "carry out trials and test new remedies . . . Before starting treatment, classify the illness and specify the various names under which authors had described it . . . List different opinions on its cause and healing, and devise a treatment plan suitable for the age, temperament, sex etc., determining the reasons on which it is based and the purpose it should serve . . . In order to verify the effectiveness of any remedy, take account of its natural history and pharmaceutical preparation, and show the grounds for trying it, and whether or not the persons present expect good or bad results from the tests . . . Take notes on all this, adding day by day changes experienced by the patient, and if the treatment is changed in any way record the reasons for doing so . . . General comparison to decide for or against the use of such a remedy in such an illness." Cf. José Antonio Alzate (1790), "Carta de Edimburgo, 10 de Mayo de 1786," *Gacetas de literatura de México*, 4 vols. (Puebla, 1831), vol. 1.

[19] Speech by Montaña in 1802. Cf. Luis José Montaña, *Anales de ciencias naturales de Madrid*, 1802, 6:214–21.

[20] José J. Izquierdo, *Montaña y los orígenes* (cit. n. 16), p. 201.

[21] From a speech made by Mociño in 1801, we know that Mexican native plants tested in the hospitals were divided into fifteen groups according to their principal healing quality. For example, among the astringents we find Texcalamatl (*Ficus nymphaelifolia*) and the celebrated Ezpatli, or blood medicine (*Lignum nephriticum*) (Kircher, Grimaldi, and Newton experimented with color diffraction of this plant.) Of the narcotics, Picietl or tobacco and the dangerous Toloatzin (*Datura stramonium*). And of the Diaphoretics, Guayacán or *Lignum vitae*. Cf. Mariano Mociño (1801), *Ensayo para la materia médica mexicana* (Puebla, 1832), pp. 98–101. The herbalist tradition had been revived by the exiled Jesuit Javier Clavijero in his *Historia antigua de México*, published in 1780–1781 (Mexico City: Porrúa, 1945). Embroiled in the debates about the New World's "inferior" nature since his exile in Italy, in this important work Clavijero unleashed not only the great chroniclers and jurists of Renaissance and baroque New Spain, such as Las Casas, Acosta, Sahagún, Solórzano, García, Sigüenza, etc., but the encyclopaedic knowledge of natural and moral history of Francisco Hernández, Philip II's physician-in-chief for Spanish America. Cf. Jaime Vilchis, "Recepción y mundialización de la historia natural de Francisco Hernández: S.XVI–XVIII," in *The World of Francisco Hernández*, eds. Simon Varey and Rafael Chabrán (Stanford Univ. Press, in press); A. Gerbi, *La disputa del Nuevo Mundo* (México: FCE, 1982).

Figure 2. *José Antonio Alzate y Ramirez (Ozumba, 1737–México, 1799).*

tem of Carl Linnaeus. He had been commissioned to impose it upon the viceroyalty by Gómez Ortega (director of Madrid's Botanical Gardens) and, at the latter's behest, he fulfilled his task swiftly.[22] What, however, was initially a legitimate difference in scientific opinion would soon become a deep political and institutional clash of interests after Cervantes's involvement in creating the botany chair in Mexico's newly founded Real Jardin Botánico.[23]

Local journalism fueled the deployment of arguments based upon empirical and utilitarian convictions. The journalists were already mature enough to divide readers into apparently irreconcilable camps. Polemicists, boasting of their persuasive skills, created a newsprint "drama" that, although cloaked in scientific language, was actually personal. On one side, members of the expedition played the role of supporters of Linnaeus; on the opposite, a sect of rough and pedantic quack doctors appeared: it was Cervantes vs. Alzate, Linnaeus vs. Hernández, New Spain vs. Mexico, science vs. experience, botany vs. natural history. In sum, there were Manicheans in search of rhetorical legitimacy.[24] Anyone wishing to savor a perfect example of incisive, scathing, and insightful prose (baroque in style but enlightened in content) should read the works reflecting this literary and scientific antagonism. It would, however, be inappropriate to make too much of it, as the antagonists coincided in their mockery of and diatribes against the futile egotism of the academics. Moreover, they also agreed in their patriotic and patrician zeal. Something, however, separated one from the other: Creoles studied the new botanical knowledge in the crucible of the Mexican medical corpus. So Alzate exclaimed ironically, "What is the use of including this or that plant in such-and-such a genus, in such-and-such a species, if it has properties very different from those that, because of their similarities, must be included in a given category? In Europe, they conduct ill-fated experiments as a result of which parsley and hemlock are viewed as similar as far as their organisation is concerned. In New Spain, on the other hand, we eat plants and fruits reputedly poisonous, if Linnaeus' botanical laws were true. . . ."[25]

[22] The great reforming and reorganizing impulse encouraged by King Carlos III (1716–1788), aimed at the recovery of colonial power and the appropriation through science of Spanish and American territory, converted Madrid's Royal Botanical Garden into a center for the organization of American botanical expeditions in the latter decades of the eighteenth century. Politician-scientist-in-waiting Casimiro Gómez Ortega served as the garden's director. See Xavier Lozoya, *Plantas y luces en México* (Barcelona: El Serbal, 1984); Javier Puerto, *Ciencia de cámara: Casimiro Gómez Ortega (1741–1818): El Científico cortesano* (Madrid: CSIC, 1992); and Miguel A. Puig-Samper, "Difusión e institucionalización del sistema linneano en España y América," in Lafuente, Elena, and Ortega, *Mundialización de la ciencia* (cit. n. 1), pp. 349–59.

[23] This was not the only battlefield on which the struggle between scientists from the metropolis and those in the colony took place. Cf. Renán Silva, *Saber, cultura y sociedad en el Nuevo Reino de Granada: Siglos XVII–XVIII* (Bogotá: Universidad Pedagógica Nacional, 1984); José Torre Revello, *El libro, la imprenta y el periodismo en América durante la dominación española* (Mexico City: Universidad Nacional Autónoma de México, 1991); Juan J. Saldaña, "Ciencia y felicidad pública en la ilustración americana", in Saldaña, *Historia social de las ciencias* (cit. n. 8), pp. 151–207.

[24] Rogers McVaugh, "Botanical Results of the Sessé & Mociño Expedition (1787–1803)" (pt. 3, "The Impact of This and Other Expeditions on Contemporary Botany in Europe"), *Contributions of the University of Michigan Herbarium*, 1987, *16*:155–71; José L. Peset, *Ciencia y libertad: El papel del científico ante la independencia americana* (Madrid: CSIC, 1987); Roberto Moreno, *La primera cátedra de botánica en México* (Mexico City, 1988); Graciela Zamudio, "El Jardín Botánico de la Nueva España y la institucionalización de la botánica en México", in *Los orígenes de la ciencia nacional*, ed. Juan José Saldaña (Mexico City: Cuadernos Quipu 4, 1992), pp. 55–98.

[25] Alzate (1788), *Gacetas de literatura de México* (cit. n. 18), vol. 1, pp. 20–2. On Alzate, see Roberto Moreno, *Un Eclesiástico criollo frente al estado Borbón* (Mexico City: Academia Mexicana

Right from the beginning, the dispute (sprinkled with references to the new chemical nomenclature) was tainted with an unpleasant antimetropolitan aftertaste or, as revolutionary leader José María Morelos pointed out, with nationalist feeling. The dissemination of new values shaping what Clifford Geertz has called an "integrating ideology" was the most important event. It helped to create a self-governing political entity endowed with authority and legitimacy; it combined the political virtues of the Aztec "civilization" and the juridical possibilities spawned by the "sovereignty of sovereignties" of the baroque period.[26] It was a return to primeval antiquity, an attempt to recreate an individual identity combining Catholic orthodoxy with the omens of astrology, healthy herbalism, and the Anáhuac region's fertile agriculture. Summing up, a new rhetoric was developed which, in contrast to that employed by Fausto Elhuyar, Vicente Cervantes, or other missionaries from the metropolis, did not limit itself to serving as propaganda on the usefulness of modernization: it was also aimed at blending the new motherland territory with the protagonists' cultural heritage so they could be taken as a single reality.

We have alluded to the priest Alzate, not because he was the wisest or most conspicuous of Creole polemicists, but rather because he assumed the role of spokesman. As a whole, the ideas published by Juan Díaz Gamarra, Antonio León y Gama, Juan Santalices, José Bartolache, Nicolás Guadalajara, Francisco Gamboa and others worked a sort of Orphic charm, a persistent and melodious murmur capable of filling the gap between sages and laymen and of mixing old knowledge with new disciplines—ethnography, archeology, medical geography, cartography, and anthropology—in pursuit of a national memory that was soon to become nationalist.[27]

The concessions made by both metropolitan and Creole scientists were astonishing. For instance, in the Real Jardín Bótanico school's graduation examinations of 1793, Cervantes maintained that a student, apart from distinguishing plants by judging from their gender, must also "describe their virtues and . . . uphold the opinion, in opposition to the claims of Linnaeus and others, that accurate data on their properties can be obtained through well-supervised chemistry." Alzate, in his attempt to reach a new consensus, acted in the same way with regard to Lavoisier's nomenclature. In 1791, he admitted without hesitation (even using the same arguments previously employed by Cervantes himself), "I would not dare describe [Lavoisier's] system as entirely false."[28] Did he perhaps notice the similarity between the nomenclature of chemistry (the name assigned to a substance describes its process of synthesis) and that of the Nahuatl native medical tradition (the name of the medicine describes its healing properties)? Perhaps, indeed, in both cases we are dealing with

---

de la Historia, 1980). It is interesting to note that Alzate was not only the focus of a stream of public opinion in New Spain but also the center of a network of international correspondence.

[26] Clifford Geertz, *The Interpretation of Cultures* (New York: Basic Books, 1973), pp. 190–218.

[27] James R. Jacob, "*Por encanto órfico*: La ciencia y las dos culturas en la Inglaterra del siglo XVII," in *La ciencia y su público*, eds. Javier Ordoñez and Alberto Elena (Madrid, CSIC, 1990), pp. 43–69. See J. A. Ruedas de la Serna, *Los Orígenes de la visión paradisíaca de la naturaleza mexicana* (Mexico City: Universidad Nacional Autónoma de México, 1987), and José Luis Peset, "La Naturaleza como símbolo en la obra de José Antonio de Alzate," *Asclepio*, 1987, *39*, 2:285–95.

[28] Referring to Lavoisier's statement that vitriolic ether cannot be manufactured in places where the barometer reads between 20 and 24 inches, Alzate points out his error by citing that his adversary Vicente Cervantes manufactured and sold the ether at his pharmacy in Mexico City, where the barometer shows 21 inches. See Alberto Saladino, "La Química divulgada por la prensa ilustrada del Nuevo Mundo," in *La Quimica en Europa y América (siglos XVIII y XIX)*, ed. Patricia Aceves (Mexico City, Universidad Autónoma de México, 1994), pp. 177–99.

a nomenclature based on what is considered intrinsic to the nature of a given substance.[29]

The mixture of old local wisdom with new European knowledge (not in spite of, but actually thanks to the imperial dynamics of Bourbon despotism) favored the emergence of a social movement that advocated more autonomy and the spread of a type of creativity that envisioned itself as sovereign.[30] Creole initiatives towards assimilation (without giving up a sense of belonging to tradition), and towards re-creation of scientific knowledge and procedures, refuse to conform to the pattern of simple automatism characteristic of diffusionism, nor to that of inevitable antagonism between Europeans and Americans.

Thus, in Mexico, Creole scientists tried to merge the native scientific heritage and European practices and knowledge into a single discursive reality. In this way, the motherland would be politically categorized as a defender of tradition as well as of modernity. As we shall see in the next section, Creole scientists in New Grenada were less emphatic about their glorious native past. All their efforts went to show that their country was a promised land that would offer up its treasures only when it had been measured, inventoried, and mapped.

NEW GRENADA: MUTIS AND HIS CRITICS

Colombian historiography has traditionally interpreted the Enlightenment in New Granada as a continuous thread, beginning with the arrival of José Celestino Mutis (1732–1808) in Santa Fé in 1760, ending with the political clampdown by General Pablo Morillo, and reaching its climax at the turn of the century.

Historians have pointed to Mutis's influence on the cultural life of the colony. Arriving in Bogotá in 1760 as physician to the new viceroy, he soon found favor as a practical administrator and a symbol of Bourbon reformism. His ambition increased as he received support from the authorities, and reached its height with the approval of his proposal to organize an expedition to survey the mineral and vegetable wealth of the viceroyalty of New Granada. Such was Mutis's influence that from 1783, there was no scientific or cultural project that did not pass through his hands, and he was called the "Oracle of New Granada."

Mutis's death in 1808, however, heralded a reappraisal of the aims of the Royal Botanical Expedition as commanded by Mutis, as well as of its alleged achievements.[31] By then, conditions had changed radically: now, criticism of Mutis came

[29] *Exercicios Publicos de Botánica . . . presidiéndolos Don Vicente Cervantes catedrático de Botánical . . .* (Mexico City: Herederos de Don Felipe de Zúñiga y Ontiveros, 1793), pp. 8–9.

[30] The contemporary literary essay is where the Creole concept has been most clearly and brilliantly defined. After the influence—intangible but powerful—of Jorge Luis Borges in *El tamaño de mi esperanza* (1926) and of José Vasconcelos, who wrote *Ulises criollo (1935)*, the post-1960s brought a crescendo of other bold meditations on the Creole issue. We note the most important: Lezama Lima in *La Expresion americana* (cf. *idem, Confluencias* [La Habana, 1988], pp. 263–77) elevates "Creole" to a hermeneutic category of American modernity. Octavio Paz in his essay *Sor Juana Inés de la Cruz o las trampas de la fe* (Barcelona: Seix Barral, 1982, pp. 55–67) sees the awakening of the Creole spirit in the syncretism that brings indigenous culture into the mainstream. And more recently, Victor Farías in *La Metafísica del arrabal* (Madrid: Muchnik, 1992, pp. 125–43) sees in the work of the young Borges the urgent and yearning expression of a "Creole *logos*" that comes to form part of the human heritage in its own right.

[31] An indispensable reference on Spanish botanical expeditions to America is Javier Puerto, *La Ilusión quebrada: Botánica, sanidad y política científica en la España ilustrada* (Barcelona: El Serbal/CSIC, 1988). See also Antonio González Bueno and Raúl Rodriguez Nozal, *Plantas americanas*

Figure 3. *José Celestino Mutis y Bosio (Cádiz, 1732–Bogotá, 1808).*

from moderns, and no longer from reactionaries. One of his chief critics was self-taught scholar Francisco José Caldas (1768–1816).[32] In September 1808, after examining the writings left by Mutis in the *House of Botany*, Caldas wrote, "now that I have managed to detect the lacunae and gaps it contains ... I wish to save at least my botanical writings on the Southern part of the viceroyalty from the ruin that

---

para la España ilustrada: Génesis, desarrollo y ocaso del proyecto español de expediciones botáni-cas (Madrid: Editorial Complutense, 2000). Great efforts have been made to present a balanced picture of the successes and failures of Mutis and his milieu. Among them, we would recommend José Antonio Amaya, *Celestino Mutis y la Expedición Botánica* (Madrid: Ediciones Debate/Itaca, 1986). This book excuses Mutis's failures on the grounds of "the uncertainty of the environment in which he planned his mission as spreader and ideologist of science, rather than as a systematic scientist" (p. 66). Nevertheless, the man whom Alexander von Humboldt called the "Father of the Botanists" showed shortcomings in his traditionalism (p. 46) and paternalism (p. 42), his weakness for iconography and contempt for systematic botany—his "iconism" (pp. 42, 43, 53, 56), his authoritarianism (p. 42), ambition for wealth (p. 37), and his mediocre scientific training (p. 66).

[32] Caldas is a prime example of Creole science. Brought up in the colonies (Popayán, Colombia) and frustrated by his isolation, he asked Mutis to include him in the expedition he was leading through the New Kingdom of Granada. Caldas discovered that the boiling point of water changed with altitude, a discovery that Humboldt acknowledged in 1801 during his stay in the province of Quito. After Humboldt's refusal to include Caldas in his American journey, and in view of the difficulties he had with Mutis's circle, Caldas's politics became more radical. He openly championed a

Figure 4. *Francisco José de Caldas (Popayán, 1768–Bogotá, 1816).*

threatens the Flora of Bogotá. . . ."[33] This criticism of the expedition was compelling, since it not only bemoaned the lack of attention to the vital work of classification but also criticized the poor relationship between the viceregal territory and the regions whose plant life had been studied. For Caldas, territory—in the sense of a homeland, the land supporting a civilization—was the sole object of study, with all

culture and science rooted in native traditions and committed to the interests of the colony. His patriotic preaching led him to defend the originality of American culture and to support biogeography as a field that both encompassed the virtues of European science and addressed the urgent economic policy needs of New Granada. In his work *Del influjo del clima sobre los seres organizados* (1808), Caldas argued that regional variations in cultures originated from climatic differences, a hypothesis that gave scientific backing to his proposals for the federalist organization of the American states after their independence.

It is very difficult to find texts dealing with the reported student-teacher relationship between Caldas and Mutis, or the differences that led them to fight on opposite sides—Caldas as a federalist and Mutis's nephew Sinforoso Mutis, Jorge Lozano, and Antonio Nariño as centralists—during the so-called First Republic (1810–1816). However, there is a steady trickle of publications talking of tensions between factions of the Creole elite and even among royalist officials. Contrary to the usual laudatory propaganda, the commitment of the insurgents (some of whom came from Mutis's circle) to science is neither obvious nor automatic. Proof of this, for example, is the suspension of the 1812 expedition by the Constitution of Cundinamarca, or the gradual disappearance of instruments and maps from the Bogotá Observatory, which were dispersed to military commanders for war purposes. A balanced assessment of the involvement of scientists in the independence movement can be found in Thomas F. Glick, "Science and Independence in Latin America (with Special Reference to New Granada)," *Hispanic American Historical Review*, 1991, *71*:306–34.

[33] It is worth quoting the reasons for Caldas's pessimism: " . . . now that I have seen that there [are] not even two or three palm trees, that the section on cryptogamy is almost entirely blank, that the unnumbered pages, in no order, have not a single copy; that more than half of the sketches for the engravings are missing; that many anatomies are missing; that the manuscripts are in total disorder; that they are no more than inkblots; that the whole wealth of the Flora of Bogotá consists of 48 small notebooks: that the rest of the works that he has undertaken during his life have been mere summaries; that the treatise on cinchona has not been completed, save for the medical section; that the

due respect to far-ranging geographic or botanical explorations. Unlike Mutis, he felt that science should be subordinate to political interests.[34] In fact, Caldas, José Pombo, Juan Valenzuela, Jorge Lozano, and later Zea (all important members of the intellectual elite of the colony) would submit their own separate expeditionary projects as pragmatic alternatives to the academic strategy adopted by Mutis. Mutis's more or less theoretical or speculative approach was not the only subject under discussion, but also (and above all), as Restrepo has remarked, the public or private nature of scientific enterprise.[35] Numerous testimonies show the difficulties Mutis experienced when trying to determine just where his vision of the expedition as a personal endeavor should end (with regard, for example, to the Casa de la Bótanica he established in Mariquita, or to the appointment of expedition members), and where the scientific mission, financed with viceregal or metropolitan funds, should begin.

By the end of the eighteenth century, a new form of patriotism began to take shape, giving priority to the useful and public dimensions of knowledge. Priorities shifted in the scientific disciplines: from astronomy to geodesy, from botanical taxonomy to the geography of plants, and from exploration of territories to regional econometrics. In short, having identified New Grenada with a new technical standard, the Creole elite fought to substitute more integrated views for classificatory practices; to replace observation with measurement, and the university chair with

---

descriptions of these important plants are in deplorable scrap-books. . . . " Quoted in Santiago Diaz-Piedrahita, "Caldas y la historia natural," in *Francisco José de Caldas* (Bogotá: Molinos Velásquez Editores, 1994), pp. 111–23, quotation on p. 115. However, this condemnation should be qualified, since in fact Caldas changed his opinion after 1812, when he chose to collaborate with Sinforoso Mutis, who became the new expedition director after the death of his uncle, and to contribute to spreading the work done in the preceding two decades.

[34] In order to understand this difference, a special point should be made about the scathing criticisms that Caldas, Lozano, and Zea made of the way in which Mutis handled the affairs regarding his expedition, including the building of the Bogotá Observatory itself. In the latter case, Mutis's arrogance drove him to commission a building that, in imitation of the outdated models first erected in Greenwich and Paris, was already obsolete by the time of its construction. It seemed as if Mutis, who had served as an adviser on so many and such different matters, was too proud to ask for advice and thus erected a monument to the past, rather than a useful building for the present. He ordered the construction of a building whose internal layout of empty spaces and observation rooms so closely followed the European originals that it proved totally useless in Bogotá, since celestial bodies do not reach their highest point in the same position in tropical areas that they do in the northern hemisphere. Caldas's complaints show that the error cannot be put down to "the tyranny of distance." David Wade Chambers, "Does Distance Tyrannize Science?" in *International Science and National Scientific Identity*, eds. Roderick W. Home and Sally Gregory Kohlsted (Boston: Kluwer, 1991), pp. 31–3; *idem*, "Locality and Science: Myths of Centre and Periphery," in Lafuente, Elena, and Ortega, *Mundialización de la ciencia* (cit. n. 1), pp. 605–17. Arias de Greiff ridicules this case of "inadequate technology transfer," since the observatory was designed with side openings "in order to observe from within, through the windows, the progress of the bodies of the solar system from their rising in the East to their setting in the West, with their maximum height or apogee not overhead but towards the South, when they could be seen through the windows on this side. But this is not the case in equatorial regions, where these stars reach their apogee almost at the zenith. Moreover, there is not even a window facing South, since this is where the stairs are located." Jorge Arias de Greiff, "Caldas: Inquietudes, proyectos y tragedias," in *Francisco José de Caldas* (cit. n. 33), pp. 37–54, quotation on p. 40.

[35] Olga Restrepo Forero, "El tránsito de la historia natural a la biología en Colombia (1784–1936)," *Ciencia, Tecnología y Desarrollo* (Bogotá), 1969, *10:*181–275. This was later expanded and published in "Naturalistas, saber y sociedad en Colombia," in *Historia social de la ciencia en Colombia*, eds. Olga Restrepo, Luis Carlos Arboleda, and José A. Bejarano, 9 vols. (Bogotá: Colciencias, 1993), vol. 3, *Historia natural y ciencias agropecuarias*, pp. 13–327. On Mutis's secretiveness, see p. 101 *et seq.*

the laboratory.[36] Such drastic changes were implemented in a way that would have been unthinkable a decade earlier. To illustrate this point, it will suffice to quote part of Caldas's article in the *Semanario del Nuevo Reino*, of which he was editor between 1808–1811: "To observe the skies for the sake of observation would be a legitimate activity, but it would be nothing but a fruitless activity . . . This observer would be useless and the Motherland would regard him as a consumer from whom nothing is expected. *We do not wish to play this role in society*: We want our astronomical studies to improve our geography, our roads and our commerce."[37] Doubtless, this and similar texts from which we may quote hinted at the need to strengthen the role of local government in deciding political and scientific priorities. Drawing a distinction between the two types of knowledge—one, that of the "republic of letters," more susceptible to logic, and the other responding to the needs of the "civil republic"—entails a new configuration of what Bourdieu calls the "scientific field." The Creole elite aspired to occupy a central place in the public arena, arguing that science was subordinate to politics, as history was subordinate to geography: novelties that contributed noticeably to the substitution of the Kantian "rhetoric of emancipation" for the "rhetoric of patriotism" of the American and French Revolutions.[38]

Such ideas contained a fundamental criticism of the Mutisian model of science. Other criticisms had appeared earlier in the *Proposal for the Re-organisation of the Botanical Expedition* (1802), written by Zea during his exile in Europe.[39] The new

[36] This entailed a program of action that paralleled the language they chose to use. Along with conventional references to public welfare, the program called upon the elite to put its knowledge of the geography and environment of New Granada into quantitative terms. Caldas complained that the inhabitants of New Granada knew more about the situation in China than about their own, since "we do not know the size of the country where we were born." Quoted in Hans-Joachim König, *En el camino hacia la nación: Nacionalismo en el proceso de formación del estado y de la nación de la Nueva Granada, 1750–1856* (Bogotá: Banco de la República, 1994), p. 91. Caldas identified the new patriotism with the desire to "measure how far we are from prosperity." *Ibid.*, p. 88.

[37] Quoted in Luis C. Arboleda, "Ciencia y nacionalismo en Nueva Granada en los albores de la revolución de la Independencia", in *Francisco José Caldas* (cit. n. 33), pp. 139–46, quotation on p. 142. The text of Caldas, to which we have added italics, was published in the *Semanario del Nuevo Reino de Granada* in 1811, the first year of independence. There are many texts advancing this new ethos; in 1809, also in the *Semanario*, Caldas attacked the hermetic academicism that was surreptitiously attributed to Mutis when he stated, "How can a nation which has no roads, whose agriculture, industry and trade are in their death throes, spend its time in brilliant ventures, most of which are imaginary? The cultivation of a plant, a good road . . . are more important matters than all those much talked-about projects where it can show off its genius, its erudition and its eloquence" (*ibid.*, p. 145). An analysis of the thoughts of the elite on the situation and needs of their American motherland (as they always called it, rather than "New Granada motherland") is to be found in König, *En el camino* (cit. n. 36); also in José Luis Peset, *Ciencia y libertad: El Papel del científico ante la independencia americana* (Madrid: CSIC, 1987).

[38] Caldas seemed to have no doubts on this point, and in his 1807 article "Estado de la geografía del Virreinato de Santafé" he states that "Geography is the basis for all political speculation." Quoted in Restrepo, "Naturalistas" (cit. n. 35), p. 110.

[39] Francisco Antonio Zea, born in Medellín, Colombia, was a student of the learned Creole José Francisco Restrepo in Popayán. He joined Mutis's expedition as a botanist in 1791. Since he had been involved in a conspiracy against the metropolis led by Antonio Nariño in 1795, he had to go into exile in Spain. Once there, he worked at the Royal Botanic Garden in Madrid with its director Antonio José Cavanilles, whom he succeeded upon his death. In 1816 Zea sided with Simon Bolivar and became vice-president of Colombia. He died in England while on a diplomatic mission. On Zea and his plans for the renewal of colonial science, see Luis C. Arboleda, "La Ciencia y el ideal de ascenso social de los criollos en el virreinato de Nueva Granada," in Lafuente and Catalá, *Ciencia colonial en América* (cit. n. 14). This was also published in volume 3 of the *Historia social de la ciencia en Colombia* (cit. n. 35), pp. 329–59.

Creole elite could not accept that the work of plant collection within the territory should be entrusted to village scholars (local priests or herbalists) working in the Casa de la Botánica in Mariquita. Nor were they happy that the head of the expedition should concentrate upon strictly taxonomic work, or upon supervising botanical illustrations. This pointed to a need to take greater advantage of native knowledge of the use of local flora. No doubt Mutis's correspondents consulted with the native population, but he was nonetheless branded as a European elitist and denounced for his apparent inability to relate to colonial society outside of his circle of loyal supporters. Mutis's quest for intellectual recognition in Europe and his tendency to shut himself away in an ivory tower were seen as slights to the scientists of New Granada.

These criticisms of the leader of the expedition cast doubt not only on his prestige. As the Creoles advocated a new scientific ethos, they also tended to depict Mutis as an imperial agent, selfish and despotic, rather than as a patrician sensitive to local traditions and in favor of developing New Grenada, the territory they considered their motherland. Political events would not allow the translation of these ideas into action until after Mutis's death, when they eventually prevailed in the negotiation for independence forced upon the viceregal authorities by the new generation of Creoles.

LOCAL BACKGROUND AND IMPERIAL DYNAMICS

In the foregoing pages, we have sketched some of the effects that the Hispanic world's inclusion in an imperial structure had upon the globalization of science. We have taken into consideration the two extremes that imperial reality set in motion: the metropolis and the colony. Indeed, the striving for aggiornamento from the end of the seventeenth century onwards in the Iberian Peninsula, stepped up after the accession of the new Bourbon dynasty (1700) to the throne, would soon be subject to tensions that would forge an institutional identity and a professional ethos. The equation, in its Spanish version, is easily stated: a country, opting for regeneration at the expense of old institutions and with new protagonists (scientists), whose social rise was based on talent rather than lineage, and who derived the usefulness of knowledge from its quantifiability, activated a powerful machine for the social validation of scientific activities. We can see that, on the one hand, science crystallized into an essential part of imperial structure. On the other, we find scientific practices informed by the demands of the metropolis and, at the same time, by the patriotic ambitions of the colony.

From the point of view of the colony, there was not just one scientific center but many. Is this a sign of the failure on the part of the ruling elites to develop one consultative body, an academy at the top of the pyramid responsible for decisions on scientific and technical matters? Does this also show scientists' incapacity to gain legitimacy in an autonomous and self-accounting professional environment? The answer is yes. But it is also true that scientific models set up during the seventeenth century, which were easily emulated in such fields as language, history, and fine arts, adapted badly to the requirements imposed by the structuring of the empire (both in the peninsula and in America) in the course of "colonial reconquest." During the eighteenth century, science and technology were envisioned as privileged tools of the new state and imperial policy. Pressing needs on the scientists' part turned them into government agents, thus providing a foretaste of the strong links between scientists and politicians that are common in our century and were widely forged during the revolutions in the American colonies and in neighboring France.

CONCLUSION

By the end of the 1700s, the conviction that the empire was not viable spread among Hispanic elites. As in the case of many long marriages, it seemed as though the different kingdoms that formed the monarchy could not live with each other, but nor could they bring themselves to separate. Whether the flood of complaints was balanced or not, the end of a world seemed imminent. Accusations of cultural incompatibility were added to the reproaches heaped upon the metropolis for its cruel exploitation of the colonies. In Mexico City and in Bogotá, the modernizing capacity of the communication channels established by the empire began to be questioned. It is undeniable that the metropolis was responsible for the introduction of Newton, Linnaeus, and Lavoisier to the colonies, a success achieved not in spite of, but rather thanks to the imperial link. Nevertheless, Latin American intellectuals who criticized the obsolete nature of most of the institutionalized ideas did emerge. They were partly right since, if such ideas were modern in contrast with the scholasticism pervasive a few decades earlier, they were approaching their expiration date in contemporary Europe. It would be unfair to overlook this point and fail to recognize the justification for part of Alzate's support of "natural" classificatory systems and Caldas's vindication of botanical geography. In short, to reduce the nature of the controversy to a mere confrontation between Creoles and metropolitans is, after all, a contagious simplification.

The scientific credibility of the metropolitan adversaries was not the only question raised. Criticisms also abounded about holding values opposed to those locally prevalent. Suffice it to recall how the metropolis presented the public as opposed to the private, the theoretical as opposed to the pragmatic, the paradigmatic as opposed to the local, academic interest as opposed to patriotic interest, or cabinet study as opposed to field work. The elite proposed not an alternative to modernity, but an alternative way of understanding modernity. If we were to single out a value that encapsulated the new ethos, it would be pragmatism. Creoles did not hesitate to manipulate public opinion to gain social legitimacy, in order to counter the veracity of the ideas endorsed by European scientific academies. Thus, leading members of the colonial elite tried to transfer epistemological validation to Latin America; they did not give up the experimental verification of principles, but rather adopted new principles claiming that this would result in better social outcomes. They introduced, likewise, epistemological criteria of a lesser rank, incompatible therefore with those fostered by the metropolis.

Mociño, Caldas, and Alzate were not part of a backward cultural elite. Certainly they were not the most advanced scientists of their time, but within a few years they had managed to convince themselves that they formed part of the army of modernity. And yet they did not act like converts, but rather presented themselves as critics of the knowledge brought by the expeditionaries from the metropolis. For these Creoles to argue about science was more than merely to question models or experiments. As well as promoting scientific ideas, they wanted to introduce a different way of appropriating Nature. They wanted knowledge more relevant to themselves.

Perhaps they did not fully understand the importance of what they were saying, but, with the help of local journalists, they won the battle for public opinion. And it was a resounding victory, since to the public they won for science, what they said seemed obvious: science and patriotism were inseparable. The Creoles did not promote a nonscientific ideology; rather they wanted it to be accountable and politically committed. That is to say, they globalized their problems, because they managed to express them in the cosmopolitan language of modern science. The decisive thing is not the struggle between tradition and modernity, but the reshaping of tradition in light of the communications received from outside that are considered to be most effective. It can be said that the imperial structure within which colonial science developed offered a clear option to the elites of New Spain and New Granada: alternative science, not an alternative to science.

Enlightened Mineralogists:
Mining Knowledge in Colonial Brazil, 1750–1825

*Silvia Figueirôa and Clarete da Silva**

ABSTRACT

This case study of the intellectual and professional trajectories of two Brazilian colonial scientists illustrates the constitution of a local scientific context, its characteristics, and its limits. In Brazil, colonial scientists saw themselves as inextricably linked to the Portuguese metropolis; their science was not seen as an instrument of colonial resistance, but rather as an improving factor, mobilizing natural resources for the achievement of economic and intellectual goals set by both imperial and colonial interests. The development of science in Brazil was not so much a process of transmission as a part of the same process that occurred contemporaneously in Portugal.

T O SPEAK OF THE "PORTUGUESE ENLIGHTENMENT" SEEMS TO many a contradiction in terms. Because of Portugal's Roman Catholic tradition and its alleged backwardness with respect to science, some authors have concluded that science was irrelevant or even nonexistent in Portugal and its colonies during the Enlightenment.[1] Xavier Polanco's model of world-science, with its multiple and discipline-specific centers, provides an appropriate framework for reassessing this issue.[2] The spread of science in the world, far from being a uniform phenomenon, was the result of a complex interaction among different, frequently contradictory,

---

* Institute of Geosciences, State University of Campinas (UNICAMP), P.O. Box 6152, 13083-970, Campinas-SP, Brazil.

[1] See George Basalla, "The Spread of Western Science," *Science*, 1967, *156*:611–22, and the well-known and widely quoted works by Fernando de Azevedo, *A cultura brasileira: Introdução ao estudo da cultura no Brasil* (Rio de Janeiro: Instituto Brasileiro de Geografia e Estatística [IBGE], 1943); Simon Schwartzmann, *Formação da comunidade científica no Brasil* (São Paulo: Nacional / Rio de Janeiro: Financiadora de Estudos e Projetos [FINEP], 1979); and Nancy Stepan, *Beginnings of Brazilian Science* (New York: Science History, 1981).

[2] See Xavier Polanco, ed., *Naissance et développement de la science-monde* (Paris: Découverte / Conseil de L'Europe / UNESCO, 1990); *idem*, "World-science: How is the History of World-science to be Written?" in *Science and Empires: Historical Studies about Scientific Development and European Expansion,* eds. Patrick Petitjean, Catherine Jami, and Anne Marie Moulin (Dordrecht: Kluwer, 1992), pp. 225–42.

and intimate dynamics, involving the construction of local scientific contexts.[3] Science in Portugal took its own, distinctly Portuguese form. In the past two decades our understanding of the situation in the Portuguese colonies has also improved,[4] thanks to new theoretical work on the history of science in Latin America as well as to new developments in the field of colonial science studies.[5] There was in fact an important place for scientists and science in Portugal's metropolitan world, within which Brazil prominently featured. The movement to "civilize" the colonies by promoting science took a different character in each colony, however, since they were at various stages of development and differing levels of importance within the kingdom. This chapter focuses on one of the most important colonies, Brazil, and on its leading industry, mining.

Brazil was an integral part of a vast kingdom, and given the prevailing "integrated development" ideology,[6] according to which colonial primary commodities were exchanged for manufactured goods from the metropolis,[7] Brazil received particular attention. Its exploitation was not unusual but indeed similar to the experience of other European colonies overseas. Two centuries later, the British Empire still embraced the concept of "complementary development," in which colonies exchanged raw material and agricultural products for manufactured goods from Britain.[8]

PORTUGAL AND REFORM

The first phase of the Industrial Revolution, beginning in Britain in the late eighteenth century, and the rise of capitalism wrought changes not only in Europe but also in the European world overseas. Portugal introduced a series of reforms to bring itself and its colonies within the configuration of the new political-economic order.

[3] See Patrick Petitjean's introduction, "Les sciences hors de l'occident au XXe siècle," in *Les Sciences coloniales: Figures et institutions*, ed. *idem* (Paris: Organization pour la Recherche Scientifique des Territoires d'Outre-mer, [ORSTOM], 1996), p. 9.

[4] See, for example: Maria Amélia M. Dantes, "Fases da implantação da ciência no Brasil," *Quipu*, 1988, *5*, 2:265–75; Maria Heloísa B. Domingues, "A idéia de progresso no processo de institucionalização nacional das ciências no Brasil: A Sociedade Auxiliadora da Indústria Nacional," *Asclepio*, 1996, *48*, 2:149–62; Márcia Helena M. Ferraz, "As ciências em Portugal e no Brasil (1772–1822): O texto conflituoso da Química" (Ph.D. diss., Pontifícia Universidade Católica, São Paulo, 1995); Silvia F. de M. Figueirôa, "German–Brazilian Relations in the Field of Geological Sciences during the Nineteenth Century," *Earth Sciences History*, 1990, *9*, 2:132–7; *idem, Ciências geológicas no Brasil: Uma história social e institucional, 1875–1930* (São Paulo: Humanismo, Ciência e Tecnologia [HUCITEC], 1997); Maria Rachel F. da Fonseca, "Ciência e identidade na América Espanhola (1780–1830)," in *História da ciência: O mapa do conhecimento*, eds. Ana M. A. Goldfarb and Carlos Maia (São Paulo: Univ. São Paulo, 1995), pp. 819–36; Maria Margaret Lopes, *O Brasil descobre a pesquisa científica* (São Paulo: HUCITEC, 1997); Vera R. Beltrão Marques, "Do espetáculo da natureza à natureza do espetáculo: Boticários no Brasil setecentista" (Ph.D. diss., Instituto de Filosofia e Ciências Humanas, UNICAMP, Campinas, 1998); and Oswaldo Munteal, "Todo um mundo a reformar: Intelectuais, cultura ilustrada e estabelecimentos científicos na América portuguesa, 1779–1808," *Anais do Museu Histórico Nacional*, 1997, *29*:87–108.

[5] An illuminating review is Juan José Saldaña, "Nuevas tendencias en la historia de la ciencia en América Latina," *Cuadernos Americanos*, 1993, *2*, 38:69–91. Recent syntheses include Roy MacLeod, "Reading the Discourse of Colonial Science," in Petitjean, *Les Sciences coloniales* (cit. n. 3), pp. 87–96, and Antonio Lafuente and Maria Luiza Ortega, "Modelos de mundialización de la ciencia," *Arbor*, 1992, *142*:93–117.

[6] Fernando Novais, *Portugal e Brasil na crise do sistema colonial, 1777–1808*, 6th ed. (São Paulo: HUCITEC, 1995).

[7] José Jobson de A. Arruda, *O Brasil no comércio colonial* (São Paulo: Ática, 1980), p. 675.

[8] Michael Worboys, "British Colonial Science Policy, 1918–1939," in Petitjean, *Les Sciences coloniales* (cit. n. 3), pp. 99–111.

Itself on the semiperiphery of science, Portugal was a political center but not a scientific center. If in some respects the metropolitan discourse was seen as a "civilizing mission," Portugal had first to promote a scientific "civilizing" crusade within its own frontiers and then subsequently in its colonial territories overseas.[9]

In the second half of the eighteenth century, the Portuguese elite drew upon Enlightenment ideals to reform sectors of the polity considered fundamental to economic growth. The adoption of Enlightenment ideals in Portugal—and consequently in Brazil—assumed a peculiar character. Previously, the accepted view of historians has been that Portugal imported science uncritically from abroad, imitating or "mirroring" science from other countries. Recently, however, Franciso Falcon has asserted that although Portugal did indeed accept an ideology that was structured elsewhere, this by no means amounted to uncritical imitation. It was, rather, a new reading, a function of local conditions. Portugal wished to feel itself "contemporary" without abandoning the sense of being "different" or giving up valued traditions.[10] The modernization undertaken by the Portuguese government looked towards updating current knowledge and practice rather than towards structural transformation. Portugal adopted (and adapted) aspects of the Enlightenment in an attempt to "reform to preserve," a policy of "conservative modernization."[11] It drew selectively from the models available in Europe, taking and adapting only those features that fit with its distinctively Portuguese vision.

This pragmatic policy of selecting abroad only what was considered suitable for use at home accounts for the eclectic character of Portuguese reform, which combined a variety of approaches and traditions.[12] The government searched widely in contemporary Europe for models. Official documents explicitly cite the experience of France, the German states, the Netherlands, and Britain, mentioning models for scientific education (new types of courses and schools); research centers (scientific academies and societies, museums, libraries, botanical gardens); new technology and industries; and laws and administrative structures designed to promote modern industry.[13]

The reform programs, which began during the reign of King José I (1750–1777) and continued under Queen Maria I (1777–1792), were led by the prime minister, Sebastião José de Carvalho e Melo, marquis of Pombal. Pombal performed a delicate balancing act, attempting to satisfy both the conservatives, who wanted to continue to support stagnating, unproductive agricultural sectors, and the reformers, who wanted modernization. A compromise was forged, by which scientific inquiry was accepted insofar as it had practical application—or better, would bring economic returns to the crown.[14]

A major change was the introduction of science into the university curriculum, especially at the University of Coimbra. Between 1768 and 1772, courses in modern

---

[9] Francisco J. C. Falcon, *A época pombalina (política econômica e monarquia ilustrada)* (São Paulo: Ática, 1982).

[10] *Ibid.*, pp. 196–7 and 487.

[11] *Ibid.*

[12] See especially the works by Falcon, *A época* (cit. n. 9), and Novais, *Portugal e Brasil* (cit. n. 6).

[13] Concerning reform in the industrial sector, one reads that Pombal "kept before his eyes the example of the Netherlands" (File "Cartas," Carta 4ª, 20 Feb. 1777, Instituto de Estudos Brasileiros [hereafter IEB] Archives). In education, "the Minister kept the old laws he found pertinent and added others based upon the models of the universities of England, France, and German states" (File "Cartas," Carta 9ª, 31 Mar. 1777, IEB Archives).

[14] Munteal, "Todo um mundo" (cit. n. 4), pp. 88–9.

scientific subjects ("natural philosophy," added to the philosophy curriculum) and mathematics were introduced.[15] According to José Pita, "[I]f . . . Pombal's reformism sent waves through all of Portuguese society, teaching was undoubtedly where Pombal carved deeply innovative directions."[16]

The Royal Academy of Science (Real Academia de Ciências de Lisboa) was founded in Lisbon in 1779. Its members were widely read and traveled and corresponded with their peers in Europe. The academy became the focus of a far-reaching program to promote the more rational exploitation of nature within Portugal and the colonies. The movement reflected a broad awareness of science, conceived as systematic, practical, and useful inquiry.[17] The academy developed programs designed to improve agriculture through applied science. A similar initiative to improve one of Portugal's most valued industries, mining, was begun. Information on the theoretical sciences related to mining, on modern mining techniques, and on the related industries of ore extraction and metallurgy was gathered from throughout Europe. An intense, regular process of surveying Portuguese territories for mineral resources was launched and coordinated by the academy. Colonial territories in Asia and Africa were methodically investigated, and the exchange of natural products was instituted.[18] But Brazil, because of its natural and intellectual resources, was by far the most important colony.

The reform program was to have special importance for Brazil. Pombal addressed colonial governors and captains general on the need for surveys of flora and natural products of commercial value in Brazil. One response was the well-known *Flora fluminensis*, by the Franciscan friar José Mariano da Conceição Vellozo. Begun in 1779, by order of the viceroy of Brazil, D. Luiz de Vasconcellos e Souza, it took twelve years to complete. The same viceroy established the Casa de História Natural (Natural History Museum), known as the Casa dos Pássaros (Bird Museum), in Rio de Janeiro. There, from about 1784, natural products from Brazil were collected, preliminarily classified, and shipped to the Real Museu de Ajuda (Royal Museum of Ajuda) in Portugal. Earlier, in 1772, with the aim of promoting in the colony the study of the natural sciences, physics, chemistry, agriculture, medicine, and pharmacy, the crown had approved the establishment of the Academia Científica do Rio de Janeiro (Academy of Science of Rio de Janeiro). With the patronage of the Academia, memoirs and articles were published in Lisbon. The Academia existed until 1794, when members were blamed for political conspiracy. Portugal also benefited from the voyages of the naturalist Alexandre Rodrigues Ferreira to the Amazon region. Departing in September 1783, he spent nine years traveling in the provinces of Pará, Rio Negro, Mato Grosso, and Cuiabá, collecting mineralogical, botanical, zoological, and anthropological material and forming rich collections that were sent to the metropolis.

---

[15] Maria Odila L. da S. Dias, "Aspectos da ilustração no Brasil," *Revista do Instituto Histórico e Geográfico Brasileiro*, 1968, *278*:105–70, on pp. 115–16.

[16] José R. Pita, *Farmácia, medicina e saúde pública em Portugal, 1772–1836* (Coimbra: Minerva, 1996).

[17] The concept was present in many countries, even in contemporary Japan. See Togo Tsukahara, "The Dutch Commitment in Its Search for Asian Mineral Resources and the Introduction of Geological Sciences as a Consequence," in *The Transfer of Science and Technology between Europe and Asia, 1780–1880*, ed. Keiji Yamada (Kyoto: International Research Center for Japanese Studies, 1992), p. 199.

[18] Munteal, "Todo um mundo" (cit. n. 4), p. 89.

Brazil had no university until 1920, though technical schools (engineering and medicine) were founded from 1808 onward. This fact compelled Brazil's elite to study abroad, mainly at Coimbra, where Enlightenment principles "in the Portuguese way" were absorbed. Hence the local Brazilian context constituted itself as a mirror of the metropolitan context, preferring this to reaction against external cultural domination. This does not mean that colonial scientists were without a "native consciousness": in various ways they confronted current debates about the alleged inferiority of the Americas.[19] Many times, José Vieira Couto, one of the two representative Brazilian scientists whom we take as our examples of "scientists in action," supported the American side of the dispute. Nevertheless, both he and Manuel da Câmara, our other exemplary scientist, were loyal supporters of the crown and saw no distinction between the interests of Portugal and those of Brazil.

SCIENCE AND THE COLONIES

The special character of the Portuguese Enlightenment reflected Portugal's position as the metropolis for a declining mercantilist empire. Although in the late eighteenth century Portugal was not yet faced with independence movements in the colonies (as was Spain), it nevertheless was aware of its diminished position in the world.[20]

Once wealthy, the empire was now impoverished, and the decline of mining in Brazil in the late eighteenth century had badly damaged the economy of the metropolis. From roughly 1700 until 1775, gold and diamonds from Brazil had brought Portugal great wealth. The discovery of gold, around 1693–1695 in the state of Minas Gerais, had provoked a substantial gold rush.[21] So much gold was drawn from mines by means of slave labor that during the first seventy years of the eighteenth century Brazil produced more than 50 percent as much gold as had been produced in the entire world in the sixteenth and seventeen centuries combined. Later in the eighteenth century, however, production declined, from about fifteen tons a year (around 1750) to less than five tons a year (around 1785), and the mines seemed to have been depleted.[22] Diamonds, discovered in Minas Gerais and also in the states of Bahia, Goiás, and Mato Grosso around 1720, were the mineral product of second importance, but by 1800 their production had declined more dramatically than that of gold, and the known diamond fields appeared to have been fully exploited.

Contemporaries held varying opinions about the decline. Father José Joaquim de

[19] A comprehensive analysis of this debate is found in Antonello Gerbi, *O Novo Mundo: História de uma polêmica, 1750–1900* (São Paulo: Companhia das Letras, 1996). A Latin-American response is found in Maria R. F. da Fonseca, "'A única ciência é a Pátria': O discurso científico na construção do Brasil e do México, 1770–1815" (Ph.D. diss., Faculdade de Filosofia, Letras e Ciências Humanas, Univ. São Paulo, 1996).

[20] These statements are based upon current studies of Portugal and Brazil. Further inquiries may refine the picture and require revision, especially in relation to other Portuguese colonies.

[21] According to the eyewitness Antonil, "[W]e can hardly estimate how many people presently live there. . . . From cities, villages, and the backcountry of Brazil come white, mulatto, black, and many Indian people whom the *paulistas* [Brazilians born in São Paulo] took advantage of. The mixture comprises all conditions of people: men and women, young and old, poor and rich, noble and plebeian, laic and religious men from the various institutions, many of them without a convent or house in Brazil." Antônio J. Antonil, *Cultura e opulência do Brasil por suas drogas e minas* (Belo Horizonte: Itatiaia; 3rd ed., São Paulo: Univ. São Paulo, 1982), p. 263. The state of Minas Gerais remains an important mining area today.

[22] Virgílio N. Pinto, *O ouro brasileiro e o comércio anglo-português* (São Paulo: Nacional, 1979), p. 115.

Cunha de Azeredo Coutinho, a religious leader in Brazil and founder of an outstanding educational institution, the Olinda Seminary, held that "gold mines are deleterious to Portugal."[23] As a physiocrat, he regarded mining—especially gold mining—as of secondary importance to agriculture and thought its disappearance would be no bad thing for either Portugal or Brazil.[24] Others, like D. Rodrigo de Souza Coutinho, a Portuguese minister, saw practical reasons for the lowered production. They realized that in many old mines the easily accessed ore had already been removed. Few technological improvements in mining or ore extraction had been made, and since slaves were easily obtained, interest in new technology or highly trained miners was low.

Mining in Brazil had given so much wealth to Portugal in the past that the Portuguese government had no intention of giving up on the declining industry now. They saw that in Europe the mining industry was gradually being revolutionized by science and becoming more profitable. New mines were opened as better knowledge of the earth, obtained from the developing sciences of geology, mineralogy, and chemistry, changed prospecting into an exact art. New techniques of metal extraction were developed, so that metals could now be obtained more easily from high-grade ores and also from low-grade ores previously thought useless. Advances were made in metallurgy, and more metal products (especially iron, though not yet much steel) were being manufactured. Professional scientific training for mine managers was now available in countries such as Saxony and France, and these centralized states had also provided legislative models for state support and professionalization of the mining industry.

In keeping with its policy of conservative modernization, the Portuguese government set out to revitalize the industry, particularly in Brazil, by selectively introducing modern scientific theory and technology related to mining.[25] Gold mining retained a special allure for the Portuguese elite, but those with technical knowledge were aware of the rising industrial importance of iron, coal, and other mineralogical products. In the movement begun by Pombal, concerted efforts were made to gather, with the help of the Academy of Science, information from scientifically advanced European countries about the latest techniques for mineral prospecting, mining technology, ore extraction, metallurgy, mine management (including efficient state administration of mining, practiced in France, Sweden, and some of the German provinces), and mining legislation. Improved prospecting methods might uncover new mineral deposits both in familiar areas and in the vast unexplored territories of Brazil and lead to the opening of new mines. Modern mining technology and mine administration might improve the yield of mines now being worked with outmoded techniques and slave labor. Many of the old mines in Brazil had been considered exhausted, but with modern extraction methods low-grade ore might give good yields. Scientific methods of metallurgy might lead to more efficient production of metals such as gold and iron.

---

[23] José J. da C. A. Coutinho, *Discurso sobre o estado actual das minas do Brazil* (Lisbon: Imp. Régia, 1804), p. 18.

[24] A physiocrat is a member of a school of political economists founded in eighteenth-century France and characterized chiefly by a belief that land and agriculture are the source of all wealth (cf. *Webster's New Collegiate Dictionary*).

[25] For a detailed analysis of Brazil's importance within the Portuguese Empire, see the following books: Pinto, *O ouro* (cit. n. 22); Novais, *Portugal e Brasil* (cit. n. 6); Arruda, *O Brasil* (cit. n. 7).

A certain amount of information could be gathered through reading and correspondence. (By 1803, two manuals to improve mining instruction had been printed in Lisbon.[26]) Obviously it was also desirable to send Portuguese and Brazilian scientists abroad to learn for themselves. In 1790 three recent graduates of the University of Coimbra were sent on an extended study tour of the leading scientific and mining centers of Europe.

One of the tour members, a Brazilian, Manuel Ferreira da Câmara (1764–1835), and another Brazilian scientist of the same era, José Vieira Couto (1752–1827), will be our examples of "scientists in action" in a colonial context. Their lives illuminate the complex interactions that resulted in the construction of a local scientific context.

MANUEL DA CÂMARA

Manuel Ferreira da Câmara Bethencourt e Sá was born in the Diamonds District of Minas Gerais, almost certainly at Santo Antonio da Itacambira, around 1764. His wealthy family owned gold mines.[27] When his parents moved to the state of Bahia, he and a brother remained behind and were raised by an aunt at Caeté–Minas Gerais, a gold-mining town.[28] The interest in mining that he acquired here would last throughout his life.

As young men, Câmara and his two brothers were sent to Portugal to study at the newly restructured University of Coimbra. He enrolled in law in 1783 and in philosophy (which now included science courses) in 1784 and received his degrees by 1788.[29] The following year he lived in Portugal with another young Brazilian Coimbra graduate, José Bonifácio de Andrada e Silva, later to become a well-known mineralogist.[30]

The two graduates must have been seen as promising young scientists. In 1789 both Câmara and Andrada were elected to membership in Lisbon's Royal Academy of Science. In 1790, Câmara was commissioned by Queen Maria I to lead a crown-subsidized study tour of Europe for himself, Andrada, and another recent Coimbra graduate, Portuguese-born Joaquim Pedro Fragoso de Sequeira. The three traveled together in Europe for almost a decade. Their itinerary was prescribed by detailed royal instructions (published on 31 May 1790), which stipulated that they were to sit in on courses in mineralogy, chemistry, metallurgy, and other "mining arts" at universities and technical schools in Freiberg (Saxony) and Paris. They were also to study mining operations in Europe's greatest mining districts: Saxony, Bohemia,

---

[26] De Gensanne, *Mineiro do Brasil melhorado pelo conhecimento da mineralogia, e metallurgia, e das sciencias auxiliadoras* (Lisbon: Ofic. Antônio Rodrigues Galhardo, 1801), illustrated, 135 pp.; and Le Febvre, *Mineiro livelador ou hydrometra* (Lisbon: Ofic. Antônio Rodrigues Galhardo, 1803), 100 pp. (First names are not given.)

[27] Unless otherwise stated, biographical details are drawn from Marcos C. de Mendonça, *O intendente Câmara: Manuel Ferreira da Câmara Bethencourt e Sá, intendente geral das minas e diamantes, 1764–1835* (São Paulo: Nacional, 1958).

[28] Silva, 1844, cited in Manuel S. Pinto, "A experiência européia de Manoel Ferreira da Câmara e seus reflexos no Brasil—Algumas notas," in *Geological Sciences in Latin America: Scientific Relations and Exchanges*, eds. Silvia F. de M. Figueirôa and Maria Margaret Lopes (Campinas: Instituto de Geociências, UNICAMP, 1995).

[29] Pinto, drawing upon primary sources, has shown that Câmara obtained both the *bacharel* and *formado* (undergraduate and graduate) degrees. "A experiência" (cit. n. 28), p. 246.

[30] *Ibid.*, pp. 252–3.

Hungary, Russia, Sweden, Norway, and England.[31] This journey displays the hallmark of the Portuguese Enlightenment in its pragmatic and eclectic character—searching selectively abroad for reforms within a framework of conservative modernization. The choice of places, with its emphasis on both theoretical and applied science, suggests that Portugal was attentive to contemporary advances in science and technology. As E. P. Hamm has said, "[M]ining, especially in German-speaking central Europe, helps us to make sense of enlightenment mineralogy in general. . . . Mines provided much more than economic ground for studying the earth; they were the place, both intellectual and social, for the making of knowledge about the earth."[32]

When Câmara's journey began, Freiberg was Europe's leading center for the study of mining science, which included mineralogy, geology, chemistry, engineering, mining technology, mine management, ore extraction, metallurgy, and mining legislation. "The centralized state bureaucracies of the Enlightenment recognized that mining officials required . . . an understanding of the natural history of the earth" and also "legal and engineering knowledge."[33] At the universities, the close connection of mineralogy with chemistry and geology was weak, and links with practice were missing: "[I]f university textbooks were the basis for a history of eighteenth century mineralogy the story would be tidier and simpler."[34] The Bergakademie Freiberg embraced both theoretical research and practical mining. Neptunism and the new scientific discipline of geognosy flourished under the influence of Abraham Gottlob Werner (1749–1817). Students visiting Freiberg spread Werner's ideas to other countries. Câmara and Andrada were both influenced by him, and Neptunist ideas are present in their writings.[35] There were also working mines and metallurgical works in Saxony to observe.

Câmara, who came from the great gold-mining district in Brazil, had always had a special interest in gold mining. As he writes in an early memoir, he had tried to learn more about it in all of his studies:

> [I]t was rather natural that any native of a country whose main mining product is gold would do what I did. When I studied natural history . . . , my particular attention was drawn to gold veins. . . . Studying chemistry, I took the opportunity of reading what writers had said about separation, fusion, and refining of gold, to the profit of my country, in particular, and the state, in general; I attempted . . . to accommodate and simplify ideas.[36]

As his later career in Brazil was to show, however, his interests went far beyond gold and embraced subjects such as ore extraction in general and metallurgy. Iron

---

[31] Mendonça, *O intendente* (cit. n. 27), pp. 26–7.

[32] E. P. Hamm, "Knowledge from Underground: Leibniz Mines the Enlightenment," *Earth Sciences History*, 1997, *16*, 2:77–99, on pp. 77 and 79.

[33] Martin Guntau, "The Natural History of the Earth," in *Cultures of Natural History*, eds. Nicholas Jardine, James A. Secord, and Emma C. Spary (Cambridge: Cambridge Univ. Press, 1996), p. 215.

[34] Hamm, "Knowledge" (cit. n. 32), p. 92.

[35] Câmara published two articles espousing Neptunist views. See Manuel F. da Câmara, "Über das Verhalten des Obsidians vor dem Löthrohre (aus dem Französischen Übersetzt)," *Bergmanniches Journal*, 1794, *6*, 1:280–5; *idem*, "Schreibes von Herrn da Camera de Bethencourt an Herrn Hawkins: Einige Versuche mit dem Obsidiane betreffend," *Bergmanniches Journal*, 1794, *6*, 2:239–49.

[36] Manuel F. da Câmara, "Observações físico-econômicas acerca da extração do ouro nas minas do Brasil" (1789), cited in Pinto, "A experiência" (cit. n. 28), p. 247.

production, which was becoming, as the Industrial Revolution advanced, increasingly important for industry, was another of his special interests.

In the preface to his book *Rapport des résultats des expériences chimiques et métallurgiques faites dans l'intention d'épargner le plomb dans la fonte des minérais d'argent* (published in 1795, while he was still on the tour), Câmara cited his teachers in chemistry and metallurgy at Freiberg and described how Saxony had profited from modern mining and extraction methods. He also cited French, German, and Swedish techniques of mineral exploitation and mining legislation as examples to be followed.[37]

Given his preparation, Câmara was ideally placed to launch a program of "enlightened" science in Brazil. The tour was successfully completed around 1789. In November 1800 he received his reward. He was named to one of the highest mining offices in Brazil, becoming "General Intendant of Mining, in the regions of Minas Gerais and Serro do Frio," where he was to supervise important mines for the crown. In the same year his Brazilian friend and tour companion Andrada was rewarded with the highest mining office in Portugal, becoming "General Intendant of Mining for the Portuguese Kingdom." The two remained in close touch over the years, but, despite their high positions, were to experience great resistance to change from mine owners, and many of the reforms that they desired were never implemented.

In 1800 Câmara returned to Minas Gerais, where he encountered opposition from conservatives from the beginning. Despite his high government appointment and backing by the crown, he was not even able to take up his office until seven years later, on 27 October 1807. Once in, however, he retained his position until 6 April 1822. His ideas for reforming mining were opposed by powerful local mine owners, members of the elite with great influence in the government of Minas Gerais. Câmara's ideal of government embraced an absolutist state, exemplified by European countries where the administration of mines was "trusted to men of all ranks, but trained since their childhood in the art of metal extraction and smelting."[38] The wealthy mine owners, however, had been operating their mines themselves for many decades without government interference, except for taxation. Although production was low, they had prospered, and they did not want to hand over control of their mines to professional mining engineers or accept strict government regulation. Because cheap slave labor was available, they had little interest in modern technology.

An early effort to achieve some of his scientific goals was legislation that Câmara proposed to the Brazilian Assembly supporting a "general system of mineral economy." When the bill was passed into law on 15 March 1803, however, he had not attained his goals. As he reported to the Portuguese minister of foreign and colonial affairs, Luiz de Sousa Coutinho, a key figure in the Portuguese government and a tenacious advocate of the mining sector, the bill had "suffered more discussion than any other ever presented at the Assembly, and in the end was so [changed and] disfigured that he could perceive only the faint outline of his ideas."[39]

Câmara had proposed, for example, that a school "similar to those of Freiberg

[37] Manuel F. da Câmara, *Rapport des résultats des expériences chimiques et métallurgiques faites dans l'intention d'épargner le plomb dans la fonte des minérais d'argent, réduites à un ordre systematique & précedées d'observations sur la fonte des minérais d'argent et de plomb. Adressé au Conseil des Mines de Son A. S. Monseigneur l'Electeur de Saxe* (Vienna: Imp. de Patzowsky, 1795).
[38] Mendonça, *O intendente* (cit. n. 27), p. 53.
[39] Cited in *ibid.*, pp. 34–5.

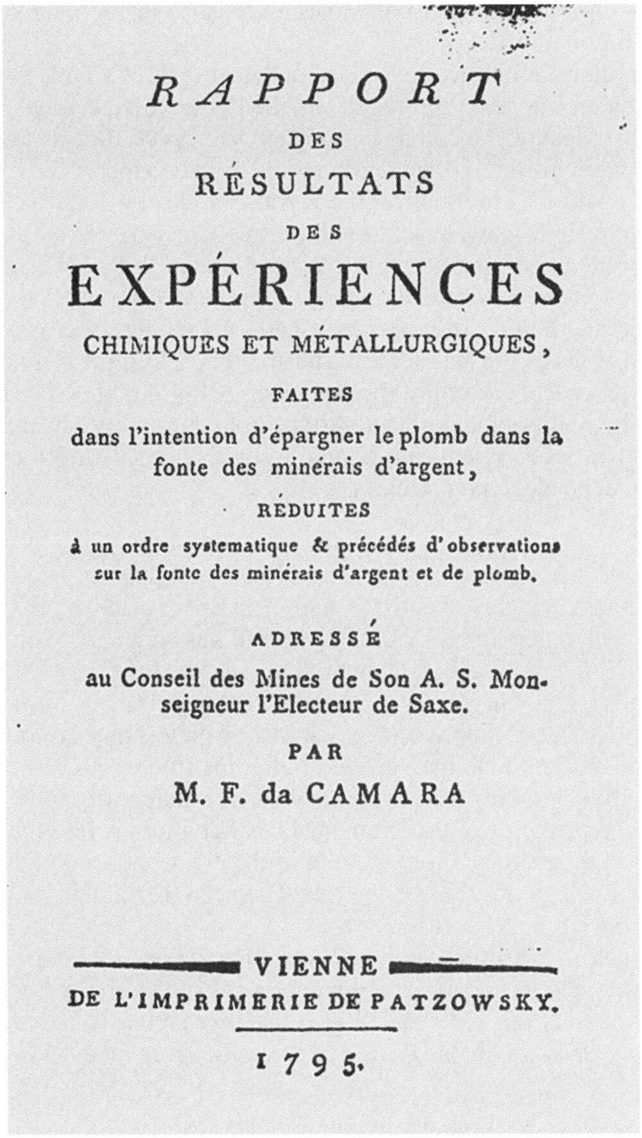

RAPPORT

DES

RÉSULTATS

DES

EXPÉRIENCES

CHIMIQUES ET MÉTALLURGIQUES,

FAITES

dans l'intention d'épargner le plomb dans la
fonte des minérais d'argent,

RÉDUITES

à un ordre systematique & précédés d'observations
sur la fonte des minérais d'argent et de plomb.

ADRESSÉ

au Conseil des Mines de Son A. S. Monseigneur l'Electeur de Saxe.

PAR

M. F. da CAMARA

VIENNE
DE L'IMPRIMERIE DE PATZOWSKY.

1795.

Figure 1. *Frontispiece of Manuel Ferreira da Câmara's 1795 book reporting some of his chemical and metallurgical experiences while studying in the mining academy of Freiberg (Saxony). (Library of the Bergakademie Freiberg, Germany).*

and Chemnitz [also in Saxony], which have brought to those countries such highly praised advantages,"[40] be established in Brazil, but that was not to happen for many years. Even after Brazil became independent in 1822, Câmara, then sitting as a representative of Minas Gerais, was unable to get the Constitutional Assembly to establish a mining academy (Bergakademie). Only many decades later, long after his

[40] *Ibid.*, pp. 119–20.

death, was the Ouro Preto School of Mines established in 1876, on a very different (French) institutional model.[41]

In 1808, an unusual political event strengthened one of Câmara's metallurgical projects. Napoleon conquered Portugal, and the Portuguese government retreated to Brazil. Strictly speaking, the colony became the metropolis. Because the Portuguese high officials were in Brazil, they could more strongly support their scientific programs. Câmara was able to establish the Royal Iron Works (Real Fábrica de Ferro) at Gaspar Soares (in Minas Gerais). (This was certainly one of the first iron works in Brazil, although credit is often given to the German Wilhelm L. von Eschwege, who established one in Minas Gerais at the same time.) Iron works were also built in the state of São Paulo. The enterprise had only limited success, owing to the poor technical skills of the smelter artisans and the lack of good-quality ore, but it established a precedent for future development. All of his life Câmara tried to put into practice the philosophical guidelines acquired during his education in Europe: a strong belief in science, education, and a powerful state, and a conviction that Brazil's future depended upon mining.

JOSÉ VIEIRA COUTO: TRAVELS IN MINAS GERAIS

Our second illustration of a "scientist in action" in Brazil on behalf of the Portuguese metropolitan government is José Vieira Couto. He was born in 1752 in Minas Gerais, at Arraial do Tijuco (now called Diamantina), and died there in 1827.[42] We know little about his parents, but it is likely that they belonged to the upper classes of the region. He studied at Coimbra and graduated in philosophy and mathematics in 1778.[43] Several authors state that during or after his studies at Coimbra he traveled in Europe. Although there is no evidence that he studied medicine,[44] it is certain that he practiced medicine in the Diamonds District after his return to Brazil.[45] His expertise was in mineralogy. From the turn of the eighteenth century, Couto and his brothers occupied high positions at the Real Extração (Royal [Diamond] Quarry) in their hometown.[46]

In 1798, as part of its effort to uncover new mineral resources in Brazil, the crown commissioned Couto to undertake a mineralogical survey in the Serro Frio region of Minas Gerais. Between 1799 and 1805, he surveyed mines in Serro Frio and other areas, including Sabará and Vila Rica, and went almost as far as Goiás, traveling up

[41] The model was that of the Ecole de Mines de Saint-Etienne.

[42] According to the baptismal records of Arraial do Tijuco, 1740–1754. Box 297, p. 133, Palácio Arquiepiscopal de Diamantina.

[43] Francisco de Moraes, "Relação dos 1.242 alunos brasileiros que freqüentaram a Universidade de Coimbra de 1772 (reforma pombalina) a 1872," Anais da Biblioteca Nacional, 1940, 62:137–305.

[44] Ferraz, "As ciências" (cit n. 4), and Pita, Farmácia (cit. n. 16), inform us about the required courses that had to be taken before a student could enroll in the medical course at Coimbra. Couto must have taken them all, because they were part of the natural philosophy curriculum. According to the university's regulations, in order to enroll in medicine the student must first have studied natural philosophy, learning chemistry, physics, and natural history. The latter comprised the study of the three kingdoms of nature (animal, vegetable, and mineral), which had various medicinal properties. Besides these disciplines, mathematics was also mandatory.

[45] We can deduce this from an official document, "Term of declaration, delivery of three diamonds made by Dr. Jozé Vieira Couto in the return of his mineralogical and metallurgical expedition," where he is mentioned as a physician. Manuscript 14,4,22. Manuscripts Section, Biblioteca Nacional do Rio de Janeiro (hereafter BNRJ).

[46] His brothers were Manoel Vieira Couto, Joaquim José Vieira Couto, and Antonio Vieira Couto.

and down the São Francisco River in the so-called Abaete backcountry, in the far western part of the state. He faced, according to the account he wrote, inhospitable deserts and unexplored regions lacking roads or even trails and often had to cut paths through the jungle. In addition to the mineralogical survey, as a well-rounded scientist he also made observations on geography, agriculture, and demography.

Couto's travel logs eventually became published memoirs.[47] The first, *Memoir on the State of Minas Gerais, Its Territory, Climate, and Mineral Products*,[48] was written in Serro Frio in 1799.[49] The second, *Memoir Concerning the Mines of the State of Minas Gerais, Their Description, Assays, and Location in Guise of an Itinerary*,[50] describes a long journey in Minas Gerais from Serro, through the cities of Mariana and Sabará, to the Abaete backcountry. The third, *Memoir on the Various Natural Mineral Chloride Mines of Monte Rorigo*, was written in 1803.[51] Another text attributed to him is entitled "Journey to Indaiá, Together with a Memoir on the Abaete Mines."[52] His last work, "Memoir on the Cobalt Mines of the Minas Gerais State," was written in 1805.[53]

Couto's travels and memoirs reveal the extent of his participation in an important European tradition of scientific investigation. Travel literature, which often included stories of searching for valuable metals, gems, and minerals, played a significant role in the construction of Enlightenment natural history. Johann-Gottlob Lehman, an author whose works were emphasized at the Bergakademie Freiberg, wrote technical books on mining but also published "Instructions" for "mineralogizing" travelers. As Hamm notes, in this era "scientifically inclined travellers rarely limited their observations to one particular field and they frequently made anthropological, political and cultural observations alongside botanical, zoological and mineralogical remarks."[54] Besides mineralogical observations, Couto's *Memoirs* give a general description of the geography, climate, and population of Minas Gerais. In his opinion, the regions into which Minas Gerais was subdivided were fertile, and agriculture, now subordinated to mining, could flourish there.

At the same time, there were, of course, rich metal veins that made mining as important a resource as agriculture. Gold, he wrote, is a "common metal in these territories, and its matrix is constituted by pure quartz or mixed with iron ores, espe-

---

[47] All titles are in Portuguese, here presented in translation.

[48] The complete title is *Memoir on the State of Minas Gerais, Its Territory, Climate and Mineral Products: The Necessity of Reestablishing and Supporting the Declining Mining Industry in Brazil; Trade and Export of Metals and Royal Interests, with an Appendix on Diamonds and Natural Azote.*

[49] Manuscript 11,933-1,1,5, Coleções Especials, BNRJ. This was published in the *Revista do Instituto Histórico e Geográfico*, 1848, *11*. (The published book is cited hereafter as 1799 *Memoir*.)

[50] The complete title is *Memoir concerning the Mines of the State of Minas Gerais, Their Description, Assays, and Location in Guise of an Itinerary, with an Appendix about the New Diamond Region, Its Description, Mineralogical Production, and Resources Which This Land May Yield*, published in *Revista do Arquivo Público Mineiro*, 1905, *10*:56–166.

[51] The complete title is *Memoir on the Various Natural Mineral Chloride Mines of Monte Rorigo and the Manner of Improving Them with the help of Artificial Ones, Purification of Potassium Nitrate or Saltpeter*, published as a book by the Imprensa Régia, Rio de Janeiro, 1809. It was republished in the magazine *Auxiliador da Industria Nacional* in 1840.

[52] The "Journey to India" belongs to the 1801 *Memoir*, since it makes literal transcriptions of part of that text and adds detailed analyses of minerals collected by Couto. Excerpts from this text were published in the magazine *Recreador Mineiro*, 1845, *2*; n. pp.

[53] Ref. 988, Cota 54-v-12 (3), Biblioteca da Ajuda, Lisbon. Dr. Manuel Serrano Pinto, geologist and teacher at Aveiro University, Portugal, kindly sent the authors a copy.

[54] Hamm, "Knowledge" (cit. n. 32), p. 87.

cially the kind [described as] *specularis* by Wallerius and *hoematites* by Linnaeus."[55]
When Couto classified the metallic ores he found in the region, he employed the
systems of both Johann Wallerius (1709–1785) and Carl Linnaeus (1707–1778).
Wallerius used a chemical model and emphasized mineral structure, whereas Linnaeus classified according to external character, applying to minerals the same methodology that he used to classify plants. The parallel use of different classification
systems reflects the influence of the eclectic and pragmatic approach of the Portuguese Enlightenment, fully adopted at Coimbra, where Couto studied.

In Couto, we find an acceptance of the close interdependence between science,
mining education, and the use of modern technology. His model for mining education was certainly the Bergakademie Freiberg. Interestingly, he never mentioned
(nor did Câmara, later) the Real Seminario de Minería (Royal Mining School) in the
Spanish colony of New Spain (Mexico), the first mining school in the Americas,
which had been founded in 1792 on the model of Freiberg. In this period, contacts
between scientists from Spanish America and Portuguese America barely existed:
all eyes turned rather to Europe.

Couto, like Câmara, emphasized "the need for a national metallurgical art" in
Brazil and for an iron industry. Like Câmara, he believed that state initiative was
vital for the development of mining and industry in Brazil. The colony was rich in
resources but underdeveloped, and "even today our mining consists only of gold
mining." Some argued that "everything is already explored and depleted." Others
feared that with the decline of gold mining, the economy would collapse and both
mining and agriculture would disappear. Consistent with Enlightenment views, he
argued that obsolete techniques had been responsible for the decay of Brazil's mining: "[W]e Portuguese, despite having very rich mines, . . . still have not yet taken
the first steps in mining."[56] Couto proposed to raise gold production to "heights of
grandeur . . . it [had] never reached before" by surveying new territories and using
modern techniques to extract valuable metals from mountains once explored superficially but "intact at their core."

The kingdom as a whole, he believed, would benefit from this exploitation in the
colony. Couto's introduction to his 1799 *Memoir* demonstrates that he had absorbed
the ideology of integrated development. In his text, the metropolis–colony dichot-
omy disappears, and there is no place left for an independence movement. The interests of the Brazilian people and the crown are those of the same family, with the
same concerns:

> I flew to the peaks of the mountains, I descended to the interior of caves, and returned
> from my peregrinations with samples of almost all metals, which I exposed at the foot
> of the throne. I spoke about the royal interests, which I could never distinguish from
> those of the people—and how could one ever separate the interests of the same family?
> Between father and son?[57]

This strong belief led Couto and Câmara, as well as a significant number of other
colonial scientists of this period, to perform the assignments they were given by the
crown, which sometimes stretched the very limits of personal ability and political
co-optation.

[55] 1799 *Memoir.*
[56] *Ibid.*
[57] *Ibid.*

Figure 2. *Cover sheet of José Vieira Couto's Memoir of 1799 discussing natural resources and mining in Minas Gerais, Brazil. (Arquivo Histórico Ultramarino, Portugal, box 147, doc. 1, code 11326).*

CONCLUSION

What can we learn from these case studies? Portuguese mercantilist practice in the sixteenth and seventeenth centuries saw colonies as territories for natural exploitation *tout court*. From the mid–eighteenth century, however, Portugal introduced "enlightened" colonial policies aimed at the scientific exploration and use of natural resources. Câmara and Couto are good examples of enlightened Portuguese and Brazilians who believed that improvements in mining and agriculture would bring economic development.[58] In this they had the support of the metropolis. Within the Portuguese Empire, there was great continuity of policy between the metropolis and the colonial scientists, but with certain distinctions among the colonies. The goal was to combine scientific inquiry with policies fostering economic exploitation and political action.[59] In contrast to Spanish America, where political independence was combined with efforts to build national science, Brazilian colonial scientists never placed themselves outside the Portuguese Empire. On the contrary, they incorporated an "ideology of integrated development," adapted to the specific reality of Brazil.[60]

Both Couto and Câmara followed "common-sense" geology in terminology and methodology. Their intent was to locate, describe, and classify mineral ores. A descriptive natural history style prevailed—not surprisingly, as nature still had much to reveal, especially in the New World. However, both men also revealed a distinctively Brazilian tendency to appropriate, amalgamate, and employ the various scientific traditions they had encountered while abroad—including French chemistry, German metallurgy, and the different systems of Wallerius and Linnaeus. Their experience argues against the validity of linear diffusionist models of the spread of scientific knowledge and also against the arguments of scholars such as Fernando de Azevedo, Nancy Stepan, and Simon Schwartzman, who have described scientific practice in Brazil before the twentieth century as a simple reflex, an importation of theories created elsewhere.[61] Colonial science developed in Brazil not so much as a consequence of transmission via the metropolis but rather as part of the same contemporaneous process that implemented a modern scientific milieu in Portugal.

The eclectic and pragmatic view typical of the Portuguese–Brazilian Enlightenment comes alive in the texts of these two scientists. They show themselves as enthusiasts of science. But the science that attracts them leads to the resolution of practical problems: useful knowledge, with implied economic value for the state. Instruction in science is the privileged path. However, as members of an elite aware of its privileges in a country still dominated by slavery (not abolished until 1888), the Enlightenment ideology they preferred was closer to that of Voltaire and Didérot than to Rousseau and the French Revolution. When Câmara wrote that the applications of physics would serve to "apply machines to gold extraction and with them replace the feeble force of miserable slaves,"[62] this was not humanitarian reasoning but the reasoning of efficiency. Mining, agriculture, and their technical problems constituted

---

[58] Novais, *Portugal e Brasil* (cit. n. 6), p. 251.

[59] Francisco J. C. Falcon, "Da ilustração à revolução: Percursos ao longo do espaço-tempo setecentista," *Acervo*, 1989, *4*, 1:63–87, on p. 80.

[60] As demonstrated by Dias in her article "Aspectos da ilustração" (cit. n. 15).

[61] See, for instance, Azevedo, *A cultura* (cit. n. 1); Stepan, *Beginnings* (cit. n. 1); and Schwartzmann, *Formação* (cit. n. 1).

fields in which "scientists in action" could put "science into practice." But in Brazil, as in Portugal, science, as the "light of civilization," was conceived more as a means of increasing power and privilege and helping to keep public order than as a means of liberating "people" into "citizens."

[62] Câmara, "Observações" (cit. n. 36), p. 247.

Racism and Medical Science in South Africa's Cape Colony in the Mid- to Late Nineteenth Century

*Harriet Deacon**

ABSTRACT

Racism has been a particular focus of the history of Western medicine in colonial South Africa. Much of the research to date has paradoxically interpreted Western medicine as both a handmaiden of colonialism and as a racist gatekeeper to the benefits of Western medical science. This essay suggests that while these conclusions have some validity, the framework in which they have been devised is problematic. Not only is that framework contradictory in nature, it underplays differences within Western medicine, privileges the history of explicit and intentional racial discrimination in medicine, and encourages a separate analysis of racism in law, in the medical profession, and in medical theory and practice. Using the example of the Cape Colony in South Africa, this paper shows how legislation, class, institutional setting, and popular stereotypes could influence the form, timing, and degree of racism in the medical professional, and in medical theory and practice. It also argues for an analytical distinction between 'racist medicine' and 'medical racism.'

INTRODUCTION

IN RECENT YEARS, CONSIDERABLE ATTENTION HAS BEEN PAID TO the interplay between racism, medicine, and empire.[1] This work has concentrated on the history of racial inequality or discrimination in medical professionalization, medical institutions, public policy, and in the incidence, perception, and treatment of specific diseases.[2] It is not surprising that racism has been a key issue in the

---

* Heritage Department, Robben Island Museum, Robben Island 7400, South Africa.

[1] Key texts include Sander Gilman, *Difference and Pathology: Stereotypes of Sexuality, Race and Madness* (Ithaca: Cornell Univ. Press, 1985); selected papers in Roy MacLeod and Milton Lewis, eds., *Disease, Medicine and Empire* (London: Routledge, 1988) and David Arnold, ed., *Imperial Medicine and Indigenous Societies* (Manchester: Manchester Univ. Press, 1988); Megan Vaughan, *Curing their Ills: Colonial Power and African Illness* (Cambridge: Cambridge Univ. Press, 1991); Mark Harrison, *Public Health in British India: Anglo-Indian Preventive Medicine, 1859–1914* (Cambridge: Cambridge Univ. Press, 1994); Warwick Anderson and Mark Harrison, "Race and Acclimatization in Colonial Medicine," *Bulletin of the History of Medicine*, 1996, 70:62–118; selected papers in Dagmar Engels and Shula Marks, eds., *Contesting Colonial Hegemony* (London: British Academic Press, 1994) and in Andrew Cunningham and Bridie Andrews, eds., *Western Medicine as Contested Knowledge* (Manchester: Manchester Univ. Press, 1997).

[2] In African (other than South African) medical history, significant work on race and colonial medicine includes Adell Patton, *Physicians, Colonial Racism and Diaspora in West Africa* (Gainesville:

Osiris, 2001, 15:00–00

relatively new field of South African medical history, given the significance of racism in the country's history.[3] However, since the 1980s, revisionist medical history has concentrated mainly on the politics, rather than the ideology, of race in Western medicine—asking whether medicine, like the missionary endeavor, was a handmaiden of colonialism.[4] Historians have been interested in the practitioners of Western medicine and the extent to which they aided colonial domination by initiating and implementing the racist medical policies of colonial governments or industries.[5] A small but growing body of research examines indigenous medical traditions and practitioners in South Africa.[6]

Many South African medical historians have adopted the dual framework of colonial and underdevelopment theory. This means that they have been trapped in a "catch-22," arguing *(a)* that Western medicine was a detrimental agent of colonialism, but also *(b)* that black people were disadvantaged because they did not have equal access to its practitioners and therapies. Racist theory and practice in Western medicine certainly encouraged both *(a)* the treatment of black patients as inferior and different and *(b)* the limitation of black patients' access to its services.[7] As Roy MacLeod has noted, "the political uses of medical knowledge are not unambiguously one-sided; its effects are not simple."[8] While the theoretical framework of

Univ. Press of Florida, 1996); Megan Vaughan, "Idioms of Madness: Zomba Lunatic Asylum, Nyasaland, in the Colonial Period," *Journal of Southern African Studies*, 1983, *9*, 2:218–38; Philip Curtin, "Medical Knowledge and Urban Planning in Tropical Africa," *American Historical Review*, 1985, *90*, 3:594–613; John Cell, "Anglo-Indian Medical Theory and the Origins of Segregation in West Africa," *Amer. Hist. Rev.*, 1986, *91*, 2:307–35; Jock McCulloch, *Colonial Psychiatry and the "African Mind"* (Cambridge: Cambridge Univ. Press, 1995); Heather Bell, *Frontiers of Medicine in the Anglo-Egyptian Sudan, 1899–1940* (Oxford: Oxford Univ. Press, 1999).

[3] The first serious historical analysis was Edmund Burrows, *A History of Medicine in South Africa up to the End of the Nineteenth Century* (Cape Town: Balkema, 1958).

[4] One of the texts that set the parameters for this research was Shula Marks and Neil Andersson, "Issues in the Political Economy of Health in Southern Africa," *J. Southern African Stud.*, 1987, *13*, 2:177–86.

[5] On the racialization of the nursing profession in South Africa, see Shula Marks, *Divided Sisterhood: The Nursing Profession and the Making of Apartheid in South Africa* (London: Macmillan, 1994). On the Western medical profession, see Harriet Deacon, "Cape Town and Country Doctors in the Cape Colony During the First Half of the Nineteenth Century," *Social History of Medicine*, 1997, *10*, 1:25–52. On the relationship of Western medicine to the colonial state, see, for example, Elizabeth van Heyningen, "Agents of Empire: The Medical Profession in the Cape Colony, 1880–1910," *Medical History*, 1989, *33*:450–71; Maynard Swanson, "'The Sanitation Syndrome': Bubonic Plague and Urban Native Policy in the Cape Colony, 1900–1909," *Journal of African History*, 1977, *18*, 3:387–410 and "The Asiatic Menace: Creating Segregation in Durban 1870–1900," *International Journal of African Historical Studies*, 1983, *16*, 3:401–21; Randall Packard, *White Plague, Black Labor: Tuberculosis and the Political Economy of Health and Disease in South Africa* (Pietermaritzburg: Univ. of Natal Press and James Currey, 1989); Elaine Katz, *The White Death: Silicosis on the Witwatersrand Gold Mines, 1886–1910* (Johannesburg: Witwatersrand Univ. Press, 1994).

[6] See, for example, Harriet Ngubane, *Body and Mind in Zulu Medicine* (London: Academic Press, 1977); Catherine Burns, "Louisa Mvemve: A Woman's Advice to the Public on the Cure of Various Diseases," *Kronos: Journal of Cape History*, 1996, *23*:108–34; David Gordon, "From Rituals of Rapture to Dependence: The Political Economy of Khoikhoi Narcotic Consumption, c. 1487–1870," *South African Historical Journal*, 1996, *35*:62–88; Harriet Deacon, "Understanding the Cape Doctor Within the Context of a Broader Medical Market," in *The Cape Doctor: A History of the Medical Profession in the Nineteenth-Century Cape Colony*, eds. Harriet Deacon, Elizabeth van Heyningen, and Howard Phillips, forthcoming.

[7] This is an interesting contrast to the feminist critique of Western medicine for treating women as inferior but *over*medicating them. See Elaine Showalter, "Victorian Women and Insanity," *Victorian Studies*, 1980, *23*, 2:157–81.

[8] Roy MacLeod, "Introduction," in MacLeod and Lewis, *Disease, Medicine and Empire* (cit. n. 1), p. 11.

colonialism and underdevelopment can explain why Western medicine in the colonial context was often racist in conception and application, it cannot explain how Western medicine could at the same time be politically, economically, and culturally loaded, and also useful.

In South African medical history as a whole, as perhaps elsewhere, there has been a tendency to see racism occurring mainly where white colonists met black indigenes.[9] This perspective constructs racism as an issue that attained its full expression in the European colonies: racism is the one topic about which colonial historians feel they can write authoritatively. This essentialist view of race has been challenged by recent work on the (re)construction of whiteness in European colonies, and on the importance of understanding racism through filters of gender and class.[10] There are other problems, too: within the history of medicine, as in other branches of South African history, the intellectual history of racism stands strangely neglected in contrast to the history of racial discrimination.[11] Even in charting the history of racial discrimination, we have documented official pronouncements, policies, and legislation at the expense of understanding local practices. Little has been written on the relationship between racist theory and practice, diverse forms of discrimination in different contexts, and the correlations between racism and other types of discrimination.[12] Indeed, in spite of recent interest in the history of racist medical and scientific theories, we are still far from understanding the trajectory of racism in colonial medicine.[13]

In order to understand the variations, trends, and contradictions within colonial Western medicine, we require a more nuanced analysis of its relationship to racism. We must explore differences in the application and elaboration of racist ideologies in various colonial contexts.[14] At the same time, Western medicine cannot be treated as a homogenous entity, as doctors from different European medical traditions practiced in different political, economic, cultural, and institutional situations in the colonies.[15] In order to understand racism in Western medicine, we should combine analyses of racialization within the profession, discrimination in medical practice, and

[9] In this framework, racism is "about black people" in the same way that gender is "about women."
[10] See for example, Ann Laura Stoler, "Sexual Affronts and Racial Frontiers: European Identities and the Cultural Politics of Exclusion in Colonial Southeast Asia," and Lora Wildenthal, "Race, Gender and Citizenship in the German Colonial Empire," in *Tensions of Empire: Colonial Cultures in a Bourgeois World*, eds. Frederick Cooper and Ann Laura Stoler (Berkeley: Univ. of California Press, 1997). See also Anne McClintock, *Imperial Leather: Race, Gender and Sexuality in the Colonial Context* (London: Routledge, 1995).
[11] Saul Dubow, *Scientific Racism in Modern South Africa* (Cambridge: Cambridge Univ. Press, 1995), p. 3.
[12] For a comparison of discrimination in hospitals and prisons, see Harriet Deacon, "Racial Segregation and Medical Discourse in Nineteenth-Century Cape Town," *J. Southern African Stud.*, 1996, 22:287–308. On the relationship between racial and gender discrimination in medicine, see Sally Swartz, "Colonialism and the Production of Psychiatric Knowledge in the Cape, 1891–1920" (Ph.D. diss., Univ. of Cape Town, 1996).
[13] See, for example, Paul Rich, "Race, Science and the Legitimization of White Supremacy in South Africa, 1902–1940," *Inter. J. African Hist. Stud.*, 1990, 23:665–86; Dubow, *Scientific Racism* (cit. n. 11); Andrew Bank, "Liberals and their Enemies: Racial Ideology at the Cape of Good Hope, 1820 to 1850" (Ph.D. diss., Univ. of Cambridge, 1995).
[14] Bank, "Liberals and their Enemies" (cit. n. 13), suggests that European settlers elaborated different types of racist ideologies in the eastern and western parts of the Cape Colony.
[15] For example, Deacon, "Cape Town and Country Doctors" (cit. n. 5) argues that Continental and British educational and professional backgrounds produced doctors with different approaches to professionalization at the Cape in the early nineteenth century.

racism in medical theory. This essay explores this idea by considering the dynamics between racism and Western medicine in the Cape Colony during the mid- to late nineteenth century.

The Cape Colony, which now comprises most of the western, northern, and eastern Cape Provinces of South Africa, was under British rule from 1795–1803 and from 1806–1910, when it became part of the Union of South Africa. The inhabitants of the Cape at this time consisted mainly of indigenous Africans, including Nguni speakers in the northern and eastern areas, Khoekhoen (then called "Hottentots"), and San or Bushmen; slaves (mainly from the East Indies) and their descendants; Dutch-Afrikaners descended from Dutch and German settlers; and British settlers. As this essay suggests, professional and institutional factors acted as key influences on the emergence of a highly differentiated, although broadly racist, field of colonial medicine in the Cape Colony. It also suggests that, in understanding racism in colonial medicine, we should differentiate between racist medicine (the institution of discriminatory practices in medicine based on broader social discrimination) and medical racism (the application of racially discriminatory practices in medicine justified on medical grounds).

SCIENCE AND RACISM IN SOUTH AFRICA

The relationship between racism and science is hotly contested terrain. Some have argued that Western science and medicine were never free of racial content. Indeed, Keenan Malik and others have argued that the liberalism of the European Enlightenment, which "introduced a concept of human universality which could transcend perceived differences," made the modern concept of race possible.[16] The fact that (racial) difference was recognized within liberal discourse created a space for the paternalism of "civilization"; the construction of an egalitarian moral economy created the opportunity for exclusions from it.[17] On the other hand, others have implied that racism, when present, was just a temporary and undesirable adjunct to scientific theory.

It is not, of course, an easy matter to decide whether racism in science is best identified by examining explicit theoretical content, or the way in which theories were used or applied. It is hard to separate racist theory from practice because both emerged out of broader, often unarticulated, racist discourses, only some of which were arbitrarily elevated by contemporary science to the status of theories. In understanding the trajectory of racism within science it is, however, useful to differentiate (if only in degree) between "scientific racism," which created scientific justifications for racist practice, and "racist science," which incorporated elements of popular racism in the theory and practice of an otherwise universalist science. (In the same way, medical racism can be differentiated from racist medicine.) This distinction can help us to understand the degree to which notions of race became an essential part of the way in which the scientific enterprise was defined.

In the eighteenth century, a biologically based concept of race gained currency in

---

[16] Keenan Malik, *The Meaning of Race* (Basingstoke: Macmillan, 1996), p. 42.
[17] David Goldberg, *Racist Culture: Philosophy and the Politics of Meaning* (Oxford: Blackwell, 1993), p. 6.

Europe and became central to the articulation of a bourgeois identity.[18] Although Europe remained the dominant partner in the formulation of racist doctrines, the colonial contribution was significant. Not only did empire provide an important political motivation and ideological touchstone for the elaboration of racial difference as a fixed biological reality, but it was a key testing ground for racist theory and practice, some of which was then assimilated into popular and scientific discourses in Europe. Metropolitan and colonial science had common origins and enjoyed frequent crossfertilization, but they were formulated and practiced in different political and ideological contexts. They were also shaped in diverse ways by the patchy and delayed transfer of ideas and people. Scientific theory and practice, and its racism, thus exhibited different features in different contexts.[19]

The relationship between metropole and colony was, of course, by no means equal. The development of an independent colonial science was thus not just a matter of time, as Basalla's work seems to suggest.[20] Colonial scientists often "functioned as collectors of facts, while metropolitan scientists acted as theorists or gatekeepers of scientific knowledge."[21] The development of independent, national scientific communities was thus difficult, and perhaps impossible, to achieve in European colonies.[22] Like colonial historians on racism, colonial scientists, especially in Africa, felt more confident in espousing theories about racial difference considered specific to the colonies, than in competing with metropolitan scientists' theories considered applicable to both colony and metropole. Even compared with other settler colonies, colonial South Africa was slow to develop scientific institutions and a home-grown scientific identity. The South African Association for the Advancement of Science was not founded until 1903, and the first signs of a confident "South Africanized" (albeit not perhaps independent) science emerged only in the late 1920s.[23]

Even though racist practice was commonplace, scientific racism was slow to develop in South Africa. A kind of professional schizophrenia characterized the Cape medical profession, which in practice was deeply rooted in popular colonial racism but in theoretical orientation remained largely universalist and metropolitan.[24] Doctors oscillated between asserting the differences between black and white patients and their essential sameness. South African discussions of eugenics were less intense and less nuanced or consistent than elsewhere.[25] Rich has suggested that

[18] Ann Laura Stoler and Frederick Cooper, "Between Metropole and Colony: Rethinking a Research Agenda," in Cooper and Stoler, *Tensions of Empire* (cit. n. 10), pp. 2–3.

[19] For a Latin American example, see Nancy Stepan, "Race, Gender and Nation in Argentina: The Influence of Italian Eugenics," *History of European Ideas*, 1992, *15*:749–56, p. 749. On South Africa, see Dubow, *Scientific Racism* (cit. n. 11).

[20] George Basalla, "The Spread of Western Science," *Science*, 1967, *156*:611–22.

[21] Susan Sheets-Pyenson, *Cathedrals of Science: The Development of Colonial Natural History Museums during the Late Nineteenth Century* (Kingston: McGill-Queen's Univ. Press, 1988), p. 15.

[22] For a review of the debates on the development of a national scientific community in Australia, see Jan Todd, *Colonial Technology: Science and the Transfer of Innovation to Australia* (Cambridge: Cambridge Univ. Press, 1995), pp. 7–8. For a general overview of the comparative literature, see Sheets-Pyenson, *Cathedrals of Science* (cit. n. 21), pp. 12–15.

[23] Dubow, *Scientific Racism* (cit. n. 11), pp. 12–13.

[24] Harriet Deacon, "Racial Categories and Psychiatry in Africa: The Asylum on Robben Island in the Nineteenth Century," in *Race, Science and Medicine*, eds. Waltraud Ernst and Bernard Harris (London: Routledge, forthcoming).

[25] Dubow, *Scientific Racism* (cit. n. 11), pp. 16–17.

> Until the rise of anthropological research in the 1920s and 1930s, there was little attempt to back up propositions [in the South African debate on racial segregation] with systematic evidence. Arguments often fell back upon well-worn stereotypes [couched] in a 'scientific' vocabulary.[26]

It was only in the 1920s that scientific racism would fully catch up with discriminatory practice as South Africa sought to present itself as a "human laboratory" for "race relations."[27] Even then, "the newer scientific discourse . . . ended up perpetuating older conceptions of African society inherited from nineteenth-century travelers and missionaries."[28]

Before the 1880s, few Cape doctors styled themselves as "scientists." Andrew Sparrman, a Swedish doctor and scientist who visited South Africa in the late eighteenth century, was disappointed in the standard of Western doctors practicing at the Cape, and suggested that they were more interested in making money than in science or healing.[29] Sparrman, a student of Linnaeus, was unusual in documenting his investigation of racial differences between patients. In 1775, he visited the Caledon Warmbaths where he helped to treat a Madagascan slave from Cape Town who was suffering from an ulcerated leg. His comments on the case indicate he expected to find that racial differences were more than skin deep, but he could not support that assumption through his observations of the black man's wound:

> Being curious to examine a negro's flesh, I had for some time undertaken to look after the sore myself . . . The raw flesh appeared exactly of the same colour with that of an European . . . [As] the ulcer began to heal, [it threw out] fresh fibres in the same manner as ours do, with something whitish on the side of the skin, which otherwise was of a dark colour.[30]

Undaunted by finding similarity where he sought difference, in November of the same year Sparrman treated several farm workers (slaves and Khoisan) and a settler girl for "bilious fever." He treated them all with a strong mixture of tobacco, water, and alcohol. He gave racial explanations for the differences he observed in the patients' symptoms, disease progress, and reactions to his treatment. He explained these differences in cultural rather than biological terms, suggesting that those indigenous Khoekhoen ("Hottentots") who had recently "made too sudden a transition from their strange manner of living" responded less quickly to the medication than settler patients and Khoekhoe servants brought up with the family.[31] At this time in Europe, the idea of race was still linked to culture as well as to biology. By the 1880s, Cape doctors had begun to develop cultural explanations for racial difference in the field of psychiatry, a shift in perspective that would soon characterize other fields of medicine, too.[32]

At least until the end of the century, the British model of "gentlemanliness" was

---

[26] Rich, "Race, Science and White Supremacy" (cit. n. 13), p. 667.

[27] Dubow, *Scientific Racism* (cit. n. 11), p. 14.

[28] Rich, "Race, Science and White Supremacy" (cit. n. 13), p. 667.

[29] Andrew Sparrman, *A Voyage to the Cape of Good Hope towards the Antarctic Polar Circle and round the world . . . from the year 1772 to 1776*, 2 vols., 2nd ed. (London: Robinson, 1786), p. 48.

[30] *Ibid.*, p. 143.

[31] *Ibid.*, pp. 351–3.

[32] Deacon, "Racial Categories and Psychiatry" (cit. n. 24).

more important in the self-definition of the South African medical profession than was the idea of a colonial or national scientific community. Sparrman found few medical colleagues in the Cape Colony with whom he could discuss his ideas on racial difference. It was not that Cape doctors were antiracist, but that they did not put their racist ideas into scientific terms within a medical discourse subject to experimental proof. Well into the twentieth century, the colony lacked the resources, libraries, and universities needed to encourage the development of a scientific community. Earlier, some doctors had contributed to scientific discourse generated abroad (e.g., the trial of anaesthesia in Grahamstown in 1847), but there was little scientific innovation within the colony itself.[33] The two professional medical organizations established before the 1880s were both short-lived.[34] The *Cape Town Medical Gazette*, which began in 1847, also had a short life, and its editor's calls for a medical society went unheeded.[35] In the 1850s, Collis Browne, a military surgeon quartered at the Cape Town Castle and a keen inventor, complained that military doctors had failed to keep up with the "progress of modern science" in the medical field.[36] It was little better among civilian doctors.

THE RACIALIZED MEDICAL GENTLEMAN

By the beginning of the nineteenth century, when Britain had just taken over the Cape from its Dutch colonizers, immigrant medical practitioners in the Cape, as elsewhere, had begun to aspire to being "medical gentlemen," although there was always a gap between the gentlemanly ideal and the reality of colonial life.[37] The notion of the medical gentleman was racialized, gendered, and class specific, circumscribing a community of middle-class, European-trained doctors who were almost all white men. The Cape medical profession situated itself in the strange half-light between European identity and colonial reality. While the doctors did come into contact with black patients and indigenous ways of healing, they practically ignored the scientific study of indigenous medical pharmacology, although the Western Cape boasts a unique floral kingdom.[38] Cape doctors were more likely to look to their European (mainly British) colleagues and institutions for professional affirmation than to each other. This particularly intense form of "colonial cringe" led to a much greater delay in the establishment of colonial medical schools at the Cape than in India, Canada, and Australia.[39]

The spread of the gentlemanly ideal was associated with medical professionalization. In the eighteenth century there was little legal control over who could practice medicine. Indigenous and slave healers were left to their own devices, and Western

[33] On the Grahamstown trial, see Burrows, *A History of Medicine* (cit. n. 3), p. 170.

[34] *Ibid.*, p. 133.

[35] *Cape Town Medical Gazette*, 1847, *1*, 3:32. See also Burrows, *A History of Medicine* (cit. n. 3), p. 350 and C. Blumberg, "The South African Medical Society and its Library," *Cabo*, 1978, *2*, 4:18–25.

[36] Thomas Lucas, *Camp Life and Sport in South Africa* (1878; reprint, Johannesburg: Africana Book Society, 1975), pp. 33–5.

[37] On medical gentlemen elsewhere, see Penelope Corfield, *Power and the Professions in Britain, 1700–1850* (London: Routledge, 1995); Robert Gidney and Winnifred Millar, *Professional Gentlemen: The Professions in Nineteenth-Century Ontario* (Toronto: Univ. of Toronto Press, 1994).

[38] Deacon, "Understanding the Cape Doctor" (cit. n. 6).

[39] Howard Phillips, "Medical Education," in Deacon, Van Heyningen, and Phillips, *The Cape Doctor* (cit. n. 6).

doctors were not organized into a profession with strict rules governing membership and practice. In 1807, however, the autocratic military government in Cape Town tried to impose price and quality controls on the sale of imported medicines. Western-trained doctors used this as an opportunity to seek government regulation of the whole medical profession. Not surprisingly, this won for the doctors government endorsement and a legal monopoly over professional medical practice at the Cape.[40] Indigenous healers were prohibited from charging for their services, and female midwives had to come under the control of licensed male doctors.[41]

Almost all Cape doctors were European settlers, and many were recent immigrants.[42] Almost no black or female Western doctors were in practice; a handful were licensed late in the century. Although there was no racist legislation, the Cape profession excluded most black and female practitioners in practice by defining itself in masculine, Western terms and by insisting on a European medical training. Most colonial families could not afford to send their sons to study overseas. The European medical schools did not usually admit women, and moreover, few Cape and European families wanted to go to the expense of educating their daughters at a time when women were supposed to become housewives. It was also difficult to overcome official, medical, and settler prejudice against black people becoming Western doctors. Most black communities consulted indigenous healers by preference, and there were few opportunities for black doctors outside missionary hospitals. The overwhelming number of white immigrant medical practitioners was also an important factor. Many more black doctors were trained in the Western tradition in West Africa—"the white man's grave"—where from midcentury on there was an Anglicized black elite whose members sought posts in a government service shunned by white doctors.[43]

By the early nineteenth century, the Cape medical profession was thus highly racialized in terms of its professional identity and racial composition, in spite of the absence of formal legal barriers to the entry of black doctors.

RACIST MEDICINE

Within different institutions and areas of medical specialization, racism was expressed in different ways and the timing of segregation varied. This was only partly due to the interplay between racist medicine and medical racism; disease profiles, patterns of institutionalization, and popular stereotypes also played a role. How doctors viewed medicine was closely connected to their conception of their status and responsibility as settlers and medical gentlemen.[44] They were, however, obliged to make a living by selling medical services. Their relationships with black patients were thus influenced by the unequal relations of colonialism, but also by class relations in the context of their consultations (private, charitable, or public). Moreover,

---

[40] In the other British colony in South Africa, Natal, indigenous practitioners were allowed a limited practice.

[41] Deacon, "Cape Town and Country Doctors" (cit. n. 5).

[42] *Ibid.*, p. 33; Van Heyningen, "Agents of Empire" (cit. n. 5), p. 452.

[43] Patton, *Physicians, Colonial Racism and Diaspora* (cit. n. 2).

[44] John Harley Warner has suggested that even in fiercely Republican antebellum America, doctors' professional identity was predicated partly on "moral character" or social respectability. John Harley Warner, *The Therapeutic Perspective: Medical Practice, Knowledge and Identity in America, 1820–1885* (Princeton, N.J.: Princeton Univ. Press, 1986), pp. 1, 15–16.

the extent and timing of racial segregation in medical and other institutions was influenced by site-specific functional and professional factors.[45] These circumstances can also help explain variabilities in the theory and practice of racism in colonial medicine.

Given a white male profession whose members aspired to be "gentlemen," as well as the racism of settler society, one might expect both the theory and practice of medicine at the Cape to have been racist. This is broadly true, but we have to be very aware of the ways in which we measure the form and extent of this racism. Many historians have focused on documenting overt or covert racism in medical legislation and the racist implementation of "color-blind" medical legislation in South Africa by the end of the nineteenth century.[46] This focus obscures some key issues, however: *(a)* that private black patients may not have experienced the same degree of racist practice as institutionalized black patients, and *(b)* that circumstances within institutions played a key role in shaping the extent and timing of racial discrimination.[47]

The profile of black patients attended by colonial doctors was by no means homogenous, and neither was their treatment.[48] Black patients went to Western doctors for a few specific ailments; some slaves or workers went at the behest of their employers; and many sought free medical care, food, and lodging in colonial hospitals because they were destitute or abandoned. Some black patients actually chose Western medicine in preference to other types of care. Others were treated under duress: Western doctors were often called in by slave owners to ensure that slaves could continue working. Doctors were also used to testify in favor of an owner accused of maltreating a slave.[49] Since doctors identified with and depended on settlers as clients, they seldom asserted the rights of sick slaves against the demands of their masters. When they did, it caused great friction between doctors and the settler community.[50]

Given the lack of detailed records, our understanding of the size and nature of private practice serving black clients is limited. However small, this group of clients is important. The records of Peter Chiappini, an Edinburgh-trained doctor practicing in Cape Town in the 1840s, suggest that he had a varied clientele, ranging from wealthy white settlers to poor black laundresses. Some of his poor black clients consulted him frequently and paid bills as private patients, although on the lowest

[45] Deacon, "Racial Segregation" (cit. n. 12).

[46] Swanson, "The Sanitation Syndrome" (cit. n. 5).

[47] Some work has begun to consider these issues. On psychiatric theory and practice see, for example, Swartz, "Colonialism and Psychiatric Knowledge" (cit. n. 12); Harriet Deacon, "Madness, Race and Moral Treatment at Robben Island Lunatic Asylum, 1846–1910," *History of Psychiatry*, 1996, *7*:287–97. On institutional contexts, see Deacon, "Racial Segregation" (cit. n. 12).

[48] White patients (particularly Dutch-Afrikaans settlers) also sometimes consulted black medical practitioners, whether midwives or traditional healers. See Harriet Deacon, "Midwives and Medical Men in the Cape Colony before 1860," *J. African Hist.*, 1998, *39*:271–92.

[49] Cape slavery is often described as relatively "mild," but this does not mean that slaves escaped severe (even fatal) beatings, malnourishment, and psychological abuse at the hands of their masters.

[50] For example, Daniel O'Flinn, surgeon at Stellenbosch, suffered financially because of his "kindness and attention" to slaves ("Report on O'Flinn," *Confidential Reports on Civil Servants, 1843–51*, Cape Archives, Cape Town, CO 8551). Occasionally, a slave owner was charged with murder as a result of a doctor's report: see, for example, the case of the contradictory postmortem reports on a dead slave by Drs. Shand and Tardieux in Burrows, *A History of Medicine* (cit. n. 3), pp. 56–7, 87.

scale.[51] Western medical care was probably more expensive than other options, so we might assume that only wealthier black clients could afford it. Yet Chiappini had black clients from the lowest income groups. Chiappini's black clientele may have been introduced to Western medical care in the hospitals, at the weekly free clinics some urban doctors offered, or by former employers and slave owners. Given the history of abuse of female slaves by male settlers, it is particularly interesting that not only were some of Chiappini's private black clients women, but that his specialty for Muslims was in an area of gynecology (removal of the afterbirth).[52] Chiappini probably attracted private black patients because he spoke Dutch-Afrikaans and practiced close to the tenements of Cape Town's burgeoning urban working class. (Afrikaans is a Creole language, derived from Dutch, Malay, and African languages, that was spoken by some settlers of continental European origin, slaves, and Khoek-hoen, hence the term Dutch-Afrikaans to designate people who were identified with the white settler community.)

In rural areas with few hospitals and Western doctors, and great distances between farms and settlements, black people consulted Western doctors for specific ailments only. Blacks with eye problems came from afar to consult Dr. John Fitzgerald at the Kingwilliamstown hospital in the eastern Cape, for example.[53] Although hospitals and free clinics were not always available, there were other ways of getting access to Western medicine. Missionaries often freely administered it to potential converts. Apothecaries and traveling salesmen sold medical advice as well as drugs, the most popular of which were patent medicines made in Halle. Similar remedies, marketed as "Dutch medicines," are in widespread use among black South Africans even today.

The willingness of black clients to use Western medicine does not mean that it was free of racial bias. Nor was racism in Western medicine by any means homoge-nous, since Western doctors came from different traditions of racism (ranging from liberal to openly racist), and since settler stereotypes of Khoekhoe, Muslim, and African patients varied. The extent to which Western doctors could and did discrimi-nate against their patients was also influenced by legislation, existing practices among their peers, and the financial aspect of the doctor-patient relationship. Some black patients paid for consultations themselves (reimbursement was often scaled to income), but most did not. In general, there was greater leeway for systematized discrimination and the application of racist policies in institutions where patients were not paying for medical services.

Given the constraints on private practice, government and missionary hospitals were the main point of contact between black patients and white doctors. These colonial institutions were conducive environments for the development of racist the-ory and practice, as their custodial overtone lent itself to discrimination. In rural

---

[51] Unfortunately, we do not yet know what the comparative cost of consulting slave or indigenous practitioners might have been in Cape Town during the 1840s, but we can assume it was fairly low, and even a visit to the local apothecary would probably have been cheaper than Chiappini's lowest rate.

[52] On abuse of women slaves, see Pamela Scully, "Rape, Race and Colonial Culture: The Sexual Politics of Identity in the Nineteenth-Century Cape Colony, South Africa," *Amer. Hist. Rev.*, 1995, *100*, 2:335–59. On Chiappini, see Deacon, "Midwives and Medical Men" (cit. n. 48), p. 287, n. 86.

[53] See, for example, Burrows, *A History of Medicine* (cit. n. 3), p. 182.

areas the town jail was often used as a "hospital" or holding place for the homeless and destitute, and hospitals were often established in old barracks or prisons. Most hospital patients were poor, and many were black. Treatment was generally free or nearly so. Patients often had nowhere else to go; some were forcibly institutionalized. Their interaction with doctors was thus less individualized, encouraging discrimination against black patients.

Hospital care in the nineteenth-century Cape (as elsewhere) was not particularly effective, and was shunned by most middle-class patients.[54] Racial discrimination was almost inevitable when a doctor had to divide his time and resources between white and black: if the doctor did not prioritize white patients, they often protested. Once separate institutions for black and white patients had been established in the 1890s, discrimination was entrenched, but at the budgetary, staffing, and policy level rather than at the discretion of institutional staff.

Despite similarities between hospitals, prisons, and other colonial institutions, and between hospitals of different sorts, they presented different patterns of racist practice.[55] At the Robben Island leper, mental, and chronic sick hospitals, for example, racial segregation emerged at different times. The timing depended partly on the numerical balance between black and white patients and the group's class profile. The social and medical stereotypes of the diseases treated also affected patterns of segregation. But medical racism—racist medical theories of insanity and leprosy—did not materially influence the timing of racial discrimination. These theories did little more than justify differential treatment based on race and the creation of separate asylums for black and white mental patients.[56]

MEDICAL RACISM

Racist medical theory did not always accompany racist medical practice. Racism in medicine could arise from circumstances other than the elaboration of formal medical arguments for racial difference in patient diagnosis and care. Colonial doctors often used general justifications for racial discrimination (social entitlement) rather than specifically formulated medical reasons. At the Cape, racist medical theories were slow to develop because of conditions in separate black institutions, a desire to underplay differences between metropolitan and colonial science, and an unquestioning acceptance of racism in colonial society. In Cape psychiatry, for example, doctors treated black and white mental patients differently from the beginning, but only in the 1880s and 1890s did they advance theoretical arguments to justify this, suggesting that black patients responded better to physical rather than psychological therapy because of their supposedly lower developmental status.[57] Such racist theories emerged from doctors' experiences with large numbers of black patients in insti-

[54] Exceptions were the curative wards of the New Somerset Hospital and middle-class lunatic asylums, such as those in Grahamstown or Valkenberg.

[55] Deacon, "Racial Segregation" (cit. n. 12).

[56] The exception to this rule was the reversal of some forms of racial discrimination in the 1860s at the Robben Island Lunatic Asylum with the application of "moral management" treatment, which assumed that insanity was "everywhere the same" (Deacon, "Racial Categories and Psychiatry" [cit. n. 24]).

[57] Deacon, ibid.

Figure 1. *A leprosy patient at Robben Island c. 1900.*

tutional settings where their prejudices were often confirmed. These theories were often simply tacked on to pre-established racist practices.[58]

The case of leprosy provides an important example of the emergence of racist medical theory from a popular stereotype. Portraying the leper as black was central to medical theory and public health policy in the Cape. Of course, most leprosy

---

[58] A similar pattern can be observed in British psychiatry, which did not use the "scientific" evidence collected to produce insights into the types, causes, and possible results of insanity during the nineteenth century. See Richard Russell, "The Lunacy Profession and its Staff in the Second Half of the Nineteenth Century," in *The Anatomy of Madness III: Essays in the History of Psychiatry*, eds. William Bynum, Roy Porter, and Michael Shepherd (London: Routledge, 1988), p. 298.

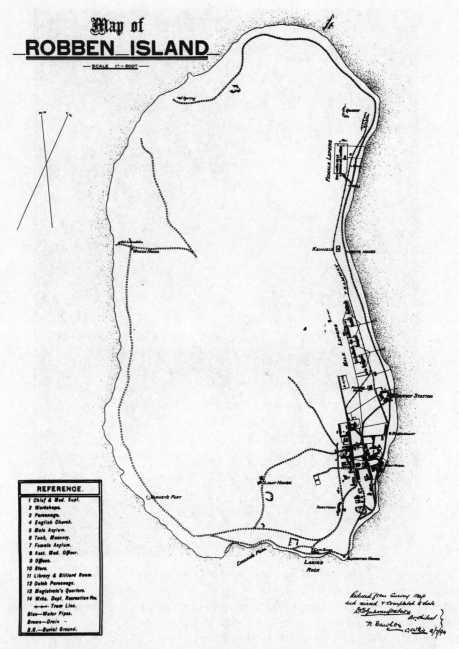

Figure 2. *Robben Island, 1894. The leprosy hospital wards for men and women were separated from the main village area in 1891.*

Figures 3 and 4. *As these pictures show, most leprosy patients at Robben Island were black, while most staff members were white.*

patients in Cape institutions were black. Although it is possible that different populations may have had different levels of generationally acquired immunity to the disease, poverty—more widespread among blacks—may also have influenced its spread and the pattern of institutionalization.[59] But both medical and popular explanations for differential rates of institutionalization for leprosy focused on existing settler stereotypes of black culture, rather than on socioeconomic conditions.

[59] Packard, *White Plague, Black Labour* (cit. n. 5), p. 31.

From the beginning of the nineteenth century, for example, it was popularly believed that Khoi ("half-castes") and later "Malays" (Muslims) and Africans were
particularly susceptible to leprosy.[60] The negative stereotype of the Hottentot, which
stressed idleness, unpleasant smell, promiscuity, "dirty" habits, and nomadic lifestyle, was translated into a popular aetiology of disease.[61] Once the *lepra* bacillus
had been identified as the causal agent during the 1870s, specific black groups
("mixed breeds" and the highly visible small trader, often a Muslim, selling vegetables or retailing other food items), rather than black "habits," were identified as a
source of the disease.[62] After 1883, officials, doctors, and colonists often voiced fears
about leprosy being transmitted from black farm servants to white farmers or their
children.[63] In 1896, Dr. Samuel Impey attributed the historical spread of leprosy in
South Africa almost entirely to various black groups, especially "Bushmen."[64] By
the 1890s, both medical theory and popular discourse attributed blame for the contraction and spread of leprosy to the supposedly "pathogenic" culture of black
people in general.[65]

This led to a particularly harsh approach to the management of the disease. Unlike
many other countries that passed public health regulations to control leprosy, the
Cape, after almost a decade of concern that the disease was spreading and affecting
whites, opted for compulsory segregation of all lepers under the Leprosy Repression
Act of 1891. Officials and doctors believed that only a limited number of people
would be affected, and that the disease could be stamped out using quarantine techniques like those used for smallpox or plague. But because leprosy was also perceived to be a black disease, it was particularly easy to apply procedures that in-
fringed basic civil liberties. At the time, blacks in general were losing civil rights
and being shifted out of the public sphere in the colony.[66]

The racist stereotyping of the leper as black, and therefore the selection of harsh
public health legislation, created problems for Cape institutions that had to impose

[60] On Khoi, see Minutes of Evidence, "Report of the Commission of Inquiry into the General
Infirmary and Lunatic Asylum on Robben Island," *Cape Parliamentary Papers*, G31-1862, p. 161.
Especially Muslims with low morals were suspect; see Joseph Matthews, *Incwadi Yami or Twenty
Years' Personal Experience in South Africa* (New York: Rogers and Sherwood, 1887), p. 350.

[61] Minutes of Evidence, "Report of the Commission of Inquiry" (cit. n. 60), p. 161.

[62] Minutes of Evidence, "Report of the Select Committee on the Spread of Leprosy," *Cape Parliamentary Papers*, A23–1883, pp. 11, 33; "Report of the Select Committee Inquiring into the Spread
of Leprosy," *Cape Parliamentary Papers*, G3-1889, p. x.; Minutes of Evidence, "Report of the Select
Committee . . . 1883," *Cape Parliamentary Papers*, A23-1883, p. 22; Minutes of Evidence, "Report
of the Select Committee . . . 1889," *Cape Parliamentary Papers*, G3-1889, p. 4.

[63] "Abstracts of Replies to Interrogations," "Precis of Report of the . . . Leprosy . . . Commission,"
Cape Parliamentary Papers, G4-1895, p. 89; Arnold Simons, Minutes of Evidence, "Report of the
Select Committee . . . 1889" (cit. n. 62), p. 8. After 1882, a related contraction scenario was commonly presented for syphilis in which a black nursemaid transmitted the disease first to the child and
then, through the child, to the whole family (Aberdeen and Caledon, "Reports of the Civil Commissioners, Resident Magistrates and District Surgeons for 1882," *Cape Parliamentary Papers*, G91-
1883; "Reports of the Civil Commissioners, Resident Magistrates and District Surgeons for 1883,"
Cape Parliamentary Papers, G67-1884).

[64] Samuel Impey, *Handbook on Leprosy* (London: Churchill, 1896), chap. 1.

[65] See, for example, "Leprosy Notification," *South African Medical Journal*, July 1898: 63–7, p.
63, and Hutchinson to Black, 6 Feb. 1902, Cape Archives, Cape Town, Health Branch, Letters re
Leprosy Commission of 1894–5, CO 7663. Hutchinson's views were echoed in the Indian Medical
Service for much of the nineteenth century, where these views had roots in indigenous knowledge
(Mark Harrison, pers. comm., 30 Jan. 1994).

[66] Harriet Deacon, "Leprosy and Racism at Robben Island," in *Studies in the History of Cape Town*,
ed. Elizabeth van Heyningen (Cape Town: Univ. of Cape Town Press, 1994), vol. 7, pp. 45–83.

these regulations on whites. In practice, black and white leprosy patients were treated differently within institutions, and campaigns were started to highlight the plight of white patients who were forced to stay there. A system of "home segregation" was set up to cater almost exclusively to white patients whose families could support them in a separate dwelling. This was gradually followed by medical rationales for differences between black and white leprosy sufferers. By the 1940s, a visiting leprosy specialist speculated that the reasons for a higher incidence of the more severe lepromatous leprosy in whites were (a) that Europeans had less close contact with each other and thus only the more susceptible contracted it in a severe form, and (b) that Europeans were at a climatic disadvantage in South Africa, or (c) that the disease produced in Europeans more mental depression and therefore more physical weakness.[67] The colonial stereotype of the leper had attained a "scientific" gloss, in which racial or cultural differences were linked to clinical pictures.

CONCLUSION

The introduction to this paper pointed out the contradictions inherent in arguing (a) that Western medicine was an agent of colonialism and therefore evil in its effects, and also (b) that black people were disadvantaged because they did not have equal access to Western medicine. It suggested that we seek a more nuanced approach to understanding the relationship between racism, medicine, and empire.

To some extent, both (a) and (b) are true. Given the social profile of the medical profession at the Cape and the close relationship it enjoyed with the colonial state and the settler community, Western medicine inevitably shared many of the aims and assumptions of colonialism. Doctors acted as agents of the colonial state when they represented its authority in courtrooms or enforced its legislation in their practices. They rejected integration with indigenous healing systems, pathologized indigenous cultures, and deferred to European professional standards. Racism in medicine grew out of racism in colonial society. These attitudes were not only reproduced at the individual level, but they played a role in the construction of the colonial medical profession and the definition of the ideal "medical gentleman." Black people had restricted access to Western medicine because it was expensive and because its curative energies and expenditures were mainly targeted at white settlers. Although hospitals admitted large numbers of black patients, they were still underrepresented relative to the general population. Racial discrimination within hospitals meant that black patients received less and lower-quality medical care than did white patients. Racist medical theories helped to entrench and justify such practices.

To describe Western medicine as just an agent of colonialism, however, oversimplifies its form and impact, assumes that colonized peoples were no more than victims, and underestimates differences among practitioners and patients. The colonial state apparatus (and its medical wing) was neither homogenous nor omniscient, and it could not fully predict or control the consequences of Western medical interventions.[68] Western hospitals were used to remove the homeless, insane, or contagious

[67] Ernest Muir, "Report on Leprosy in the Union of South Africa," *Leprosy Review*, 1940, *11*:43–52, p. 43.
[68] Engels and Marks, *Contesting Colonial Hegemony* (cit. n. 1).

from colonial cities and settler farms. Some patients, many of them black, were forced to stay against their will in hospitals such as the leper institution on Robben Island. But other patients, both black and white—and relatives who wished to be rid of them—used illness as an opportunity to obtain free food and lodging. While colonial doctors often discriminated against black patients who were not paying for their own treatment, private practice serving black clients may have encouraged a more equitable approach. The fact that some black patients sought assistance from Western medicine in certain areas (such as surgery or placental delivery) but not others suggests that they were able to shop for what was perceived to be the most effective medical care without necessarily buying into the overall philosophy of any one system.

We cannot assume without comparative evidence that the primary discrimination experienced by black patients, even in the racist environment of the colonial hospitals, had to do with race rather than with gender or type of disease. If some of what Western medicine could offer was useful to black patients, and it probably was, this was not necessarily an unintended consequence of colonialism, nor a kind of false consciousness. Doctors were supposed to cure people, and colonial subjects as well as doctors recognized this as a key element of professional identity. In comparing how doctors treated black and white patients we can see, however, clear evidence of discrimination, the effects of which need to be documented and assessed. We can also note the discrepancy in the proportions of white and black people treated by Western medicine, but it would be foolish to ascribe this solely to restricted access. Some doctors actively sought black clients but were unable to persuade black communities of their worthiness.

Western medicine in the colonies was certainly racist in its professional makeup and in the construction of its theory and practice. Yet this generalization tells us little that is interesting about racism or, indeed, about Western medicine. This essay suggests that, in seeking to understand the trajectory of racism in colonial medicine, we should distinguish between rich and poor black patients and between hospital and private practice. We should contextualize racist practices within the framework of different professional formations, types of disease, stereotypes of disease, and medical theories. We cannot understand the form, timing, intention, and function of racism in colonial medicine without considering these factors together.

The Colonial World as Mission and Mandate: Leprosy and Empire, 1900–1940

Michael Worboys[*]

ABSTRACT

The history of medicine in twentieth-century empires has been dominated by studies of "imperial tropical medicine" (ITM) and its consequences. Historians have
been fascinated by the work of medical scientists and doctors in the age of high
imperialism, and there are many studies of medicine as a "tool of empire." This
paper reviews work that explores colonial medicine as a broader enterprise than
ITM in three spheres: missionary activity, modernization, and protection of the
health and welfare of indigenous peoples. To illustrate the themes of mission and
mandate, it discusses the development of policies to control leprosy in the tropical
African and Asian colonies of Britain in the first half of this century, especially
the work of the British Empire Leprosy Relief Association (BELRA). Although
BELRA's efforts did little to change imperial medical and health agendas, they
had an important impact locally and ideologically, and show how closely interwoven the themes of Christian caring, medical humanism, colonial development,
and welfare policy had become by the outbreak of the Second World War.

INTRODUCTION

THE HISTORY OF COLONIAL MEDICINE IN THE TWENTIETH century
has predominantly focused on what John Farley has termed "imperial tropical
medicine" (ITM) and its consequences. Practitioners of ITM concentrated on the
control of a group of vector-borne parasitic diseases, principally malaria, sleeping
sickness, smallpox, yellow fever, bilharzia, and hookworm. Their main approach
was through so-called vertical programs, which targeted specific diseases one by
one, attacking pathogens or their vectors. Farley characterizes ITM as a triad of
"definition, imposition and non-involvement," where problems and solutions were
"defined and imposed by practitioners of Western-style medicine without involving
the indigenous populations."[1] Such programs began in the late nineteenth century
and were first promoted by scientists and colonialists as having the potential to stop
tropical colonies from being the "white man's grave," and hence to consolidate impe-

---

[*] History, Sheffield Hallam University, Collegiate Campus, Sheffield, S10 2BP, United Kingdom.
[1] John Farley, *Bilharzia: A History of Imperial Tropical Medicine* (Cambridge: Cambridge Univ.
Press, 1991), p. 293.

rial rule and promote trade. There is no doubt that ITM was one of the most powerful tools of empire, and it helped shape the style and aims of colonial rule.

In the last ten years, however, historians have begun to investigate medicine in the colonies as an enterprise broader than ITM. Looking at the period 1900–1950, scholars have extended work in three directions: first, by considering the place of medicine in the spread of Western Christianity through the efforts of medical missionaries; second, in exploring the deployment of medical ideas, practices, and institutions as forces in the modernization of colonial societies and indigenous cultures; and third, in recognizing the work that state and voluntary institutions undertook, albeit unevenly, to protect the health and welfare of indigenous peoples.[2] Thus, the picture of medicine in the colonies that is now emerging shows its development as both mission and mandate: the missions of spreading Christianity and introducing modern ("scientific") rationality, and the mandate of consolidating colonial rule and promoting material development, while at the same time protecting the welfare and health of indigenes. On each front historians have also begun to explore the responses of indigenous populations to colonial medicine; hence the growth of interest in the interactions between Western and indigenous medical systems.

This essay discusses the attempts to devise and implement policies to control leprosy in Britain's tropical African and Asian colonies from 1900 to 1940, especially the work of the British Empire Leprosy Relief Association (BELRA).[3] Leprosy has received relatively little attention from historians of colonial medicine, who have tended to follow the disease agenda of ITM and focus on the control of tropical-parasitic diseases that most affected European health, or on the major campaigns against diseases that were politically, economically, or socially important. Yet, in the first half of the century, leprosy became an archetypal tropical disease, associated with backwardness and "dark races," and seen as being especially communicable in hot climes and requiring Western know-how to combat. The history of leprosy control illuminates the complex relations among the individuals and agencies that promoted various health projects, including the changing tensions and alliances between colonies and across empires. However, this essay does not discuss subject peoples' experiences of leprosy control measures. Instead, it focuses on shifts in perceptions and policies.

MISSIONS

The first dedicated provision of modern Western medicine to the indigenous peoples in European colonies was by Christian missionaries.[4] How many people were treated by medical missionaries, and with what impact, is the subject of debate. Doubts about the impact of colonial medical practice reflect a growing trend in colonial history to question the extent of European influence and to emphasize the continuing

[2] Marcos Cueto, ed., *Missionaries of Science: The Rockefeller Foundation in Latin America* (Bloomington: Indiana Univ. Press, 1994), p. xvii.

[3] Zachary Gussow, *Leprosy, Racism and Public Health: Social Policy in Chronic Disease Control* (Boulder, Colo.: Westview Press, 1989).

[4] Michael Worboys, "The Spread of Western Medicine," in *Western Medicine: An Illustrated History*, ed. Irvine I. Loudon (Oxford: Oxford Univ. Press, 1997), pp. 249–63; Charles M. Good, "Pioneer Medical Missions in Colonial Africa," *Social Science Medicine*, 1991, *32*:1–10.

relative autonomy of many features of indigenous cultures.[5] In the nineteenth century, medical work was a small part of the larger missionary project. For example, between 1838 and 1914 the London Missionary Society sent 987 missionaries abroad, of whom only eighty-two were doctors (forty-eight to China, twenty to India, six to Madagascar, and eight to central Africa). Protestant missions embraced medicine earlier and more wholeheartedly than did Catholic ones, reflecting their instrumentalist rather than fundamentalist stance to conversions.[6]

Nineteenth-century medical missionaries were, according to A. F. Walls, the "heavy artillery" of missionary work, being mobilized with the aim of using the technical superiority of Western medical practice to bolster evangelism.[7] This latter point is disputable, as before the twentieth century Western medicine was not conspicuously more effective than other systems, especially outside of Europe. Thus, many mid-nineteenth-century travelers, for example David Livingstone, often preferred indigenous healers and their remedies, believing they were more suitable for local diseases. However, in the late nineteenth century the technical gap between Western medicine and the healing systems of other cultures was seen to widen with the improvement of surgery, the development of sera and vaccines, and the understanding of etiologies, giving new impetus to state and missionary medical ventures.

In 1893, the Church Missionary Society founded the Livingstone Medical College in London.[8] This school provided training in self-preservation, first aid, new medical techniques, and nursing to new lay missionaries and to old hands at home on leave. From 1900, medical missionaries were well represented in the postgraduate courses at the new schools of tropical medicine, learning the latest ideas and hygienic recommendations. The aims of medical missionary work at this time and later were to continue the work of Christ the healer, to protect the health of missionaries themselves, to provide an opening into alien cultures to facilitate conversions, and to represent the superiority of Western civilization. There was much debate in missionary circles about the conflict between saving bodies and saving souls, and it was usually incumbent upon medical missionaries to build a chapel before a hospital.[9] After 1900, the focus of medical missionary work began to swing from Asia to Africa though, as we shall see with leprosy campaigns, the missionary presence in India remained influential.

The opening that modern medicine promised to provide into indigenous cultures was paradoxical. The ideal was for indigenous peoples to try Western medicine, be impressed by its curative powers and superiority over native healing systems, and then to try the religion that had brought the means of their cure. Indeed, cured patients were possibly more effective agents of Christianity than missionaries, as R. G. Cochrane of BELRA explained in 1927:

[5] Gwyn Prins, "But What Was the Disease: The Present State of Health and Healing in African Studies," *Past and Present*, 1989, *124*:159–79.

[6] Christopher P. Williams, "Healing and Evangelism: The Place of Medicine in Later Victorian Protestant Missionary Thinking," in *The Church and Healing*, ed. William J. Shiels (London: Blackwell, 1982), pp. 271–87.

[7] A. F. Walls, "'The Heavy Artillery of the Missionary Army': The Domestic Importance of the Nineteenth Century Medical Missionary," in Shiels, *Church and Healing, ibid.*, pp. 287–98.

[8] Stanley G. Browne, Frank Davey, and William A. R. Thornton, eds., *Heralds of Health: The Saga of Christian Medical Initiative* (London: Christian Medical Fellowship, 1985).

[9] Michael Gelfand, *Godly Medicine in Zimbabwe* (Harare: Memba Press, 1988).

The evangelisation of the world depends on the indigenous people of each nation, and what better missionary could be found than the leper cured of his disease, going back to his people telling them out of the fullness of his heart, the twofold gospel of spiritual and physical healing.[10]

In Africa, the feature of Western medicine that most impressed indigenes was surgery, especially operations for the removal of tumors and cataracts. The apparent ability to make the blind see must have seemed magical, though missionaries likened it to one of Jesus's miracles. In both Africa and Asia, the indigenous population was also willing to try Western drugs, which were plant- and mineral-based like local remedies. It is interesting that few historians comment on the ways that Christian missionary medicine combined those supposed old foes: religion and science. Perhaps this is because an inevitable "warfare" between these parties is no longer assumed. However, Megan Vaughan has argued that there were important contrasts in the approaches of missionary and state medical personnel.[11] Missionaries concentrated on individual patients and their care, while civil and military doctors focused on disease agents, vectors, and populations. By this time both missionary and state medical personnel were dismissive of the health beliefs and practices of indigenous peoples.[12] When they thought about them at all, it was largely as obstacles to the introduction and adoption of Western beliefs and practices. Indeed, missionaries lived with the contradiction of criticizing "primitives" for mixing religion and healing—a combination demonized in the term "witch doctor"—while the same association was the very rationale for their work.

Of course, most medical missionaries saw Christianity as a "rational" religion whose "truths" were tried and tested, as opposed to the superstitious beliefs of "primitive" cultures. The inextricable linkage between body, mind, and spirit that indigenes expected in all medical systems, including the Western, meant that few practitioners of Western medicine expected its takeover to be rapid or easy. No doubt influenced by debates in the history of science about the incommensurability of knowledge systems, and perhaps by the growing cultural import of alternative medicine in the West, historians have recently structured their discussions of the transfer of Western medicine to colonial settings around the notion of "contested knowledge."[13] Thus, the reductionism and objectivity of Western medicine has been seen to be fundamentally different and at odds with the holism and subjectivity of traditional healing systems. However, this is too simple a contrast. Neither system was monolithic and there were major changes over time. Any "contests" were quite uneven, as the practitioners of Western medicine had, or were aligned with, the major sources of power in colonial societies. Historians have also written in terms of the "resistance" and "responses" of indigenous people, assuming they were reactive and defensive. However, the case can be made that indigenous healers and patients were pro-active, the adoption of Western medicine was usually selective and synthetic,

[10] Robert G. Cochrane, *Leprosy in India: A Survey* (London: World Dominion Press, 1927), p. 22.

[11] Megan Vaughan, *Curing Their Ills: Colonial Power and African Illness* (Oxford: Polity Press, 1991), p. 60.

[12] On earlier, less hostile attitudes, see Mark Harrison, *Climates and Constitutions: Health, Race and British Imperialism in India, 1600–1850* (New Delhi: Oxford Univ. Press, 1999), pp. 25–110.

[13] Andrew Cunningham and Bridie Andrews, eds., *Western Medicine as Contested Knowledge: Studies in Imperialism* (Manchester: Manchester Univ. Press, 1997).

and many groups and activities remained relatively unaffected by it, especially outside the major towns and commercial ventures.

MANDATES

From the late nineteenth century, doctors and scientists promised to help governments control tropical colonies and their peoples. Tropical medicine promised to explain and normalize the mysteries of exotic diseases, assimilating them to the dominant etiological and pathological models of infectious diseases, albeit in a separate specialism.[14] The most famous attempt to reorder and assimilate the tropical world to Western medicine was the chain of Pasteur Institutes that spanned France, the French colonial empire, and eventually countries beyond.[15] Founded with a mixture of voluntary, private, and state support in different locations, the institutes were established to diffuse antirabies measures and other vaccinations. Over time, staff took up work on other diseases and any opportunities available to advance the "Pasteurisation of the world."[16] Yet the traffic in ideas and techniques was never one-way. The Pasteur Institute system became polycentric, and in the interwar years some of the overseas institutes were more dynamic and successful in research than were those in France. For example, in Tunis, Charles Nicolle completed the research on typhus fever that won him the Nobel Prize in 1928 and used this to obtain a senior post back in France.[17] At the same time, other overseas Pasteur Institutes reduced their research ambitions and established an important local role, providing laboratory services for state medical departments. Thus the Pasteur Institutes, as well as showing the problems with notions of centers and peripheries, offer further evidence of the blurred boundaries between state and voluntary agencies, pure and applied research, and research and consultancy.

The growth in medical services, the development of research, and the perceived successes of specific disease-control programs led both colonial agencies and medical personnel to reflect positively on their contribution to the "civilizing mission." Pasteur Vaillery-Radot, Louis Pasteur's grandson, explained in 1938 the value of medicine to a vision of French colonialism:

> If French peace reigns over the boundless regions, if epidemics are prevented or thwarted, if sanitary reforms can be undertaken, cities built up, and harbours opened to trade, if Europeans can live safely in hostile Africa and the Far East, if morbidity and mortality decrease in a striking way in native populations, all the transformations must be attributed to colonial medicine.[18]

The role of colonial medicine in promoting the development of colonies went through three phases. Before 1914, priority was given to protecting the health of

[14] Michael Worboys, "Tropical Diseases," in *Companion Encyclopedia of the History of Medicine*, eds. W. F. Bynum and Roy Porter (London: Routledge, 1993), pp. 521–36.

[15] Anne-Marie Moulin, "Patriarchal Science: The Network of the Overseas Pasteur Institutes," in *Science and Empires: Historical Studies about Scientific Development and European Expansion*, eds. Patrick Petitjean, Catherine Jami, and Anne-Marie Moulin (Dordrecht: Kluwer, 1991), pp. 307–22.

[16] Bruno Latour, *The Pasteurization of France* (Cambridge, Mass.: Harvard Univ. Press, 1989).

[17] Kim Pelis, "Prophet for Profit in French North Africa: Charles Nicolle and the Pasteur Institute of Tunis, 1903–36," *Bulletin of the History of Medicine*, 1997, *71:*583–622.

[18] Quoted in Anne-Marie Moulin, "The Pasteur Institutes Between the Two World Wars: The Transformation of the International Sanitary Order," in *International Health Organisations and Movements, 1918–1939*, ed. Paul Weindling (Cambridge: Cambridge Univ. Press, 1995), p. 251.

European functionaries, soldiers, traders, and managers to support the political and
economic health of each colony. It was in this context that disease-specific approaches were introduced and ITM was established. After 1920, by which time mor-
bidity and mortality rates among Europeans had improved, the health of the indigenes became an issue in situations where they worked in mines and plantations,
and were potential consumers in towns. In the third phase, from the late 1920s—in
the context of the collapse of world trade and the rise of colonial nationalism—
the development and health policy agenda again changed. The welfare of indigenes
became a focus of debate, if little action.[19]

The issue of the indigenous population's health was shaped by the ideas of "trust-
eeship" and the "dual mandate" that had emerged in Britain in the 1920s.[20] This
mandate required colonial governments to try and preserve indigenous cultures,
while at the same time promoting their economic and cultural development. The
new policy in British dependent colonies was advanced by people from varying
backgrounds and with different interests. One important factor was the experience
of colonial troops in Europe during the First World War. Soldiers from Asia and
Africa experienced high rates of sickness and death from tuberculosis and other
infections that were common in Europe. While the effects of wartime conditions
were not discounted, the dominant explanation was that colonial soldiers were "vir-
gin soil" populations that had not been exposed to, and developed specific immunities to, the main Western infections. They were also were seen to lack the constitutional strength to acquire such immunities, which was often viewed as a racial
weakness that would take many generations to be "selected" out.[21]

One conclusion drawn from this experience was that certain types of colonial
development, especially urbanization and improved communication, would pose a
threat to the well-being of colonial subjects if their biological adaptation to new
social and disease ecologies lagged. In the case of tuberculosis, Lyle Cummins drew
an influential, though controversial, analogy between individual European children
and African communities, suggesting that just as most children acquire immunity
by gradual exposure to infection over their lifetimes, so Africans would only acquire
effective immunities over many generations.[22] Such ideas also resonated with the
view among psychologists that Africans had "childlike" minds, and anthropologists'
assumptions that African tribal societies were examples of human societies in their
original, primitive forms. Cummins, for example, adopted a thoroughgoing social
evolutionism, seeing African societies at the level of medieval Europe and modern
Asian societies as equivalent to Europe in the seventeenth and eighteenth centuries.
Cummins's views were challenged by another leading tuberculosis expert, Louis
Cobbett, who wanted to make a clear distinction between the acclimatization of an
individual and the racial immunity that was "deeply fixed in the blood."[23]

[19] Michael Worboys, "The Discovery of Colonial Malnutrition," in David Arnold, *Imperial Medicine and Indigenous Societies* (Manchester: Manchester Univ. Press, 1988), pp. 206–25.

[20] Michael Havinden and David Meredith, *Colonialism and Development: Britain and its Tropical
Colonies, 1850–1960* (London: Routledge, 1993), p. 312 *et seq.*

[21] Michael Worboys, "Tuberculosis and Race in Britain and its Empire, 1900–1950," in *Race, Science and Medicine, 1700–1960*, eds. W. Ernst and B. Harris (London: Routledge, 1999), pp. 144–66.

[22] S. Lyle Cummins, "Tuberculosis in Primitive Tribes and Its Bearing on the Tuberculosis of Civilised Communities," *International Journal of Public Health*, 1920, *1*:158.

[23] Louis Cobbett, "The Resistance of Civilised Man to Tuberculosis: Is it Racial or Individual in
Origin?" *Tubercle*, 1925, *6*:590.

However, the short-term policy implications of both analyses were similar: socio-economic development had to be slow, carefully managed, and not seek to emulate Western models. That said, the indigenous populations of Africa and Asia could not be left to subsistence agriculture and primitive beliefs. Their backward economic and cultural life had to be reformed and set on a more rational, ordered, and efficient course. Such medical ideas supported the dominant interwar economic policies for empire, namely, that colonial territories should remain primary producers of agricultural commodities and other natural products, while Britain supplied them with capital equipment, industrial goods, and services.

The contradictions of the dual mandate were often very evident to state, missionary, and even private medical personnel. For example, medical officers in South African mines were supporting economic modernization, while at the same time dealing with its consequences in the high incidence of tuberculosis among miners.[24] State medical officers, when faced with the problems of urban health, blamed both colonizers and colonized: the former had permitted uncontrolled growth without adequate infrastructure, while the latter compounded the situation due to their "ignorant" and "irrational" beliefs. The tensions within the dual mandate increased successively in each decade after 1920, though for a decade after 1945 it seemed that the circle could be squared with insecticides, antibiotics, vaccines, and the building of hospitals and clinics.[25]

LEPROSY

The nature of leprosy was contested in European medicine from the 1850s. Its historical and biblical associations with disfigurement, dirt, and the isolation of sufferers all testified to its contagiousness. However, in the 1860s, an enquiry by the Royal College of Physicians of London, prompted by West Indian officials, decided that the disease was hereditary, noncontagious, and did not require segregation. This conclusion was questioned, all the more so when Edward Hansen, working in the 1870s in Norway where leprosy was still rife, claimed a germ etiology for the disease. The historic decline of the condition in Europe was still explained in social evolutionary terms by the extinction of weak individuals and the rise of civilization. Such speculations were congruent with contemporary experience in which the disease was present among "primitive" races living in conditions of "dirt" and "filth" who were "ignorant" about the true nature of the disease. The relative economic backwardness of Norway could just about be made to fit this model, though it was suspected that other factors, notably a fish diet, predisposed Scandinavians to the disease. Anxieties about European vulnerability to infection in the tropics, along with the growing acceptance of Hansen's bacillus, ensured that contagionist ideas remained influential. They were endorsed by the Royal Commission on Leprosy in India in 1891, although the commission decided against strict segregation because the infectivity of the disease was relatively weak.[26] Already in India, leprosy had

[24] Randall Packard, *White Plague, Black Labour: Tuberculosis and the Political Economy of Health and Disease in South Africa* (Pietermaritzberg: Univ. of Natal Press, 1989).

[25] Jock McCulloch, *Colonial Psychiatry and "The African Mind"* (Cambridge: Cambridge Univ. Press, 1995).

[26] Sudhir Kakar, "Leprosy in British India, 1860–1940: Colonial Politics and Missionary Medicine," *Medical History*, 1996, *40*:215–30.

become a focus of medical missionary activity. Its biblical associations gave it particular significance for Christians, while the perceived risks of caring for lepers offered opportunities to show dedication to the cause, if not martyrdom. Indeed, lepers
had their own mission societies, the most influential of which was the Mission to
Lepers in India, founded in 1874.

In the 1910s, doctors in India, the Philippines, and French Indo-China began to
call for new approaches to leprosy from missionary societies and the state. This was
prompted by several changes in ideas and policy. The doctors were uneasy about the
repeal of segregation measures in the previous decade, which had fed wider fears
that the incidence of the disease would increase. Also, the biological and etiological
similarities between the leprosy and the tubercle bacilli led doctors to think that
European antituberculosis measures might be a suitable model for leprosy control.[27]
Doctors also drew parallels between leprosy and the emerging view of tuberculosis
as a disease of civilization, associated with urbanization, improved communication,
and economic development. Tuberculosis, moreover, was seen as a disease of a particular stage of civilization, a rite of passage to the modern urban, industrial, and
developed world. Edward Muir set out this view in his textbook on leprosy in 1921,
claiming its rising incidence in Asia and Africa was due to the introduction specifically of "clothes, trains, [and] brickhouses."[28] He stated that the disease "is
not common in the savage nor is it common in those who are highly civilised. It is
found chiefly in the savage who is being civilised and is part of the price that has to
be paid for civilisation." His assumption was that the indigenes were as yet biologically unadapted to wearing clothes and thus vulnerable to irritation of the skin, and
culturally unadapted as they were ignorant of hygienic regimes and lacked the
skills and resources to institute modern sanitation. Trains and other new forms of
transportation had facilitated migration to and from towns, while brickhouses had
produced overcrowded, ill-ventilated, and insanitary dwellings favorable to infection. The solution to these problems was not, of course, to return to nakedness,
settled village communities, and straw huts, but to move forward through education
to new hygienic standards that would strengthen bodies and reduce opportunities
for infection.

This was the long-term solution. In the 1910s, doctors in the Philippines and India,
led respectively by V. G. Heiser and Leonard Rogers, claimed to have found a technical fix that offered a short-term answer. They promoted a treatment that was said to
arrest the progress of the disease, if not cure it, maintaining that Africans and Asians
did not have to pay the same price for civilization as Europeans. An additional boon
was that leper hospitals and colonies might become centers of treatment and cure,
making them more attractive to patients and to potential government, Christian, and
private sponsors. Indeed, it was argued that leper hospitals, if made part of larger
public health programs, would contribute to the eventual eradication of the disease
from communities, if not from the world.

These new ideas on treatment and control centered on the use of derivatives of

---

[27] Mark Harrison and Michael Worboys, "'A Disease of Civilisation': Tuberculosis in Africa and
India, 1900–1950," in *Migrants, Minorities and Health: Historical and Contemporary Studies*, eds.
L. Marks and Michael Worboys (London: Routledge, 1997), pp. 93–124.

[28] Edward Muir, *Handbook on Leprosy: Its Diagnosis, Treatment and Prevention* (Calcutta: R. J.
Grundy, 1921), p. 1.

chaulmoogra oil.[29] This substance was an indigenous remedy that had become part of the European pharmacoepia, being already favored as a treatment for tuberculosis. Heiser and Rogers converted the oil into a modern specific remedy by isolating its active principle and using chemical derivatives. Their work shows the strength of research in the colonies, while the exchange of materials and ideas between the two men and other experts demonstrates the vitality of polycentric colonial research networks. These networks crossed political and geographical boundaries, and showed no evidence of any "tyranny of distance."

At the end of the First World War, Heiser held the grand title of director of the East in the Rockefeller Foundation's International Health Board (IHB) in New York. Rogers, then head of the Calcutta Medical School, lobbied Heiser to have the IHB take up leprosy and develop a program for its eradication.[30] This did not happen. Among many reasons were questions about the efficacy of chaulmoogra oil, the adequacy and cost of supplies, and complex local variations in the politics of hospitalization and segregation. Unable to interest either colonial governments or the Rockefeller Foundation, Rogers, on his retirement in 1921, began his own campaign. In 1923, he orchestrated the establishment of the British Empire Leprosy Relief Association—a charitable organization whose goal was to collect funds to enable and coordinate new approaches to the control of leprosy.[31]

BELRA's launch in 1923 was a financial failure. The event coincided with difficult economic and political conditions, and Rogers, who was not a charismatic figure, found it very difficult to excite metropolitan or imperial interest. Unwilling to give up, the organization set out to establish local branches across Britain and the empire, spreading the word about the disease and its control. BELRA became most effective ideologically, spreading the idea of the poor leper and the heroic leprosy caregiver. Only in India were sufficient funds collected for BELRA to institute new measures, which further benefited from close cooperation with the Mission for Lepers.[32] Indeed, this set the trend for other colonies where BELRA, rather than linking with local, state, or international medical agencies as hoped, developed close links with medical missionary organizations and other voluntary bodies.[33]

The construction of leprosy, as created and disseminated by Rogers and BELRA, was complex, as were plans for its prevention and control. Rogers had to put across the difficult point that the disease was contagious, but not highly so. There were two types of the disease, affecting either the "skin" or the "nerves." Only active skin cases, some 8 percent of the total, were highly contagious and caused major disfigurement. Those sufferers with nerve cases had an essentially internal disease and only posed a small threat to others. In the 1920s, Rogers favored the notion that the disease was a "house infection," caught only by long and intimate exposure through cohabitation over many years. The more intimate the relations between people—

---

[29] Leonard Rogers, "Recent Advances in the Treatment of Leprosy and its Bearing on Prophylaxis," *Practitioner*, 1928, *120*:209–19; V. G. Heiser, "Recent Progress in the Control of Leprosy," *Proceedings of the American Philosophical Society*, 1932, *71*:167–71.

[30] Correspondence between Leonard Rogers and V. G. Heiser, 1918–1922, Contemporary Medical Archives Centre, Wellcome Institute for the History of Medicine, London, ROG/C13/13–34.

[31] Leonard Rogers, *The Foundation of the British Empire Leprosy Relief Association (BELRA) and its First 21 Years of Work* (Watford, U.K.: Voss and Mitchell, 1945).

[32] J. Lowe, "Leprosy in India: The Present Outlook," *Indian Medical Gazette*, 1932, 67:208–10.

[33] Rogers, *Foundation of BELRA* (cit. n. 31), pp. 15–20.

particularly conjugal relations or children sharing beds, especially in conditions of poor sanitation and heat—the greater the risk of infection, especially from the more common nerve cases. Whether the term "house infection" was calculated to worry Europeans and indigenous elites into supporting BELRA is unclear, but it would certainly have concerned homes with "house boys" and other servants.

BELRA sought to advance several control policies, but all were politically sensitive, so every effort was made to present them as mere technical measures. First, it wanted all infective cases to be removed to settlements or farm colonies so the disease could be arrested and sufferers could learn hygienic living. Existing asylums were criticized for being "prison-like." BELRA officials hoped to make the new leper colonies open, humane, improving, and attractive. Rogers believed, naively, that the offer of treatment and perhaps a cure would attract voluntary patients and remove the stigma of the disease. He reasoned, rightly, that the threat of "prison" led to cases being hidden, leading to further infection and patients only coming to the attention of doctors at an advanced stage. Second, BELRA wanted to remove healthy, as yet uninfected children from parents who were diagnosed as sufferers or contacts. It was intended that such children be bought up by healthy relatives or in special homes. They would still be allowed to see their parents; the important thing was that children should not live with infected parents. Third, BELRA's staff and institutions were to become a vehicle for the development and dissemination of chaulmoogra oil therapy, which by the mid-1930s was being promoted in Britain as "magic oil." Rogers's "great discovery" for healing the sores of lepers was even compared to Wilberforce's great crusade that led to the end of slavery in the British Empire.[34]

Through the 1920s and 1930s, the most visible of BELRA's activities, both in Britain and overseas, were the visits by its roving ambassador-investigators. Three men wrote most of the reports: Robert Muir, who worked from Rogers's old base in Calcutta; F. G. Oldrieve, and R. G. Cochrane. Between 1923 and 1939 almost every African, Asian, and West Indian colony was visited. The BELRA reports catalogued local control measures and the success its agents had in persuading local officials, doctors, missionaries, and others to raise funds and adopt BELRA policies. The success of this work depended on the extent to which a local constituency for leprosy work already existed or could be created; hence, there were wide variations between colonies. For example, in Ghana (then the Gold Coast), Oldrieve's visit in 1926 stimulated the foundation of a local branch of BELRA.[35] This event prompted the medical officer in Ho, a town in the west of the colony, to start a leper settlement and the colony's medical service to appoint a leprologist, funded by BELRA. The leprologist concentrated on surveys, while small settlements were founded by missionaries and local doctors with some support from BELRA. It took a visit by Muir in 1936 to kick-start planning for institution building on a large scale, but this fell foul of the economic depression and then the Second World War. What happened in Ghana was repeated elsewhere. BELRA's overall attempt to push leprosy up the agenda of colonial public health failed, and its local achievements remained modest.

---

[34] Report of a talk by Rev. Tubby Clayton, founder of TocH, BBC, Easter 1934, Contemporary Medical Archives Centre, Wellcome Institute for the History of Medicine, London, ROG/C13/336. TocH was a society that aimed to maintain wartime comradeship and the ideals of service.

[35] Stephen Addae, *The Evolution of Modern Medicine in a Developing Country: Ghana, 1880–1960* (Durham: Durham Univ. Press, 1996), pp. 396–7.

Leonard Rogers continued to promote various treatments of the disease, arguing optimistically that leprosy was eradicable if cases were treated early and the infection of children was halted. However, the reality was that by the mid-1930s, only 1 percent of sufferers across the empire received institutional care. Also, many doctors were beginning to doubt the efficacy of chemical treatments and called for other methods of control. Indeed, when Rogers ceased to be active in BELRA in the mid-1930s, its policies changed to emphasize isolation, but only in settlements that inculcated a "leprosy-consciousness" in sufferers and that protected children in the community.[36] In selling this idea to metropolitan audiences, BELRA's leaders stressed that the modern leper colony was no longer an almshouse or lazaret, but rather it was analogous to an agricultural or industrial settlement: it was educative, self-supporting, and a model community for the outside world—an ideal first stage on the path to civilization. The markers of this were to be the abandonment of irrational views on the nature of leprosy, the adoption of hygienic behavior, plus the end of "promiscuity, general and sexual, and overcrowding." What this signaled was the decline of medical influence within BELRA and its takeover by philanthropic and Christian agencies and their ideals. BELRA had struggled on many fronts for a number of years. It was a one-issue charity, launched in a period of economic crisis, with a noncharismatic, retired, colonial doctor at the helm and an undeveloped infrastructure for collecting and distributing resources. BELRA's position had not been helped by the demise of chalmoogra oil and the absence of any new methods of prevention and treatment. Leprosy had become a much more intractable problem, unamenable to quick medical fixes. Moreover, while European workers in leper colonies did not catch the disease in large numbers, the infection of high-profile individuals, from Father Damien in Hawaii onwards, pointed to the fact that "civilized peoples" had not developed physiological immunity. All this left experts and charity workers to fall back on was the historical experience that leprosy diminished with the rise of civilization. In other words, the incidence of leprosy was determined by cultural conditions, what Robert Muir in 1938 called "mental and social immunity."[37]

CONCLUSION

This essay has argued that the history of colonial medicine in the twentieth century cannot be written solely in terms of the growing power and influence of disease-control-centred medical science and technology, as represented by imperial tropical medicine. The diverse medical efforts mounted by various agencies and individuals had many aims, methods, and outcomes. Medical missionary work continued to be important locally, with specific diseases, and ideologically. Colonial medical services certainly concentrated on public health measures, albeit with changing priorities, though we should not overlook the steady spread of Western clinical medicine through colonial hospitals and private practice, and the expectation that its individualist, interventionist style would be an effective vehicle for transforming the thought, behavior, and morality of indigenous peoples.

Although efforts to control leprosy in the British Empire had achieved very little institutionally or practically by 1939, the work of BELRA nicely illustrates the im-

---

[36] *British Medical Journal*, 1937, *i*:164.
[37] Edward Muir, "The Epidemiology and Control of Leprosy," *Brit. Med. J.*, 1938, *i*:36.

portance of non-ITM medical endeavors. BELRA certainly raised the profile of leprosy as a tropical disease linked to race and backwardness, and pushed the idea
that individual sufferers deserved compassion and care. They had less success in
promoting the view that the rising incidence of leprosy was caused by colonial "development," nor that it was a public health problem. BELRA's ideology shows how
close medical humanism and Christian caring could become, both in the initial therapeutic aims of the new leper settlements and in the late-1930s aspiration that they
become model moral communities. Indeed, the leper colonies were very similar in
aim and character to the settlements planned for tuberculosis sufferers in Britain.[38]
The work of BELRA also shows the importance of the voluntary sector in colonial
medicine, both in its strengths and weaknesses. Imperial initiatives in medicine, in
the sense of empire-wide programs, were not pursued by the British government or
its agencies to any great extent in the years after 1918. These were left to local,
regional, and special-interest groups. Thus, BELRA was important in keeping the
flag of imperial health flying and, through its fundraising activities, in continuing to
represent the dangers of tropical diseases to the British public, as well as in presenting the opportunities for medical service in terms of both mission and mandate.

[38] Linda Bryder, "Papworth Village Settlement: A Unique Experiment in the Treatment and Care
of the Tuberculous?" *Med. Hist.*, 1984, *28*:372–90.

Part IV: Colonial Science and the New World System

Locality in the History of Science:
Colonial Science, Technoscience,
and Indigenous Knowledge

David Wade Chambers * *and Richard Gillespie* **

INTRODUCTION

D URING THE SECOND HALF OF THE TWENTIETH CENTURY, THE
"colonial world" became a prominent research focus for historians of science. In the process of establishing this new subdivision of knowledge, colonial
science historians took pains to clarify their use of the term "colonial," an exercise
that helped refine the terminology of the larger colonial and postcolonial discourse.[1]
But these discussions were more concerned with the meaning of "colonial" than
with the meaning of "science," consideration of which was generally left to philosophers and sociologists of knowledge. And during this same period, philosophers,
sociologists, and a few historians (variously arrayed as positivists, realists, and con-
structivists) were indeed contending over the nature of science. It may now be seen
that constructivist approaches,[2] because they emphasize the *locally* contingent char-

* Science and Technology Studies, Deakin University, Deakin, Victoria, Australia.
** Museum Victoria, G.P.O. Box 666E, Melbourne 3001, Australia.

[1] The leading scholar in this enterprise is Roy MacLeod, who has published important theoretical
pieces providing a refreshing breadth of perspective. See, for example, Roy MacLeod, "On Science
and Colonialism," *Science and Society in Ireland: The Social Context of Science and Technology in
Ireland, 1800–1950* (Belfast: Queen's University, 1997) pp. 1–17; "Reading the Discourse of Colonial
Science," in *Les Sciences coloniales: Figures et institutions* (Paris: Editions de l'Office de la Recherche Scientifique et Technique d'Outre-Mer, 1996) pp. 87–96; and "On Visiting the 'Moving Metropolis': Reflections on the Architecture of Imperial Science," *Historical Records of Australian Science*,
1982, *5*, 3:1–16. In addition, he has produced a great range of locality case studies that range across
Australia, the United Kingdom, India, and the Pacific. Finally, he has edited many useful volumes,
such as Roy MacLeod and Richard Jarrell, eds., *Dominions Apart: Reflections on the Culture of
Science and Technology in Canada and Australia, 1850–1945* (*Scientia Canadensis*, 1994, *17*, 1 and
2); Roy MacLeod and Philip Rehbock, eds., *Nature in its Greatest Extent: Western Science in the
Pacific* (Honolulu: Univ. of Hawaii Press, 1988); and Roy MacLeod and Deepak Kumar, eds., *Technology and the Raj: Technical Transfer and British India, 1780–1945* (New Delhi: Sage, 1995).

[2] There are many possible entry points into the literature of constructivist thought. In addition to
books cited in the body of this paper, some recent titles that provide a useful overview include: Barry
Barnes, David Bloor, and John Henry, *Scientific Knowledge: A Sociological Analysis* (Chicago: Univ.
of Chicago Press, 1996); Peter Galison and David Stump, eds., *The Disunity of Science: Boundaries,
Contexts, and Power* (Stanford: Stanford Univ. Press, 1996); Jan Golinski, *Making Natural Knowl-
edge: Constructivism and the History of Science* (Cambridge: Cambridge Univ. Press, 1998); Ian
Hacking, *The Social Construction of What?* (Cambridge, Mass.: Harvard Univ. Press, 1999); David
Hess, *Science Studies: An Advanced Introduction* (New York: New York Univ. Press, 1997); Karin

acter of the knowledge-making process, held particular promise and powerful ana-
lytic consequence for the emerging discipline of colonial science history.[3] Gradually,
within the new field, explicit commitment to the prevailing positivist assumptions
gave way to implicit acceptance of constructivnist perspectives.

This move from a mainly positivist to a mainly constructivist orientation was, in
significant measure, empirically driven—not the result of considered debate over
abstractions. When an historian studies a particular *locality*,[4] by definition one would
expect that locality to become the "center" of his or her interest. Yet positivist colonial historians of, say, science in New Spain were, in reality, often writing the larger
social and intellectual history of Europe, and not the history of Mexico,[5] seeking
out local "traces" of European ideas and intellectual movements.[6] "'Europe'" says
Dipesh Chakrabarty, "remains the sovereign theoretical subject of all histories, including the ones we call 'Indian,' 'Chinese,' 'Kenyan,' and so on."[7] When historians
sought richer, deeper, "thicker" accounts of science in non-European localities,[8] they
soon became dissatisfied with analyses in which every standard of truth and rationality was set in Europe, and in which the very meaning of "rationality," "enlightenment," "progress," and "useful knowledge" had been defined on that distant continent. Thus, little by little, historians of local science sloughed off a paradigm of

---

Knorr-Cetina, *Epistemic Cultures: How the Sciences Make Knowledge* (Cambridge, Mass.: Harvard
Univ. Press, 1999).

[3] For example, Bruno Latour's writings have made a particularly useful contribution, both by insisting on eliminating the "great divide" between science and traditional modes of thought, and by
locating the power of modern science in its distinctive international network of institutions. The
workings of that network create the conditions that make legible and commensurable (for the center)
all the observations, measurements, representations, and texts produced in the various peripheries.
See especially Bruno Latour, *Science in Action: How to Follow Scientists and Engineers Through
Society* (Cambridge, Mass.: Harvard Univ. Press, 1987).

[4] In this paper I shall use the terms "local" and "locality" flexibly to indicate "places" in which
science is accomplished. A locality may be a region, country, city, or even a single institution, incorporating social, cultural, political, and economic factors and relationships, and including both centers
and peripheries.

[5] In fact, Mexican historians have been somewhat less Eurocentric than historians of science in
many other colonial localities. Nevertheless, at the first Mexican colloquium in the field (September,
1963), thirty-four of the sixty-one papers presented were part of a symposium on the European
Enlightenment in Latin America. Enrique Beltrán, ed., *Memorias del Primer Coloquio Mexicano de
Historia de la Ciencia*, 2 vols. (Mexico City: Sociedad Mexicana de Historia Natural, 1964). The
history of Mexican science has a venerable and distinguished disciplinary history with antecedents
in the nineteenth century. See Enrique Beltrán, "Fuentes mexicanas de la historia de la ciencia,"
Anales de las Sociedad Mexicana de Historia de la Ciencia y de la Tecnología, 1970, *2*:57–112; Juan
José Saldaña, "Marcos conceptuales de la historia de las ciencias en Latino América: Positivismo y
economicismo," *El Perfil de la ciencia en América* (Mexico City: Sociedad Latinoamericana de Historia de las Ciencias y la Tecnología, 1986); and Elías Trabulse, "Aproximaciones historiográficas a
la ciencia mexicana," *Memorias del Primer Congreso Mexicano de Historia de la Ciencia y de la
Tecnología* (Mexico City: Sociedad de Historia de la Ciencia y de la Tecnología, 1989), vol. 1, pp.
51–69.

[6] See for example, Roland D. Hussey, "Traces of French Enlightenment in Colonial Hispanic
America," in *Latin America and the Enlightenment*, ed. Arthur P. Whitaker, 2nd ed. (Ithaca: Cornell
Univ. Press, 1961), pp. 23–51. This book, originally published in 1942, uncovered useful material
but remains a classic example of a project in European history focused on Latin America, and is one
that helped set the agenda for writing colonial science history. All six of the distinguished contribut-
ing scholars were apparently English speaking and based outside Latin America.

[7] Dipesh Chakrabarty, "Postcoloniality and the Artifice of History: Who Speaks for 'Indian'
Pasts?" *Representations*, 1992, *32*:1–26.

[8] Clifford Geertz referred to the study of local cases as "thick description," without which more
general cultural meanings and power relationships cannot be understood. Clifford Geertz, *The Interpretation of Cultures: Selected Essays* (New York: Basic Books, 1973).

cultural *deficit*, replacing it with a paradigm of cultural *difference*. Within the "big picture" Europe was progressively "decentered,"[9] and in a very real sense, science was also decentered.

PERIPHERAL CENTERS AND CENTRAL PERIPHERIES

Because modern science arose principally in one geographic locale,[10] historians of science had taken the wheel as the metaphor for its international structure: its center was in Europe (displaced this century to the mid-Atlantic), with the rest of the world revolving around. But the metaphor of the wheel is exceedingly misleading. From the time of its cosmopolitan birth in the correspondence of Marin Mersenne (1588–1648) and Henry Oldenburg (1618–1677) and in institutions like the much neglected Casa de la Contratación in Seville (1539?), the Florentine Accademia del Cimento (1657), and the Royal Society of London (1660), modern science is better understood, both metaphorically and actually, as a polycentric communications network.[11] During the nineteenth and twentieth centuries that network was fully institutionalized, which represented a revolution in knowledge making more significant for both science and society than the theoretical advances of the seventeenth century traditionally known as the Scientific Revolution. Thus, from the very beginnings of the scientific movement,

> Centrality or peripherality was not primarily a matter of geographical location, but the combined effect of social, scientific, and—not the least—power relations. . . . Scientists, like other people, bore identities, they belonged somewhere, and they were loyal to something. Even more importantly, the daily activities of scientists were carried out in a framework of institutions, agendas, career opportunities, working language, financial support and patronage systems.[12]

This is to suggest that the idea of science having a European center and a global periphery perpetrated a confusing, and ultimately spurious, understanding of the relations of science and place. Then and now, Europe had major centers, minor centers, and peripheries; cities like London, indeed, had central institutions and peripheral institutions. Of course, progressively other localities developed scientific centers and peripheries. Furthermore, within Europe and without, centers rose and fell.

---

[9] Andrew Cunningham and Perry Williams, "Decentring the 'Big Picture': The Origins of Modern Science and the Modern Origins of Science," *British Journal of the History of Science*, 1993, *26*:407–32.

[10] "Modern science" as distinguished by its institutions, procedures, and technologies.

[11] See Latour, *Science in Action* (cit. n. 3), pp. 215–57, and Steven Shapin and Simon Schaffer, *Leviathan and the Air-Pump* (Princeton: Princeton Univ. Press, 1985). Sverker Sörlin has given a clear description of early processes of scientific internationalization: "National and International Aspects of Cross-Boundary Science: Scientific Travel in the 18th Century," in *Denationalizing Science: The Contexts of International Scientific Practice*, eds. Elizabeth Crawford, Terry Shinn, and Sverker Sörlin (Dordrecht: Kluwer, 1993), pp. 43–72. See also Lorraine Daston, "The Ideal and Reality of the Republic of Letters in the Enlightenment," *Science in Context*, 1991, *4*:367–86; and for the role of the Casa de la Contratación, see David Turnbull, "Cartography and Science in Early Modern Europe: Mapping the Construction of Knowledge Spaces," *Imago Mundi*, 1996, *48*:7–14, and J. Pulido Rubio, *El Piloto mayor de la Casa de Contratación de Sevilla* (Sevilla: Escuelade Estudios Hispano-Americanos, 1950).

[12] Sörlin, "National and International" (cit. n. 11), p. 45.

And whenever a scientific center arose within a locality, both science and the locality were changed by the event.[13]

Eurocentric explanations of the growth of science received a great boost with the appearance of historian George Basalla's widely known model describing "the introduction of modern science into any non-European nation."[14] The model predicted that localities peripheral to the European center would progressively "receive" the ideas of Western science, slowly establishing their own scientific organizations and personnel, perhaps producing along the way a few "heroes of colonial science."[15] In the final stage, after the colony had accomplished "seven tasks," a broad and "independent" institutional support base for science would have been established, thus allowing the given locality to compete scientifically in the world of nations.[16] The seven tasks, which are rarely discussed in the critical literature, included such activities as "overcoming" and eventually "eradicating" recalcitrant local "philosophical and religious beliefs," founding scientific societies "patterned after" the major European organizations, and importing European technologies. This unrelenting Eurocentrism was only one of the many reasons that the Basalla model was finally rejected by most historians.[17]

COLONIAL TO NATIONAL TRAJECTORIES

Basalla's model was initially attractive because it showed—in fact, seemed to prescribe—the straight and narrow path to national scientific development. Each locality was to rise in invariant sequence from a colonial to a national stage, from scientific dependency to autonomy. Colonial science was, in effect, considered a scientific adolescence that might eventually grow with the new nation-states into the maturity that Europe had long since achieved. In countries like Australia, where European settlers predominated, the predictive capacity of the model might, at first glance, seem reliable. In just a little over two hundred years, Australia moved from its first European scientific expedition (Cook/Banks) through a clearly "colonial" period to a remarkable degree of national scientific sophistication. Melbourne, for instance, is a locality that seemingly forms a perfect exemplar of how this can happen. The story of the city's move from scientific periphery to scientific center develops around the person of Frank Macfarlane Burnet (1899–1985), who became an outstanding

[13] Bruno Latour, "Give Me a Laboratory and I Will Raise the World," in *Science Observed*, eds. Karin Knorr-Cetina and Michael Mulkay (London: Sage, 1983), pp. 141–70.

[14] George Basalla, "The Spread of Western Science," *Science*, 1967, *15*:611–21. This paper, perhaps more than any other, set the initial research parameters for colonial science history. Patrick Petitjean amusingly and accurately describes Basalla's model as the work "le plus cité, et le plus réfuté aussi" by historians of science working in the field. Patrick Petitjean, "Sciences et empires: Un théme prométteur, des enjeux cruciaux," in *Science and Empires: Historical Studies about Scientific Development and European Expansion*, eds. Patrick Petitjean, Catherine Jami, and Anne Marie Moulin (Dordrecht: Kluwer, 1992), p. 6.

[15] Basalla, "Spread of Western Science" (cit. n. 14), p. 614.

[16] *Ibid.*, pp. 617–20.

[17] Although it has been subjected to devastating critique, Basalla's model continues to be cited long after every vestige of its explanatory power has disappeared. The "fall" of the model among historians of science has been well documented. This literature is extensively reviewed in David Wade Chambers, "Locality and Science: Myths of Centre and Periphery," in *Mundialización de la ciencia y cultural nacional*, eds. Antonio Lafuente, Alberto Elena, and María Luisa Ortega (Madrid: Doce Calles, 1993), pp. 605–18.

theoretician of virology and immunology.[18] Burnet declined chairs at Harvard and in London, as he was determined to make Melbourne an international center for medical research. He eventually attracted several future Nobel laureates to work with him. Today, Melbourne has seven major institutes for medical research and currently attracts sixty-four percent of the institutional awards from Australia's National Health and Medical Research Council.[19]

Thus, especially for the "neo-Europes" of the colonial world,[20] there might seem to be a hopeful and discerning congruity in Basalla's schema, especially in its postulation of a clear nexus between scientific activity and nation building. If the model works anywhere, one might expect it to be in those countries that had the decided "advantage" of European cultural, legal, economic, and technological frameworks—that is to say, in the colonies of those nations whose socioeconomic conditions had first given rise to modern science. This is especially true in invader/settler societies,[21] like Australia, where destruction of the indigenes and their traditional cultures had been ruthlessly accomplished, thereby effectively eliminating the need to "eradicate" and "replace" prevailing traditional philosophies, the first of Basalla's seven tasks of Europeanization. Australia, although obsessed with its great distance in kilometers from Europe, was socially, culturally, politically, economically, and racially closer to Europe than most of Europe's near neighbors (such as Egypt, Turkey, and many parts of the former Soviet Union).[22] But the apparent fit of the Basalla schema even with the Australian case lasts only through a very superficial reading; indeed, some of the model's leading critics actually use Australia as a counterexample.[23] At the very least, the Australian story is "richer and more complex" than the Basalla model allows.[24]

In Roy MacLeod's apt phrase, "science became a convenient metaphor . . . for

[18] Of Australia's six Nobel Prize winners in the sciences and medicine, all but Macfarlane Burnet spent most of their professional careers abroad.

[19] Christopher Sexton, *Burnet: A Life*, rev. ed. (Oxford: Oxford Univ. Press, 1999). For a detailed account of the Walter and Eliza Hall Institute, see Max Charlesworth, *et al.*, eds., *Life Among the Scientists* (Melbourne: Oxford Univ. Press, 1989). The Melbourne case was suggested by Barry Jones, former Australian minister for science and technology, personal communication, 1999.

[20] Alfred W. Crosby, *Ecological Imperialism: The Biological Expansion of Europe* (Cambridge: Cambridge Univ. Press, 1986). Crosby offers ecological explanations for the quick demographic dominance achieved in "neo-Europes." For a more recent and very interesting treatment of environmental history issues in which colonialism (and particularly colonial science) play a role, see two recent books by Richard Grove, *Green Imperialism: Colonial Expansion, Tropical Island Edens and the Origins of Environmentalism, 1600–1860* (Cambridge: Cambridge Univ. Press, 1995) and *Ecology, Climate and Empire: Colonialism and Global Environmental History, 1400–1940* (N.P., White Horse Press, 1998).

[21] The term "settler society" should not be used. It conveys an inaccurate picture of the European invasion and is offensive to the memories of millions who died in the peaceful-sounding process of "settlement." See Henry Reynolds, *Frontier: Aborigines, Settlers and Land* (Sydney: Allen and Unwin, 1987), pp. 192–3; A. Grenfell Price, *White Settlers and Native Peoples* (Cambridge: Cambridge Univ. Press, 1950), and Tom Griffiths and Libby Robin, *Ecology and Empire: Environmental History of Settler Societies* (Carlton: Melbourne Univ. Press, 1997).

[22] See David Wade Chambers, "Does Distance Tyrannize Science?" in *International Science and National Scientific Identity*, eds. R. W. Home and Sally Gregory Kohlstedt (Dordrecht: Kluwer, 1991).

[23] In particular, see MacLeod, "Moving Metropolis" (cit. n. 1), pp. 1–16, and Ian Inkster, "Scientific Enterprise and the Colonial 'Model': Observations on Australian Experience in Historical Context," *Social Studies of Science*, 1985, *15*:677–704.

[24] R. W. Home, "Introduction," in *Australian Science in the Making*, ed. R. W. Home (Cambridge: Cambridge Univ. Press, 1988), p. x.

what the Empire might become."[25] Indeed, for colonial scientists, science served as metaphor *and* means of legitimate colonial aspiration. Eventually, both colonizer and colonized came to believe that the promotion of science also promoted the cause of independence. For example, after losing the vast majority of her empire, Spain was not slow to sabotage local attempts to reform and modernize educational and scientific institutions in such remaining colonies as Puerto Rico and Cuba. Without a doubt, on both sides of the colonial divide, science was seen to provide a mechanism for increased colonial autonomy and self-sufficiency.[26] And we may speculate that the long-lasting popularity of the Basalla model may lie in its clear depiction of staged scientific growth moving ever nationward.

It is sometimes forgotten that Basalla's "three stage model" was deeply ensconced in the intellectual assumptions, not to mention the cold war ideological baggage, of the early theories of development. His famous essay appeared when W. W. Rostow's *Stages of Economic Growth*, published seven years earlier,[27] was at the height of its influence. Rostow's five stages precisely parallel Basalla's three stages. If Rostow's model provides the *economic* development corollary of modernization theory, Basalla's similar model plays a kindred role for *scientific* development. But there were flies in the ointment of modernization theory: some regions could not escape perpetual underdevelopment, dependency, exploitation, or cultural breakdown. In significant measure, these problems bedeviling world economic development—which have discredited "stages of economic growth" theories—also infect the international science system. In other words, many localities are held structurally in scientific underdevelopment due to such factors as brain drain, the high costs of technoscientific labs and equipment, inability to support the full range of scientific disciplines in any one locality, and a subjugated position in the institutional relations of knowledge and power.

The Basalla/Rostow approach to modernization assumes that the patterns that characterized scientific/economic development in the West provide a model for other localities around the world to follow. Without considerable modification this assumption is effectively blind to both history and culture, and is premised on the notion that "pre-scientific" localities, today, start from a position similar to Europe's before scientific take-off hundreds of years ago. Furthermore, the philosophy, religious beliefs, values, and institutions of traditional societies are considered probable obstacles, in effect, so much chaff to be blown away on the winds of scientific change.[28]

These considerations alone—without surveying the full critique that has been mounted over the last twenty years against staged, linear, and progressive models— suggest the need for a new framework for comparing histories of local science. From the extensive discussion of the Basalla model over the years, we have learned much about how such a framework ought to look. It should be symmetrical and interactive across the great divides—center/periphery, local/global, national/colonial, and tradi-

[25] MacLeod, "Moving Metropolis" (cit. n. 1), p. 244.

[26] See David Wade Chambers, James E. McClellan, and Heidi Zogbaum, "Science/Nation/Culture in the Caribbean Basin," in *Cambridge History of Science*, ed. Ronald Numbers, vol. 8 (Cambridge: Cambridge Univ. Press, forthcoming).

[27] W. W. Rostow, *Stages of Economic Growth* (Cambridge: Cambridge Univ. Press, 1960).

[28] Michael Shermer, *Why People Believe Weird Things: PseudoScience, Superstition and Other Confusions of our Time* (New York: Freeman, 1997).

Figure 1. *Joseph Dalton Hooker in the Himalayas receiving colonial tributes (in this case scientifically undescribed rhododendrons). Hooker insisted that plants be sent to Kew Gardens to be described; indigenous people and colonial botanists had only local knowledge. By William Tayler (1849). (Reproduced with permission of the Trustees of the Royal Botanic Gardens, Kew.)*

tional/modern. It should be nonlinear, nonstaged, and nonprescriptive, but it should specify a set of parameters that allow systematic comparison of the great array of independent and interdependent local histories of the production, application, and diffusion of natural knowledge. It should be dynamic and flexible and should identify vectors of communication, exchange, and control. Finally, the framework should take careful note of the social infrastructures that support knowledge work in both "Western" and "traditional" settings, without privileging one knowledge system over the other, thus allowing examination of both local and global contingencies of knowledge production and inculcation in the chosen locality.[29]

[29] Needless to say, this is a tall order. It is no wonder that some have suggested it unlikely that such a model will ever be devised, especially considering the cultural, social, and economic diversity of the cases for which the model must account! See Petitjean, Jami, and Moulin, *Science and Empires* (cit. n. 14), pp. 6–9.

SCIENCE AND PLACE

How does one articulate the *place* of knowledge or the *locality* of science? To some, even to formulate such a question is nonsense. According to intellectual legacies inherited from the Greeks and, more recently, from empiricist portrayals of scientific knowledge, "the place of knowledge is nowhere in particular and anywhere at all."[30] In other words, under the old philosophical paradigm, "the significance of place is dissolved."[31] Not surprisingly, then, historians of science have, on the whole, shown little interest in the complex interactions of science and place.[32] Exigencies of place might have been seen to present obstacles against, or encouragement for, doing or applying science, but so-called externalist explanations have been effectively isolated from the central processes of knowledge construction. On the other hand, *colonial* science historians very early began to realize that their stories were made interesting primarily by parameters of locality.[33]

Parameters of Locality

Until recently within the field, the most commonly found unit of locality was the colony or the nation-state. But to confine our interest to national cases would be arbitrary, needlessly limiting, and ultimately unsound. Localities mark the intersection of history, environment, language, and culture, and geographic boundaries are only one of the possible desiderata in defining a case study. Localities may be bounded by tangibles, such as socioeconomic circumstances, legalities, colonizing forces, topographies, and technologies; and by abstractions, such as beliefs about time, space, and progress. They may be further shaped by such factors as race and gender, ideology, and religious belief. To define a scientific locality, then, is simply to nominate a local frame of reference within which we may usefully examine the role of knowledge construction and inculcation.

[30] See Steven Shapin's delightful essay examining the social uses of solitude, "'The Mind Is Its Own Place': Science and Solitude in Seventeenth-Century England," *Sci. Context*, 1990, *4*:191–218, quotation on p. 191. Shapin also reminds us that "Most writers who insist both on the global character of mathematical and scientific knowledge and its universal application tend to overlook the immense amount of work that is done to create and sustain the artificial and formal environments in which 'application' happens," p. 209. See also Adi Ophir and Steven Shapin, "The Place of Knowledge: A Methodological Survey," *Sci. Context*, 1991, *4*:3–21; and Steven Shapin, "Placing the View from Nowhere: Historical and Sociological Problems in the Location of Science," *Transactions of the Institute of British Geographers*, n. s., 1998, *23*:5–12.

[31] Joseph Rouse, *Knowledge and Power* (Ithaca: Cornell Univ. Press, 1987), p. 77. This book is still extremely valuable for colonial science historians beginning to think about the concept of local knowledge.

[32] The ramifications of taking "place" seriously have been extensively discussed in theoretical writings in many fields, such as geography, anthropology, postcolonial studies, and feminist studies. See for example, Donna Haraway, *Simians, Cyborgs, and Women: The Reinvention of Nature* (New York: Routledge, 1991), pp. 183–202; Michael Keith and Steve Pile, eds., *Place and the Politics of Identity* (London: Routledge, 1993); Bill Ashcroft, Gareth Griffiths, and Helen Tiffen, eds., *The Post-Colonial Reader* (London: Routledge, 1995), and others cited below. For further discussion of the notion of knowledge "spaces," see David Turnbull, "Reframing Science and Other Local Knowledge Traditions," *Futures*, 1997, *29*, 6:551–62; Stanley Jeyerada Tambiah, *Magic, Science, Religion, and the Scope of Rationality* (Cambridge: Cambridge Univ. Press, 1990); Edward W. Soja, *Thirdspace: Journeys to Los Angeles and Other Real-and-Imagined Places* (Cambridge, Mass.: Blackwell, 1996); and homi bhabha, *The Location of Culture* (New York: Routledge, 1994).

[33] As we have seen, the diffusionist slant of the Basalla model allowed us to maintain the fiction that we were dealing with universal truths variously transmitted and applied.

What does this approach mean for studies of, say, the history of Caribbean science? One might define the locality geographically as the chain of islands, or as a particular island, or as the entire basin including an outer rim reaching to North, Central, and South America. Additionally, one might look at the area as a "colonial locality," within which a number of empires acted and interacted over a particular time frame. Or the colonial locality might simply be made up of the Spanish colonies. Alternatively, one might construct a "traditional knowledge locality," examining how tribal knowledge was constituted and the intellectual roles it played. In Mexico City, a university locality for the construction of knowledge might be differentiated from a mining-school locality. For some purposes, the world of medicine and health might be seen to constitute a separate knowledge space. And so on.

Such interpretive flexibility, allowing overlapping hierarchies of locality within a single geographical area, might seem daunting to the historian. Clifford Geertz comments that, in trying to explain phenomena, to turn from invoking master narratives ("grand textures of cause and effect") to providing "local frames of awareness" is to exchange "a set of well-charted difficulties for a set of largely uncharted ones."[34] But colonial historians of science have already begun mapping these uncharted localities; one might even say that their developing focus on locality is one of the field's greatest achievements within the history of science. The problem remains, however, that if we do not find a separate "other" vantage point from which to interpret and compare—whether we call it master narrative, theoretical model, or third space—we are left with the certainty of sinking into a vast sea of nativist ethnohistories.

Is this an infinite regress, leading in one direction to solipsism and in the other to a pretense of universal objectivity that hides the subjugation of local culture and local knowledge? Perhaps the best way forward, based on what we have learned, is to construct a new, more responsive, democratic, and self-questioning global discourse.[35] This process would necessarily nourish and sustain the local histories and local cultures that alone can provide "external" critique of the modernity project and the structures of power that it affords. The local and the global are a dialectical pair and must remain so in our histories.[36]

Vectors of Assemblage

In any colonial locality, vectors of assemblage encompass elements of process and of accumulation: the historical emplacement of the institutional and the physical framework for science. Telling this story has been the major work of most colonial science historians. The local scientific infrastructure is made up not only of organizations, buildings, museums, gardens, laboratories, instruments, chemicals, minerals, disciplines, schools, textbooks, and journals, but also of ideas and strategies,

[34] Clifford Geertz, *Local Knowledge: Further Essays in Interpretive Anthropology* (New York: Basic Books, 1983), p. 6.

[35] Chakrabarty calls for "a history that deliberately makes visible, within the very structure of its narrative forms, its own repressive strategies and practices." Chakrabarty, "Postcoloniality" (cit. n. 7), p. 25.

[36] See also Edward W. Said, "Figures, Configurations, Transfigurations," *Race and Class*, 1990, *32*:1–16; Katherine Hayles, *Chaos Bound: Orderly Disorder in Contemporary Literature and Science* (Ithaca: Cornell Univ. Press, 1990), pp. 213–14; and David Turnbull, "Local Knowledge and Comparative Scientific Traditions," *Knowledge and Policy*, 1993–4, *6*:29–54.

metaphors, theories and taxonomies, values, communities of trained personnel, and new socioprofessional roles for them to fill. David Turnbull usefully suggests the use of Deleuze and Guattari's term "assemblage" to denote, in his words, this "amalgam of places, bodies, voices, skills, practices, technical devices, theories, social strategies and collective work that together constitute technoscientific knowledge/ practices." The term "vectors of assemblage" suggests active and evolving practices as well as the constructed social and physical environments. For historians, the term "implies a constructed robustness without a fully interpreted and agreed upon theoretical framework."[37] In truth, there is a fine assortment of theoretical approaches that illuminate our understanding of the various elements of this assemblage of people, places, ideas, and things: biography, environmental history, medical history, cultural studies, material culture, feminist theory, etc.

In colonial localities, the vectors of assemblage sustain the imperial metropolitan connection to the science system, but if deliberately so constructed may also allow the attainment of nationalist cultural and socioeconomic objectives. Although exceedingly rare, in some cases these institutions may provide a base for the preservation of traditional local knowledge systems.[38] Recently, debates over intellectual property have recognized the value of indigenous knowledge of taxonomy in relation to health, but the resulting property laws, rather than protecting indigenous rights, have often served to transform this knowledge into commodities, profitable only to large corporations.[39] This development has now progressed to the point where any analysis of the infrastructure of late-twentieth-century science must look at the vectors of assemblage devoted to commodification in science, including such social mechanisms as copyright laws and the privatization of university research, as well as the appropriation of indigenous knowledge.

Around the globe, indigenous voices have been raised against these changes in the infrastructure of technoscience—changes that threaten the traditional social ethos and moral economy of science as much as the rights of indigenous peoples. In the words of Victoria Tauli-Corpus, "We are told that the companies have intellectual property rights over these genetic plant materials . . . this logic is beyond us . . . we indigenous peoples . . . have developed and preserved these plants over thousands of years."[40] To understand the weight of Tauli-Corpus's argument, it is useful to consider what lies behind her use of the words "developed and preserved." Aside from

[37] See Turnbull, "Local Knowledge" (cit. n. 36), p. 34.

[38] In nineteenth-century Mexico, for example, there was an attempt, especially by José Antonio Alzate, to support the indigenous natural taxonomies rather than those of Linnaeus. See, for example, Patricia Aceves Pastrana, *Química, botánica y farmacia en la Nueva España a finales del siglo XVIII* (Mexico City: Universidad Autónoma Metropolitana, 1993), pp. 55–74. In some localities traditional medicine has been partially sustained, or at least tolerated, in relation to Western medical practice.

[39] The intellectual property rights debate has now given rise to its own large literature, which cannot be reviewed here due to lack of space. But see C. Lind, "The Idea of Capitalism or the Capitalism of Ideas? A Moral Critique of the Copyright Act," *Intellectual Property Journal*, 1991, 7:70–4; E. C. Hettinger, "Justifying Intellectual Property," *Philosophy and Public Affairs*, 1989: 35; and Laurie A. Whitt, "Cultural Imperialism and the Marketing of Native America," *American Indian and Culture Research Journal*, 1995, 19:1–31.

[40] Victoria Tauli-Corpus, "We Are Part of Biodiversity, Respect Our Rights," *Third World Resurgence*, 1993, 36:25, quoted in Laurie A. Whitt, "Metaphor and Power in Indigenous and Western Knowledge Systems," International Conference on Working Disparate Knowledge Traditions Together, Deakin University, Victoria, Australia, 1994.

the fact that indigenous bodies of knowledge may often be sophisticated in content, as has been increasingly recognized in areas like taxonomy, indigenous knowledge localities employ complex vectors of assemblage, which may include maps, calendars, training of personnel, techniques, procedures, skills, manipulation of material, interpretation of results, prediction, meetings, and the preparation of texts.[41] In other words, indigenous involvement in the production of natural knowledge is neither trivial nor inconsequential. By exploring indigenous knowledge localities in the same way that we explore Western scientific localities, we attain a better position for effective comparison of these quite disparate knowledge systems.[42]

Network of Exchange and Control

As the process of assemblage develops in any locality, vital connections and linkages are made both locally and internationally. We have seen how the letter writers and travelers of the early science movement led in a direct line to the first global information network. This network, the international science system, becomes ever more polycentric and hierarchical, with major and minor centers and close and distant peripheries defined not geographically but in terms of scientific authority and social power. The network includes laboratories, journals, public and private funding agencies, museums, libraries, educational institutions, corporations, doctors' surgeries, administrative reports, and so on. It is important to keep the assemblage and the exchange network analytically separate, although both are required to participate in modern science. Other knowledge systems have their own assemblages and networks, but the fact that they are socially incommensurable may, in some cases, be more important than their conceptual differences.

   In the conglomerate vectors of assemblage that form the local infrastructure of technoscience, most people and things are tied directly into the international science system. This system does such varied work as formulate priorities for research funding, privilege certain modes of inquiry, set standards for the size of things, authorize knowledge claims, and establish regimes of cultural transmission, including education and popularization. The history of colonial science is arguably little more than the gradual connection of the locality into this global scientific communications

---

[41] Many of these elements have been entirely overlooked or underestimated as parts of knowledge production and communication. Texts, for example, may be inculcated in song, dance, architecture, and ceremonial business. Calendars have received some attention, but until very recently indigenous mapmaking traditions were completely undervalued as evidence of sophisticated natural knowledge. See David Turnbull, *Maps Are Territories: Science is an Atlas* (Chicago: Univ. of Chicago Press, 1993), pp. 19–53; David Woodward and G. Malcolm Lewis, eds., *Cartography in the Traditional African, American, Arctic, Australian, and Pacific Societies, The History of Cartography* (Chicago: Univ. of Chicago Press, 1998); Barbara Mundy, *The Mapping of New Spain: Indigenous Cartography and the Maps of the Relaciones Geográficas* (Chicago: Univ. of Chicago Press, 1996); Mark Warhus, *Another America: Native American Maps and the History of Our Land* (New York: St. Martin's Press, 1997); G. M. Lewis, ed., *Cartographic Encounters: Perspectives on Native American Mapmaking and Map Use* (Chicago: Univ. of Chicago Press, 1998); Laura Nader, *Naked Science: Anthropological Inquiry into Boundaries, Power, and Knowledge* (New York: Routledge, 1996); and David Turnbull, "Mapping Encounters and (En)Countering Maps: A Critical Examination of Cartographic Resistance," *Knowledge and Society*, 1998, *11*:15–44.

[42] Such a detailed comparison can be found, for instance, in Helen Watson and David Wade Chambers (with the Yolngu community at Yirrkala), *Singing the Land, Signing the Land* (Geelong: Deakin Univ. Press, 1989).

network, which historically was based in and controlled by the metropolitan center. In other words, this is the system that monitors, coordinates, authorizes, legitimates, classifies, and situates theoretically the flow of observational and experimental information. Perhaps the best description of these vectors is found in Latour's "centers of calculation."[43] Without this connection, a scientific locality cannot be taken seriously, no matter the perfection of its assemblage or the quality of work being done.

But this network is more than a science system, more than just an information exchange. It also enables mechanisms of social control, commodity transaction, exploitation, and appropriation. For example, Warwick Anderson suggests that "The recognition that even the most formally structured technical knowledge may be implicated in colonial accumulation and acquisition is long overdue. . . . inquiry into the textual economy of the laboratory . . . indicates an expansion of the power of the laboratory to represent and, in so doing, to constitute, regulate and legitimate colonial social realities. . . . The appropriation of colonial bodies and their insertion into a metropolitan discourse is in a sense a simulacrum of the whole colonial enterprise."[44]

INDIGENOUS KNOWLEDGE SYSTEMS

In the final section of this paper, we would like to follow a slightly different tack, introducing (though not fully arguing) the case that historians of science, in their accounts of particular localities, should be prepared to take stock of the nature, content, and role of indigenous knowledge systems.[45] There are a number of reasons why this is important. From studying "non-Western" cultures and their knowledge of nature, we contribute to our understanding, and to the conservation, of great intellectual traditions that are tens of thousands of years in the making. In doing this we enhance our understanding of the human mind, of human culture, and most especially of the noble—and sometimes ignoble—encounter of humans and nature. The twentieth century introduced the final stages of a half-millennium of global multicultural engagement, marked principally by conflict and holocaust.[46] By helping preserve the multiple varieties of human understanding of the natural world, we go to the heart of preserving cultural diversity. And perhaps we will improve the possibility of constructive cultural reconciliation in a deeply troubled world.

Finally, from a practical point of view, there is an increasing realization that indig-

---

[43] Latour, *Science in Action* (cit. n. 3), pp. 215–57.

[44] Warwick Anderson, "Where Every Prospect Pleases and Only Man Is Vile: Laboratory Medicine As Colonial Discourse," *Critical Inquiry*, 1992.

[45] This is not to suggest that all historians of science must drop their tools and start working on indigenous research projects, but we do believe that the study of science in any geographic region must include reference to indigenous knowledge systems (sometimes called IKS). Furthermore, we believe that responsible teaching will include reference to IKS in all generalist courses in history and social studies of science. See David Wade Chambers, "Seeing a World in a Grain of Sand: Science Teaching in Multicultural Context," *Science and Education*, 1999, 8:633–44; *idem*, preface to Turnbull, *Maps Are Territories* (cit. n. 41), p. v. See also the chapter on ethnoscience in Sally Gregory Kohlstedt and Margaret W. Rossiter, eds., *Historical Writing on American Science: Perspectives and Prospects*, Osiris, 1986, 1:209–28.

[46] Sadly, but not surprisingly, modern technoscience has been an active agent in the European global conquest, which has brought devastating consequences for nature and for other cultures. This fact is not lost on indigenous peoples.

enous knowledge has a crucial part to play in the preservation of biodiversity and the management of natural resources. The desire for environmentally sustainable development has prompted attempts to establish a dialogue between science and indigenous knowledge, combining the strengths and perspectives of both systems.[47] This interest in bringing disparate knowledge systems together in productive collaboration is also seen in medicine and public health.[48] Scholars have also drawn upon indigenous knowledge to counter what they see as the dangerous reductionism and culture-bound nature of Western science, its negative impact on native peoples, and its influence on the way we perceive the natural environment.[49] Indeed, some have argued for the incorporation of value systems into the sciences.[50]

Historians of science have a reasonably good record in relation to some of the more obvious traditional cultures. For instance, it is not uncommon for major general histories—in attempting to provide the big picture—to treat scientific civilizations of the Old World (China, India, Islam) and the New (Maya, Aztec, and Inca).[51] But if these major cultures are among the best known, they are by no means the only interesting indigenous bodies of knowledge available to historians.[52] It is essential that locality studies of these other knowledge traditions become incorporated into the archive of human history. Such a project, wherever carried out, must recognize the dangers of exploitation and repression that are in some measure inherent in ethnographic studies conducted from the center. For these reasons, such projects must allow the voice of the colonized and subjugated cultures to be heard in their own terms. Of course, local/global contention will not cease in this endeavor, but the local will be strengthened and the possibility of mutual exchange and contribution will be increased.

The call to recognize the intellectual stature and continuing validity of indigenous modes of thought reflects a growing international concern that has come to prominence over the last twenty years. UNESCO has commissioned a number of reports on issues relating to knowledge, culture, and development, all of which have opposed past policies of cultural assimilation (policies that have been almost

---

[47] Graham Baines and Nancy M. Williams, "Partnerships in Tradition and Science," in *Traditional Ecological Knowledge: Wisdom for Sustainable Development*, eds. N. M. Williams and G. Baines (Canberra: Centre for Resource and Environmental Studies, Australian National University, 1993) pp. 1–6.

[48] Gregory Cajete, *A People's Ecology: Explorations in Sustainable Living* (Santa Fe, N. Mex.: Clear Light Publishers, 1999).

[49] Vine Deloria, Jr., *Red Earth White Lies: Native Americans and the Myth of Scientific Fact* (Golden, Colo.: Fulcrum Publishing, 1997); Peter Knudston and David Suzuki, *Wisdom of the Elders* (St. Leonards, New South Wales: Allen and Unwin, 1992).

[50] Gregory Cajete, *Igniting the Sparkle: An Indigenous Science Education Model* (Skyand, N.C.: Kivakí Press, 1999); Gregory Cajete, *Native Science: Laws of Interdependence* (Santa Fe, N. Mex.: Clear Light Publishers, 2000); Zia Sardar, *Explorations in Islamic Science* (London: Mansell, 1989).

[51] For a recent attempt at the big picture that gives a good account of certain areas of indigenous knowledge, see James E. McClellan III and Harold Dorn, *Science and Technology in World History* (Baltimore, Md.: Johns Hopkins Press, 1999).

[52] Preliminary access to these knowledge systems has been improved by several recent publications: Helaine Selin, ed., *Encyclopaedia of the History of Science, Technology and Medicine in Non-Western Cultures* (Dordrecht: Kluwer, 1997); Douglas Allchin and Robert DeKosky, *An Introduction to the History of Science in Non-Western Traditions* (Seattle: History of Science Society, 1999); Sarah Franklin, "Science as Culture, Cultures of Science," *Annual Review of Anthropology*, 1995:163–84; David Hess, *Science and Technology in a Multicultural World: The Cultural Politics of Facts and Artifacts* (New York: Columbia Univ. Press, 1995).

universally viewed by indigenous peoples as nothing less than genocidal). For example, an influential 1981 report stated that one major international objective should be the "rehabilitation of traditional forms of knowledge and, above all, of the potentialities which have been stifled by the pressure of the dominant countries or groups."[53] And in 1995: "a culturally distinct people loses its identity as the use of its language and social and political institutions, as well as its traditions, artforms, religious practices and cultural values, is restricted. *The challenge today . . . is to develop a setting that ensures that development is integrative and inclusive.* This means respect for value systems, for the traditional knowledge that indigenous people have of their society and environment, and for their institutions in which culture is grounded."[54]

Importantly, this understanding is seen to apply to technoscientific knowledge in its relationship to indigenous knowledge systems. In 1999, in a declaration adopted by the UNESCO-sponsored World Conference on Science in Budapest, this position was developed in some detail, acknowledging

> that traditional and local knowledge systems as dynamic expressions of perceiving and understanding the world, can make and historically have made, a valuable contribution to science and technology, and that there is a need to preserve, protect, research and promote this cultural heritage. . . . Governmental and non-governmental organizations should sustain traditional knowledge systems through active support to the societies that are keepers and developers of this knowledge, their ways of life, their languages, their social organization and the environments in which they live. . . . Governments should support cooperation between holders of traditional knowledge and scientists to explore the relationships between different knowledge systems and to foster interlinkages of mutual benefit.[55]

There are many problems associated with this international call to support the study and preservation of indigenous knowledge systems (IKS). It might easily degenerate into a rush for profiteering exploitation of botanical knowledge. Furthermore, even if the IKS project is pursued with the most honorable of intentions, it is possible to view it as a lost cause. Some knowledge systems have disappeared, some are known only in fragments, some involve sacred knowledge that cannot be made public, and most can be uncovered only by learning relevant languages and by working in collaboration with native scholars, elders, and practitioners. The comparison of Western science with indigenous knowledge systems is fraught with all the difficulties associated with understanding the similarities and demarcations between markedly different cultures; these problems are compounded by looking at precisely that aspect of Western culture that is believed to provide an objective, disinterested, and non-culture-bound account of the natural world.

It is possible to conceive how a culture can accept and appreciate another culture's aesthetics—although European interest in indigenous art was a long time coming,

[53] UNESCO, *Domination or Sharing?: Report on Indigenous Development and the Transfer of Knowledge* (Unesco Publishing, 1981), p. 31

[54] UNESCO, *Our Creative Diversity: Report of the World Commission on Culture and Development* (Unesco Publishing, 1995), pp. 70–1 (italics in original); see also D. Michael Warren, L. Jan Slikkerveer, and David Brokensha, eds., *The Cultural Dimension of Development: Indigenous Knowledge Systems* (London: Intermediate Technology Publications, 1995).

[55] UNESCO, *The Declaration on Science and the Use of Scientific Knowledge: Report of the World Conference on Science* (Unesco Publishing, 1999).

and popular appreciation has sometimes involved the development of a specific product for white consumption. But the idea that very different cultures may be able to reconcile some aspects of their knowledge of the natural world has been considered an impossible project in some quarters. After all, the old paradigm argued, there is only one objective reality, and only science has developed a reliable method for describing and explaining that reality. The history of science, in such a view, is the history of pushing back the frontiers of superstition and ignorance, with religion and belief retreating in the face of superior scientific explanation.[56]

Science, like any other social activity, bears the imprint of the society of which it is part. All knowledge systems are "situated" in power relationships, value assumptions, and historical frameworks.[57] As a culturally specific knowledge system—albeit one with enormous power and one that remains a source of both good and evil—Western science, in our intellectual calculations, cannot be accorded a privileged status over indigenous knowledge. Far from being an abstract intellectual debate, this issue goes to the heart of how different cultures view one another and their ways of seeing the world.[58] Furthermore, indigenous knowledge systems demand respect as powerful cultural expressions of ways of knowing nature—ways that have clear implications for how humans should live and prosper in particular environments. The reassessment of the character of IKS in light of these findings is only just starting, and the history of science has an important role to play in this. By considering both Western science and indigenous knowledge systems as forms of local knowledge and practice, the locality approach opens up a space for more equitable comparison.

BRINGING DISPARATE KNOWLEDGE SYSTEMS TOGETHER

In the last few pages, we offer an account of taxonomy intended to illustrate some of the things that can be learned when disparate knowledge systems are brought together. Science typically is the dominant knowledge system because it resides within international networks very different from those of a politically marginalized indigenous community. For example, an elaborate system of commissions, publications, and institutions lies behind contemporary botanical and zoological classification and nomenclature; it is inconceivable that an indigenous taxonomy—no matter how internally cohesive, how comprehensive and differentiated, or even how similarly speciated—could continue to exist within that system.[59] An ethnoscientist is

[56] Mera Nanda, "The Epistemic Charity of the Constructivist Critics of Science and Why the Third World Should Refuse the Offer," in *A House Built on Sand: Exposing Post-Modernist Myths About Science*, ed. Noretta Koertge (New York: Oxford Univ. Press, 1998). Nanda makes the case that science is the "last stand" against religious bigotry and superstition.

[57] John Law, ed., *Power, Action and Belief: A New Sociology of Knowledge?* (London: Routledge and Kegan Paul, 1986); *idem, A Sociology of Monsters: Essays on Power, Technology and Domination* (*Sociological Review Monograph*, 1991); Sandra Harding, ed., *The Racial Economy of Science: Toward a Democratic Future* (Bloomington, Ind.: Indiana Univ. Press, 1993); Helen Longino, *Science as Social Knowledge: Values and Objectivity in Scientific Inquiry* (Princeton, N.J.: Princeton Univ. Press, 1990).

[58] Ivan Karp and Steven D. Lavine, eds., *Exhibiting Cultures* (Washington, D.C.: Smithsonian Institution Press, 1991).

[59] Brent Berlin, *Ethnobiological Classification: Principles of Categorization of Plants and Animals in Traditional Societies* (Princeton, N.J.: Princeton Univ. Press, 1992); Ralph N. H. Bulmer, "Why is the Cassowary not a Bird? A Problem of Zoological Taxonomy among the Karam of the New Guinea

likely to focus on how indigenous taxonomies differ from scientific taxonomy. And the history of taxonomy has much of interest to say on this issue. For example, it may be useful first to consider how scientific taxonomy emerged from earlier European folk taxonomies. Like indigenous taxonomies, those recorded by Aristotle and in early herbals of the fifteenth and sixteenth centuries listed around eight hundred taxa at the level of genus or species. Indeed, at the local level there was often no difference between genus and species, because most genera were monospecific in a given environment, and where two or more species occurred, they were often morphologically different because they were pursuing different ecological strategies.

Several technological changes transformed folk taxonomy. The printing press and woodcut permitted the printing of books that compared taxa from different regions and across time. Voyages of discovery brought back large numbers of new specimens, which were stored in herbaria, botanical gardens, and museums. Naturalists began to specialize in plants or animals, and then in more restricted groups such as birds, fishes, or insects. These huge increases in the number of recorded taxa, which were largely the result of technological change rather than intellectual breakthrough, posed practical problems of ordering and management that had to be resolved. Linnaeus began to introduce the higher categories of class and order (above genus and species).

In the late eighteenth and early nineteenth centuries, as the genus lost its place as the chief taxonomic rank, scientific taxonomy moved farther from folk taxonomy. Scientists increasingly used biological functions and anatomical structures to define species in families and other higher-order taxa. Meanwhile, genera were split again and again under the weight of newly discovered species. No plausible explanation was offered for why it was possible to sort organisms into "natural" groups based on shared characteristics until Darwin argued that the similarities between organisms were due to the "propinquity of descent." But this did not make the procedures of taxonomy any simpler; if anything, it simply raised the stakes by requiring that taxonomy also indicate evolutionary history. An obvious difference between scientific taxonomy and indigenous classificatory systems is that the latter are not based on the theory of evolution, with taxa reflecting genetic or phylogenetic groupings. However, it would be inappropriate to say that this was a way of fundamentally distinguishing scientific taxonomy from indigenous classification, or we would be forced to conclude that all taxonomy prior to Darwin was nonscientific.

A more fundamental difference lies in the social realm: scientific taxonomy seeks to create a global system of nomenclature and a hierarchical structure, based on an elaborate system of publication, formal rules, and congresses. The preamble of the International Code of Zoological Nomenclature states that "the object of the Code is to promote stability and universality," and this contrasts noticeably with the flexible use of terms in indigenous classifications, which are part of the everyday language of the community and are used in many different contexts.[60]

Clearly, classificatory systems are an integral part of culture, whether we are talk-

---

Highlands," *Man*, 1967, 2:5–25; Paul Sillitoe, *Roots of the Earth: Crops in the Highlands of Papua New Guinea* (Kensington: Univ. of New South Wales Press, 1983), chap. 8.

[60] Ernst Mayr, *Principles of Systematic Zoology* (New York: McGraw Hill, 1969); Ralph Bulmer and Chris Healey, "Field Methods in Ethnozoology," in Williams and Baines, *Traditional Ecological Knowledge* (cit. n. 47), pp. 43–55.

ing of indigenous or scientific taxonomies. With regard to indigenous knowledge, Ralph Bulmer has observed that

> ethnozoological data do not exist as a readily separable body of knowledge in traditional societies, where generally no distinction is made like that between science and other systems of knowledge in contemporary western culture. Consequently, the investigator is likely to be confronted with what may, at first sight, appear to be an unsystematic blend of detailed, credible information, mystical ideas, and superstition. It is important to realise that such a mixture of what Europeans might consider rational, empirically based knowledge and mystical or supernatural beliefs must be understood within its cultural context.[61]

There is something forced in wrenching out the zoological classification system of the Kalam of Papua New Guinea and comparing it to scientific taxonomy, because it privileges scientific knowledge and uses it to assess the "validity" of indigenous knowledge in a culturally imperialist manner. Indeed, this charge can be made against much of what is done under the label of ethnoscience. We often forget that the validity of scientific taxonomy also suffers when taken from its cultural context. If you doubt this, try to explain (without reference to evolutionary theory or the Code of Nomenclature) the reasons for the many scientific taxonomic distinctions that ignore obvious similarities or differences of structure and function.

Bulmer was always conscious of these dangers in his work, and sought to locate Kalam classifications within the whole of Kalam culture, as he understood it. But like other anthropologists, he remained the author and authority (the etymological similarity is revealing), drawing upon his Kalam informants and then ordering the material into an ethnographic narrative for others in his profession. In his later work, Bulmer formed a more equal collaboration with his major informant, Ian Saem Majnep. In their *Birds of My Kalam Country*, Majnep spoke directly to the reader (based on taped conversations with Bulmer), and the material was primarily arranged according to Majnep's cultural categories.[62]

Conversations across cultures are even more subject to all the difficulties of multiple interpretations and readings that are characteristic of conversations between individuals. Our comments above on indigenous knowledge have been constructed within a context of writing *about* indigenous peoples, not *for* indigenous peoples.[63] Written from a Western perspective, this essay is not a replacement for indigenous views of the relations between science and indigenous knowledge. On the other hand, we do hope it will usefully contest some traditional Western biases, while opening up crosscultural spaces for new research directions within the history of science in local contexts.

[61] Bulmer and Healey, "Field Methods," (cit. n. 60), pp. 43–4.

[62] Ian Saem Majnep and Ralph Bulmer, *Birds of My Kalam Country* (Auckland: Univ. of Auckland Press, 1977); George E. Marcus, "Notes and Quotes Concerning the Further Collaboration of Ian Saem Majnep and Ralph Bulmer," in *Man and a Half: Essays in Pacific Anthropology and Ethnobiology in Honour of Ralph Bulmer*, ed. Andrew Pawley (Auckland: Polynesian Society, 1991), pp. 37–45.

[63] Both of the authors have worked with indigenous people and one is of Cherokee ancestry. See Bain Attwood, introduction to *Power, Knowledge and Aborigines*, eds. Bain Attwood and John Arnold (Melbourne: La Trobe Univ. Press, 1992), pp. i–xvi.

Figure 2. *A Kob, or Papuan Lory* (Charmosyna papou). *The most important bird in the Kalam category* Yakt ok, b noman ay ayak, *or birds which men's souls can turn into. Drawing by Christopher Healey from Ian Saem Majnep and Ralph Bulmer,* Birds of My Kalam Country *(Auckland: Univ. of Auckland Press, 1977).*

CONCLUSION

For at least four hundred years, the world has witnessed the rise and growth of the technoscientific movement,[64] inculcating such enormous power and authority that it has been able to confront, overwhelm, and absorb the insights of traditional knowledge systems around the planet. This social and organizational triumph is sometimes interpreted as evidence of the universality of scientific knowledge claims. The locality approach focuses on the conditions under which this *appearance* of universality arose and is maintained. These conditions include, for example, Europe's successful politico-economic colonization of the world; the close integration of its institutions of technoscientific knowledge with its institutions of power; its unique social mechanisms of authoritative communication and intercultural exchange, employing new devices like laboratories for knowledge making and networks for retranscribing, moving, and incorporating local knowledge into the global discourse; and, finally, the enormous instrumental successes of technoscience in the manipulation of nature and in the development of technologies of control.

By understanding modern science as the spread of the ideas and institutions of rationality and progress, the old paradigm perpetuated a clear agenda for colonial historians. We searched for local "traces" of European Enlightenment. We identified when and how these ideas and institutions first appeared in a new locality, recording when the far-flung provincials finally "got it right," sometimes offering reasons for the slow uptake of European ideas. We accepted, as patterning sciences, astronomy and physics rather than chemistry and engineering; natural history rather than agronomy or mineralogy.[65] And we dutifully attempted to distinguish center from periphery, science from technology, colonial science from national, dependent science from independent, and basic science from applied, even though these categories and distinctions completely dissolved, or at least seemed less useful, when analyzed in concrete case or local context.[66]

By the 1990s, many of those questions and assumptions seemed outmoded. As historians of the colonial world embraced new perspectives—which we have here called the "locality approach"—our histories became less defensive, analytically richer, and more firmly fixed on local dimensions of the rise of the science movement. If modern technoscience was considered to be embedded in the social, cultural, and intellectual context that produced it, then the failure of European science to take hold in another locality was best explained not by seeking out backwardness or deficiency in the target culture but by uncovering the local intellectual and socioeconomic interests that stood to gain or lose by its introduction; and by understanding the structural aspects of the international science system that favored the "West" and the "North."

What, then, is the future of colonial science history as a scholarly field? To define a locality simply as a "colony" is to invite neglect of much that matters culturally

---

[64] Surely the phrase "modern science and technology" has passed its use-by date. The term "technoscience" is perhaps a little less problematic and feels especially appropriate for the "jacked in" (computerized) sciences that dominate the beginning of the new millennium.

[65] Simon Schaffer, "Field Trials, the State of Nature and the British Colonial Predicament" (manuscript, 1999). In this important paper Schaffer reminds us that the Enlightenment also set up agronomy as a model science. Had agriculture become a patterning science, needless to say, we would think about science differently in ways that especially matter in colonial and indigenous localities.

[66] Nader, *Naked Science* (cit. n. 41).

and historically. In any case, colonies and empires are primarily the products of a particular moment in history, whereas colonizing forces that dominate and exploit are always and everywhere with us. Concentrating on the multidimensional—cultural, political, and socioeconomic—*local* contexts of scientific endeavor will help to end another "great divide," the one that has seen historians of science writing about centers and historians of colonial science writing about peripheries.

Reconstructing India: Disunity in the Science and Technology for Development Discourse, 1900–1947

*Deepak Kumar**

ABSTRACT

The turn of the twentieth century saw the apogee of the British Empire in India, while at the same time the seeds of decolonization sprouted. The last decades of the Raj (1930s and 1940s) saw some flickers of "constructive imperialism," but these came too late. By then, nationalism had gathered strength. Indian leaders— including Mahatma Gandhi and Jawaharlal Nehru—and the government raced to raise development issues and debate the role of science and technology therein. By 1937, many committees had been formed and reports published, and the push was on to make India a modern nation-state. At first sight, there seemed to be unity of purpose, but in reality this was not so. As this paper shows, the thin veneer of the development discourse evaporated when put under pressure by class interests.

IN A BROADCAST TO ALL INDIA RADIO ON 2 OCTOBER 1948, a visiting British scientist, A. C. Egerton, offered the following polite suggestion:

> May I suggest, do not be too attracted by all the glamour of Western technology . . . it is wonderful, but we have in some ways industrialised too far and not made the world happier thereby. You have a chance of distilling the best out of the West and fitting it into the age-old civilization of the East. If you can improve husbandry and state of the villagers without going for too great a concentration of industry, you may in the end gain greater happiness. The key note should not be to copy and westernise, but with wisdom to fit the best of the new into the best of the old civilisation.[1]

The broadcast was commemorating the first anniversary of Mahatma Gandhi's (1869–1948) birth following his assassination in January 1948, and it naturally reflected Gandhi's concern for rural India. Over forty years earlier, Gandhi had asked his compatriots to decide if they wished to become a fifth or sixth edition of Euro-America. For him the choice was clear, and his argument for an independent national

* Zakir Hussain Centre for Educational Studies, Jawaharlal Nehru University, New Delhi, 110067, India.

This study was made possible by grants from the British Academy and the Sheffield Hallam University. Many thanks to Michael Worboys and Sanjoy Bhattacharya for care and interest in this work. The author remains grateful to Roy MacLeod for comments and corrections, as ever.

[1] Transcript of All India Radio broadcast, 2 Oct. 1948, A. C. Egerton Papers, Royal Society, London.

identity was laid out aggressively in his *Hind Swaraj*.[2] Gandhi did not suffer the crisis of identity that his preceding generation could not help but suffer. The interlocutors of nineteenth-century India had been pluralists aimed at a cultural synthesis; they wanted the best of both worlds, as did Mr. Egerton. But how to bring this about? They had pursued a great variety of strategies—imitation, translation, assimilation, "distanced" appreciation, and even retreat to isolation—without much success. Many attempts had been made, in India and elsewhere, to articulate modern scientific rationality in terms of indigenous traditions and requirements. A highly reputed Bengali laureate found new developments in science (e.g., Darwinism) consistent with the Hindu *weltanschauung*.[3] An Islamic scholar hoped that, thanks to the new interactions, "a Newtonized Avicenna or a Copernicised Averroes may spring up."[4]

Pious hopes, yet they symbolized a yearning for change—a yearning that would soon manifest itself in nationalist aspirations. An imperial rationalist discourse showed Indians how rationalism could be turned against Europeans. Rationalism came to be seen as something inherent to human nature rather than a European speciality, and as a mark of progress independent of Europeanization. One of the first to realize the necessity of rearticulating science in such national terms was a medical doctor in Calcutta, Mahendra Lal Sircar (1833–1904). In 1869, Sircar wrote "On the Desirability of a National Institution for the Cultivation of Sciences by the Natives of India," which asked for an institution that would "combine the character, the scope and objects of the Royal Institution of London and of the British Association for the Advancement of Science." He added, "I want freedom for this Institution. I want it to be entirely under our own management and control. I want it to be solely native and purely national."[5] In 1876, after much private effort and controversy, the Indian Association for Cultivation of Science was established in Calcutta. The event was no less important than the establishment in 1885 of the Indian National Congress, a political forum that would later spearhead the nationalist movement. The Association for Cultivation of Science symbolized the determination of a hurt national psyche to assert itself in an area that formed the kernel of Western superiority. The shift from colonial dependence to national independence had begun.

In recent years, several scholars have tried to portray this shift in cultural and ideational terms.[6] Earlier this was done within the impact-response framework, which to some extent was useful.[7] But the colonial encounter was so complex and intense that no single framework can suffice. New approaches, however, can be dis-

    [2] A. J. Parel, ed., *Gandhi: Hind Swaraj and Other Writings* (Cambridge: Cambridge Univ. Press, 1997).

    [3] J. C. Bagal, ed., *Bankim Rachnavali*, vol. 1 (Calcutta: Sahitya Samsad, 1969).

    [4] Maulavi Ubaidullah, *Essay on the Possible Influences of European Learning on the Mahomedan Mind in India* (Calcutta, 1877), p. 47.

    [5] Quoted in *A Century: Indian Association for the Cultivation of Science* (Calcutta: IACS Publication, 1976), p. 5.

    [6] Ashis Nandy, *Alternative Sciences* (New Delhi: Allied Publishers, 1980); Shiv Visvanathan, *Organising for Science* (Delhi: Oxford Univ. Press, 1985); K. N. Panikkar, *Culture, Ideology, Hegemony: Intellectuals and Social Consciousness in Colonial India* (New Delhi, Tulika Publications, 1995); T. Raychaudhury, *Europe Reconsidered: Perceptions of the West in Nineteenth Century Bengal* (Delhi: Oxford Univ. Press, 1984); Deepak Kumar, *Science and the Raj* (Delhi: Oxford Univ. Press, 1995); Roy MacLeod and Deepak Kumar, eds., *Technology and the Raj* (New Delhi: Sage, 1995).

    [7] B. T. McClully, *English Education and the Origins of Indian Nationalism* (New York: Columbia Univ. Press, 1940); B. B. Misra, *The Indian Middle Classes* (London: Oxford Univ. Press, 1978); R. C. Majumdar, ed., *British Paramountcy and Indian Renaissance*, vol. 10 (Bombay: Bharatiya Vidya Bhawan, 1965).

cerned in the writings of Dhruv Raina, S. Irfan Habib, Gyan Prakash, V. V. Krishna, S. Sangwan, Anil Kumar, and others.[8] They raise new questions and investigate how scientific ideas "travel," and the strategies of hegemonization and counterhegemonization. Unlike previous studies that hovered around the master-slave dialectics or the impact-response syndrome, the new microstudies emphasize the subsumed contradictions and disjunctions.[9] As a result, the canvas has now been considerably enlarged; it involves studies on the power discourse as well as textual or prosopographical analyses of cultural displacement and renegotiation.[10] The cultural interlocuters of Victorian India themselves were no less aware of the contradictions and the dilemmas that they faced. Yet they had little choice but to work for both material benefits and traditional values.[11] They wanted the best of both worlds, and in the process they strove for more autonomy and power.

WHERE TO GO, HOW TO GO

Given this cultural background, two major questions arose. First, how was the colonial government going to use scientific knowledge and techniques in the material development of the country? Second, how would Indians articulate demands for national reconstruction, and press for it?

Indians craved autonomy, no doubt, but the British scientists working in Indian establishments also wanted more independence from metropolitan London control. Independence, in addition, would give the Government of India authority to ensure more "utility-oriented" work. With the creation of the Board of Scientific Advice (BSA) by governor-general Lord Curzon in 1902, two important shifts could be seen in the government's attitude. One was the idea that science in India could and should be cultivated without supervision from London; and the second was that the country's preference for the "natural history" sciences must be replaced by public sponsorship of industrial technology.[12] This was a sort of official version of the Swadeshism (a political movement originating in Bengal and advocating home rule) then

[8] S. Irfan Habib and Dhruv Raina, "Copernicus, Columbus, Colonialism and the Role of Science in Nineteenth Century India," *Social Scientist*, 1989, *13*:3–4, 51–61; Gyan Prakash, *Another Reason: Science and the Imagination of Modern India* (Delhi: Oxford Univ. Press, 2000); V. V. Krishna, "The Colonial Model and the Emergence of National Science in India, 1876–1920," in *Science and Empires: Historical Studies about Scientific Development and European Expansion*, eds. Patrick Petitjean, Catherine Jami, and Anne Marie Moulin (Dordrecht: Kluwer, 1992), pp. 57–72; S. Sangwan, *Science, Technology and Colonization: An Indian Experience, 1757–1857* (Delhi: Anamika, 1991); Anil Kumar, *Medicine and the Raj* (New Delhi: Sage, 1998); Subrata Dasgupta, *Jagdish Chandra Bose and the Indian Response to Western Science* (Delhi: Oxford Univ. Press, 1999).

[9] Deepak Kumar, "The 'Culture' of Science and the Colonial Culture: India 1820–1920," *British Journal of History of Science*, 1996, *29*:195–209.

[10] Gyan Prakash, "Science Between the Lines," in *Subaltern Studies*, eds. Shahid Amin and Dipesh Chakrabarty (Delhi: Oxford Univ. Press, 1996), vol. 9, pp. 59–82; Dhruv Raina and S. Ifran Habib, "The Moral Legitimation of Modern Science: Bhadralok Reflections on Theories of Evolution, *Social Studies of Science*, 1996, *XXVI*:9–42; Deepak Kumar, "Unequal Contenders, Uneven Ground: Medical Encounters in British India, 1820–1920," in *Western Medicine as Contested Knowledge*, eds. A. Cunningham and B. Andrews (Manchester: Manchester Univ. Press, 1997), pp. 172–90; David Arnold, "Public Health and Public Power: Medicine and Hegemony in Colonial India," in *Contesting Colonial Hegemony*, eds. D. Engels and S. Marks (London: British Academic Press, 1994), pp. 131–51.

[11] Kumar, *Science and the Raj* (cit. n. 6), 192–5, 235.

[12] For details, see Roy MacLeod, "Scientific Advice for British India: Imperial Perceptions and Administrative Goals, 1898–1923," *Modern Asian Studies*, 1975, *9*, 3:343–84.

raging in the country. But there was no question of any talk, much less cooperation, between the official promoters of scientific independence and the Swadeshi leaders who aimed at "real" independence. The Board of Scientific Advice was a purely interdepartmental British affair; Indian scientists or leaders were given no role in it. The goals of the two groups were similar, but their methods were different. And opinions remained sharply polarized in both camps: some emphasized the glory of "pure" science, others (probably the majority) stressed the relevance of "applied" science. The Swadeshi leaders were themselves divided.[13] Government officials were even more ambivalent. The dilemma was best summed up by Thomas Holland, then former director of the Geological Survey of India, who in 1905 wrote:

> India can not perhaps afford to rank itself beside the more thoroughly developed European countries where pure science is so richly endowed; and the practical difficulty here is to discover the *profitable mean course* in which scientific research, having a general bearing will, at the same time, solve the local problems of immediate economic value.[14]

Eventually, some hesitant steps were taken by the Indian government. In 1911, an Indian Research Fund Association was created to foster medical research and public health. Three years later, an Indian Science Congress Association was formed that, at its very first meeting, urged the government to recognize the paramount claims of science upon public funds. Finally, after almost thirty years of sustained demands from Indian leaders, the government agreed to recognize India's need to industrialize. The Indian Industrial Commission was formed in 1916, with Holland as chairman. It is significant that Holland was also made president of the Indian Munition Board. For decades, the colonizers had thrived on a plantation economy; the Great War convinced them of the need to foster tertiary industrialization that would provide war-related necessities.

The euphoria generated by the Industrial Commission was short-lived, and its recommendations were shelved by the British government. Its only tangible accomplishment was the creation of a Department of Industries. But the debates that the commission provoked in India were instructive.[15] An important dissenting voice came from one of its members, Madan Mohan Malaviya, a leading nationalist. Like his cultural predecessors who had rejected rationality as a Western import, Malaviya presented a nationalist critique of British economic policies in India, and stressed that India had remained deindustrialized. The British model was inadequate. The new icons were Japan and Germany, and the new watchword was science-based technology.

---

[13] In 1905 the nationalist leaders set up two different organizations in Calcutta; one was the National Council of Education, the other the Society for the Promotion of Technical Education. The former touched on all aspects of education while the latter aimed at a technoscientific education that would accelerate industrial progress. For details, see Sumit Sarkar, *Swadeshi Movement in Bengal* (New Delhi: Peoples Publishing House, 1973).

[14] Note by Thomas Holland, Revenue Agriculture, General Branch, Proc. no. 3, Aug. 1905, file 127, National Archives of India (emphasis added).

[15] For details, see Visvanathan, *Organising for Science* (cit. n. 6), pp. 39–96.

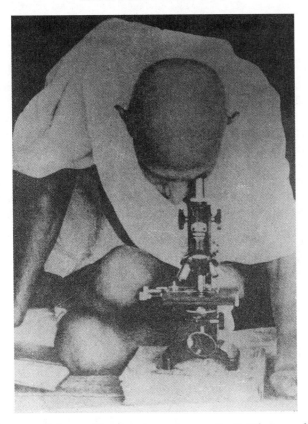

Figure 1. *Mahatma and the machine (empire under scrutiny!) (Photograph taken by S. S. Sokhey in 1930 at the Haffkine Institute, Bombay.)*

EXTRAORDINARY DISSENT: GANDHI

Amidst the growing demands for self-rule, democracy, industrialization, and development, Gandhi emerged as an extraordinary dissenter. Gandhi condemned the West for precisely those virtues in which it took pride: modernization and industrialization. Gandhi seldom used the terms "science" or "technology." His concern was with civilization and mechanization, and on these topics he talked and wrote profusely. He considered machinery "the chief symbol of modern civilisation"; "It represents a great sin," he wrote. "It is machinery (and Manchester) that has impoverished India. . . . Machinery is like a great snake-hole which may contain from one to a hundred snakes. . . . I can not recall a single good point in connection with machinery. Books can be written to demonstrate its evils."[16] Yet many times Gandhi would say he was not opposed to machinery per se:

> How can I be when I know that even this body is a most delicate piece of machinery? The spinning wheel itself is a machine. What I object to is the craze for machinery . . . today machinery helps a few to ride on the backs of millions. The impetus behind it all

[16] Parel, *Gandhi* (cit. n. 2), pp. 109–10, note 2.

is not philanthropy to save labours but greed. It is against this constitution of things that I am fighting with all my might.[17]

Machinery as an ally of capitalism was Gandhi's anathema. What about machinery under state control, as in a Communist system? Here he could foresee the state emerging as a supercapitalist and becoming equally oppressive. Means and ends both had to be "pure." Gandhi would not mind villages plying their implements with the help of electricity. But the village communities or the state must own the power houses, just as they have their grazing pastures.[18] Later, in 1946, he even agreed that "some key industries are necessary." He would not enumerate these, but preferred that they be under state ownership, provided the state professed nonviolence.[19] Gandhi was worried less about production than about distribution. He would have supported the Soviet experiment in production and distribution, had it not been based on force. He was not for mass production, but for production by the masses, using the tools that they could afford to possess and repair.[20] A simple cottage machine like the *charakha* (spinning wheel) solved the problem of distribution: there would be no concentration of wealth and power.[21] Gandhi talked of developing around the *charakha* a social program including antimalaria measures, improved sanitation, conservation, cattle breeding, etc. Gandhi's critique was thus no mere denunciation of the West; it involved the development of a whole network of what he called "ameliorative activity."[22]

Two of the few occasions on which Gandhi addressed the issue of science were his 1925 talk to students at Trivandrum and his 1927 visit to the Indian Institute of Science at Bangalore. Gandhi declared that he was not an opponent of science, but that he wanted to put certain limitations "upon scientific research and upon the uses of science." Those limitations were nonviolence, humanity, and morality. To these, he added:

> How will you infect the people of the villages with your scientific knowledge? Are you then learning science in terms of the villages and will you be so handy and so practical that the knowledge that you derive in a college so magnificently put and I believe equally magnificently equipped—you will be able to use for the benefit of the villagers?[23]

There were no easy answers. The system prepared students for an urban or industrial India, or for a position abroad. An Indian student specializing in wood chemistry in America once asked Gandhi, "Would you approve of my going into industrial enterprise, say pulp or paper manufacturer [*sic*]? Do you stand for the progress of sci-

---

[17] *Young India*, 13 Nov. 1924, pp. 377–8.

[18] *Harijan*, 22 June 1935, p. 146.

[19] *Ibid.*, 1 Sept. 1946, title page.

[20] *Ibid.*, 2 Nov. 1934, pp. 301–2.

[21] As Gandhi once beautifully explained, "No amount of human ingenuity can manage to distribute water over the whole land as a shower of rain can. No irrigation department, no rules of precedent, no inspection and no water-cess [tax]. Everything is done with an ease and gentleness that by their very perfection evade notice. The spinning wheel, too, has got the same power of distributing work and wealth in millions of houses in the simplest way imaginable." *Young India*, 27 Dec. 1923, p. 430.

[22] Gandhians find in it a concept of "technology-practice." Sunil Sahasrabudhey, "Gandhi and the Challenge of Modern Science," *Gandhi Marg*, Sept. 1983, *V*:330–7.

[23] *The Hindu*, 19 March 1925, quoted in *The Moral and Political Writings of Mahatma Gandhi*, ed. Raghvan Iyer (Oxford: Clarendon Press, 1986), vol. 1, pp. 310–15.

Figure 2. *P. C. Ray seated at center with his favorite student M. N. Saha standing at extreme left, 1916. From Jyotirmoy Gupta, ed.*, M. N. Saha in Historical Perspective *(Calcutta: Thema Publishers, 1994).*

ence? I mean such progress which brings blessings to mankind, e.g., the work of Pasteur of France or that of Dr. Banting of Toronto." Gandhi had no objection to an industrial enterprise so long as it remained humanitarian. After all, it added to the productive capacity of the country. Similarly, with his admiration for the "scientific spirit of the West" came a plea for limitations on "the present methods of pursuing knowledge." One may not agree with Gandhi's great emphasis on nonviolence and antivivisection, etc., but he definitely provided a notion of social accountability with which it is almost impossible to differ. Similarly, his search for an alternative to industrialization cannot be overlooked.

DUALISM ENDS

P. C. Ray (1861–1944), a renowned chemist, educationist, and entrepreneur, showed a greater appreciation for Gandhi than other scientists of the day. Ray had a Western education and established the first great chemical industry in India. Yet in full appreciation of modern machinery, he adopted and pleaded for *charakha*. He wrote a piece on the economics of *charakha*, and citing from Buchanan's Survey of 1800, explained how rural women generated extra income through spinning and weaving.[24] He recognized that small, self-contained villages were gone forever, never to return. As he put it, "the Ganges can not be forced back to Gangotri."[25] Yet the relevance of Gandhi's emphasis on rural values and social control was not lost on Ray, nor was the erosion of these values by the march of Western civilization. Many

[24] *Young India*, 3 Aug. 1922, p. 323. Francis Buchanan was a British official under the East India Company who conducted a survey on social practices in India.
[25] *Ibid.*, 27 Dec. 1923, pp. 429–32.

contemporaries did not agree with Ray. In particular, Rabindranath Tagore, the laureate whom Gandhi adored as *Gurudev* (teacher), felt that *charakha* was being popularized more out of blind faith in Gandhi than out of reason. He, too, was critical of the modern West, but not in Gandhian terms.[26]

The sharpest criticism of Gandhian views came from Maghnad Saha (1893–1955), a pioneer astrophysicist and an illustrious student of Ray. Saha wanted India to choose "the cold logic of technology" over the vague utopia of Gandhian economics. While he appreciated Gandhi's "genuine sympathy with the victims of an aggressive and selfish industrialism," Saha firmly refuted the claim that better and happier conditions of life could be created by "discarding modern scientific technic and reverting back to the spinning wheel, the loin cloth and bullock cart."[27]

Saha was a staunch advocate of industrial progress through careful and deliberate planning backed by scientific research.[28] His thought had two significant strands. First, he treated science and technology as distinct yet connected knowledge systems without hierarchical relations. Second, he wished to introduce changes through a mixed economy in which both state and private enterprise had significant roles.[29] Saha was suspicious of Indian capitalists and would therefore assign a greater role to the state.[30] So when M. Visvesvaraya (1860–1962), an eminent engineer, suggested that applied research be prioritized, Saha found it "rather hazardous to insist on the mere industrial aspect."[31]

Many Indian scientists and technologists sought solutions for the country's problems, but their methods differed. P. C. Ray saw the remedy in Swadeshi and *charakha*. Visvesvaraya demanded rapid industrialization through the use of Indian capital and enterprise.[32] Saha insisted on "scientific method" in every aspect of national life. S. S. Bhatnagar (1894–1955) and H. J. Bhabha (1909–1966) preferred to build centers of excellence in frontier areas of scientific research. Bhatnagar was to build a chain of laboratories under the aegis of the Council of Scientific and Industrial Research (CSIR) in 1942. Bhabha played a greater role in independent India, pioneering nuclear technology. He saw a source of power in the controlled release of fission, while Saha saw power in the huge rivers of India. Bhabha argued for the quick importation of overseas industrial models for adaptation in India, thereby gaining time and leverage. Saha believed in the development of India through a wholly independent science and technology firmly embedded in socialist economics.[33]

By the mid-1930s, the long, drawn-out freedom struggle had begun to yield sig-

[26] Ashis Nandy, *The Illegitimacy of Nationalism* (Delhi: Oxford Univ. Press, 1994), pp. 2–3.

[27] *Science and Culture*, 1935, *I*:3–4. On the significance of this journal, which Saha founded, a noted historian of science wrote, "The outstanding difficulty is not so much that only a very small percentage of the Hindu population is literate, but rather that only a very small percentage of the literate minority is able to read a journal like *Science and Culture* and to understand its rational and scientific message. The peace, security and prosperity of India will not be advanced by metaphysical and religious discussions but by the wise application of scientific and rational methods. For the sake of India's freedom, we hope that the Hindu audience of *Science and Culture* will increase considerably." George Sarton, "Science and Freedom in India," *Isis*, 1948, *38:*244.

[28] *Sci. and Cult.*, 1940, *V:*639.

[29] Dinesh Abrol, "Colonised Minds or Progressive Nationalist Scientists: The Science and Culture Group," in MacLeod and Kumar, *Technology and the Raj* (cit. n. 6), pp. 265–88.

[30] *Sci. and Cult.*, 1938, *IV:*367.

[31] *Ibid.*, 1940, *V:*572.

[32] M. Visvesvaraya, *Planned Economy for India*, (Bangalore: Bangalore Press, 1934).

[33] R. S. Anderson, *Building Scientific Institutions in India: Saha and Bhabha* (Montreal: McGill Univ. Press, 1975), pp. 91–102.

Figure 3. *Dr. S. S. Bhatnagar (in hat and dark suit), Dr. H. J. Bhabha, and Jawaharlal Nehru (with his grandchildren Sanjay Gandhi and Rajiv Gandhi) watching a balloon flight at the National Physical Laboratory, New Delhi, 1952. From A. S. Bhatnagar,* Shanti Swarup Bhatnagar: His Life and Work *(New Delhi: NISTADS Publications, 1989).*

nificant results. Provincial autonomy was introduced in 1935 and debates on planning for national prosperity began. Visvesvaraya's *Reconstructing India* of 1920 had been perhaps the first attempt to make Indians "plan-conscious."[34] In 1934, he wrote *Planned Economy for India.* The same year G. D. Birla, a major industrialist and follower of Gandhi, pleaded for planning.[35] But it was Saha who convinced the national political leadership of its necessity. In 1937, he persuaded the Congress president, Subhas Chandra Bose, to constitute a National Planning Committee (NPC). This materialized at the end of 1938 under the chairmanship of a promising politician, Jawaharlal Nehru. As many as twenty-nine expert subcommittees were formed to address different areas of national reconstruction, including agriculture, industries, population, labor, irrigation, energy, communication, afforestation, health, housing, and education.[36] Nehru was amazed at both the utility and the vastness of the planning process. He wrote:

> One thing led to another and it was impossible to isolate anything or to progress in one direction without corresponding progress in another. The more we thought of this planning business, the vaster it grew in its sweep and range till it seemed to embrace almost every activity.[37]

[34] M. Visvesvaraya, *Reconstructing India* (London: King & Sons, 1920).

[35] G. D. Birla, "Indian Prosperity: A Plea for Planning," speech delivered on 1 April 1934 at Federation of Indian Chamber of Commerce and Industry, New Delhi, later published by Leader Press, Allahabad, 1950.

[36] The Second World War disrupted the work of the NPC, which resumed at the end of the war. Before dissolving in 1949, the NPC published twenty-seven volumes of reports outlining a ten-year plan to be implemented by a government of free India. K. T. Shah, *Report: National Planning Committee* (Bombay: Vora & Co., 1949).

[37] Jawaharlal Nehru, *The Discovery of India* (Calcutta: Signet Press, 1946), p. 396.

With nationalist feelings bursting at the seams, the NPC provided an opportunity to recast the old arguments in terms of national regeneration, self-sufficiency, and all-around progress.[38] But it also laid bare inherent tensions and contradictions. Gandhians like J. C. Kumarappa and S. N. Agarwal preferred traditional technology and village industries, and attacked the NPC.[39] Even within the Congress, younger leaders leaned towards modern science, technology, and heavy industrialization. Some wanted to follow the socialist path, others favored capitalist models such as that envisaged in Purshotamadas Thakurdas's 1944 *Plan for Economic Development of India* (popularly known as the Bombay Plan). The indigenous business class was no less divided. Gandhi's politics were convenient, but his economics were not. Although socialism remained an ideal for many (including Nehru and Saha), a version of democratic socialism with a mixed economy was accepted by the NPC as the basis for future development.[40] Saha resented this dilution, and even called the Congress leaders "puppets in the hands of big industrialists." He and Nehru gradually drifted apart.[41] Later Nehru leaned more towards Bhatnagar and Bhabha, which led to government-controlled industrial and defense research. By 1945 the political leadership had made a conscious decision to modernize, and the dualism of the previous decades came to an end.

A BELATED REALIZATION

To these new nationalist demands, the British government awoke rather late. The Board of Scientific Advice was allowed to languish and disappear. The much-publicized Holland Commission Report was shelved. Meanwhile, in 1928 the Government of India preempted the question of industrial research by linking it to agriculture. The Great War and, later, the Great Depression were cited as reasons for the Indian government's inactivity, while in other countries these acted as catalysts.[42] Britain had established a Department of Scientific and Industrial Research (DSIR) in the middle of the war (1916). Similar organizations were established in Australia and New Zealand, but not in India. In 1933, Richard Gregory, editor of *Nature*, visited Indian universities and appealed to the then secretary of state for India, Samuel Hoare, to create an equivalent of the DSIR there for the development of natural resources and new industries. Hoare supported the idea, as did the provincial governments of Bombay, Madras, Bengal, Bihar, and Orissa, but the Government of India did not.[43] The finance member of the Government of India, George Schuster, did

---

[38] J. N. Sinha, "Technology for National Reconstruction: The National Planning Committee, 1938–49," in MacLeod and Kumar, *Technology and The Raj* (cit. n. 6), pp. 250–64.

[39] J. C. Kumarappa, *Why the Village Movement?* (Wardha: All India Village Industries Association Publications, 1936); S. N. Agarwal, *The Gandhian Plan for Economic Development for India* (Bombay: Padma Publications, 1944).

[40] This was done to accommodate the views of the Indian industrialists as outlined in the Bombay Plan.

[41] Anderson, *Building Scientific Institutions* (cit. n. 33), p. 29.

[42] Irrespective of the Holland Commission's recommendation to initiate new industrial research, the Government of India in 1931 proposed drastic reductions in grants even to well-established areas of research like geology, botany, agriculture, etc. This was strongly opposed by several fellows of the Royal Society who, in a resolution, condemned such action "as risking a permanent loss to India, to the whole Empire, and to the world at large." *Minutes of the Royal Society*, vol. XIII, 1931, p. 230.

[43] V. V. Krishna, "Organisation of Industrial Research: The Early History of CSIR 1934–47," in MacLeod and Kumar, *Technology and the Raj.* (cit. n. 6), pp. 289–323.

feel the winds of change and was probably the first official to talk about planning India's development. But his dangerous "Keynesian" ideas were neither appreciated nor shared by his colleagues.[44] Except for an innocuous Industrial Intelligence and Research Bureau, set up in 1934 to gather and disseminate industrial intelligence information, the Government of India dithered. It witnessed in silence the stormy sessions of the Indian National Congress, the rumblings of Indian industrialists, and the establishment of the NPC.

The onset of the Second World War changed the scenario fundamentally. The secretary of state for India suddenly realized that it was possible to outmaneuver the Congress by attempting, "regardless of conventional financial restraints," a "complete overhaul of India's national life," with the British playing the postwar role of a "bold, far-sighted and benevolent despot . . . in a series of five-year plans, to raise India's millions to a new level of physical well-being and efficiency."[45] This was not a mere reassertion of the "white man's burden." It had within it a plausible answer to Gandhi's call to the British to "Quit India." The same year, after the failure of his political mission to transfer power to the Indians, Stafford Cripps wrote "A Social and Economic Policy for India" in which he proposed giving more attention to education, population control, agricultural productivity, factory legislation, and industrialization. He also recommended the use of "modern techniques of economic planning and the modern device of the Public Corporation."[46] This was also what nationalists like Saha were demanding, yet their motives were different. From the British point of view, it was a reassertion of constructive imperialism. But the likes of Saha would not accept an "empire-driven development." They would not settle for anything less than *Purna Swarajya* (total self-rule) and complete control over the agenda and its implementation.

Schuster and Cripps were not alone in trying a different strategy. Two successive viceroys, Linlithgow and Wavell, were "sincerely concerned to see that science is properly used to give India a chance of developing on right lines."[47] Linlithgow had even written a tract on Indian peasants.[48] When the Second World War began (in September 1939 for England), a Board of Scientific and Industrial Research was established under S. S. Bhatnagar with a charter similar to that of the British DSIR. By the end of 1940, it had about eighty researchers working on the purification of Baluchistan sulphur and the development of vegetable oil blends, dyes, and emulsifiers. Impressed by its utility and success, the government of India elevated this board to the status of a council (i.e., CSIR) in September 1942. Although their man-

[44] S. Bhattacharya and B. Zachariah, "A Great Destiny: The British Colonial State and the Advertisement of Post-War Reconstruction in India, 1942–45," *South Asia Research*, 1999, *19*:71–100.

[45] L. Amery to Linlithgow, 27 May 1942, quoted in *ibid.*

[46] *Ibid.* This strategic realization soon percolated down the hierarchy. For example, Roger Thomas (an agricultural expert and adviser on postwar reconstruction in Sindh) described planning "as the essence of post-war development. . . . Rather than a reversal of *laissez faire* . . . to avoid the pitfalls of the earlier mercantile doctrine . . . " Roger Thomas Papers, MSS Eur. F. 235/46, India Office Library and Records, London.

[47] A. V. Hill (secretary of the Royal Society, London) to C. J. Mackenzie (National Research Council, Canada), 14 Nov. 1944, A. V. Hill Papers, MDA/A7, Royal Society, London (hereafter cited as Hill Papers, London).

[48] Linlithgow, *The Indian Peasant* (London: Faber & Faber, 1932). Later, in the midst of a famine in Bengal, experts discounted the possibilities of "nation building on the foundations of an ignorant, illiterate and debilated people." Roger Thomas, *Notes on the Planning for Agriculture* (Simla: Government Press, 1944).

dates were similar, in constitution and work the British DSIR and the Indian CSIR differed. The former was answerable to the British Parliament through the Privy Council and was free from bureaucratic hassles, while the latter's agenda was set by the Department of Supply and Munitions. Still, the establishment of the CSIR in the middle of the war was a remarkable development. The chemist in Bhatnagar envisioned a "spectrum of research" with pure and applied research at either end.[49] Even with a limited budget, and in the midst of political turmoil, the CSIR began work in radio research, statistical standards, building materials, and electrochemistry. But, unlike the DSIR or similar organizations in Canada and Australia, the Indian CSIR was not encouraged to contribute substantially and directly to the war effort. It remained a distant, poor cousin. All that it could boast of at the end of the war were thirty-two processes, none of which could be classified as a major breakthrough.[50]

HILL'S 1944 VISIT: A TURNING POINT

This phase of transition (1937–1947) was, however, significant in view of the ever-increasing attention being paid to development and the role of science and technology as catalytic agents. It was an intellectually stimulating period in which age-old problems were revisited and a new vision emerged. An Indian sociologist asked for "the balm of science" to restore a civilization, sick and diseased, to "health, beauty and mobility."[51] The efforts of M. N. Saha and others helped, including the British scientists A. V. Hill (1886–1977) and J. D. Bernal (1901–1971).

In early 1944, Hill (biological secretary of the Royal Society) visited India on an official mission to advise and report on the state of scientific research.[52] Hill had an excellent personal rapport with Bhatnagar and he corresponded with more than fifty other Indian scientists.[53] Hill came to India with no political views, but quickly put economic problems high on the agenda. He regretted that, since 1931, several scientific operations had been starved by "false economy."[54] He spoke of a quadrilateral dilemma, i.e., population, health, food, and natural resources.[55] To him, the fundamental problems of India were "not really physical, chemical or technological, but a complex of biological one[s] referring to population, health, nutrition, and agriculture all acting and reacting with another."[56] To represent this, he formulated $H = f(X, Y, Z, W)$, where H is total human welfare, X is population, Y is health, Z is food, and W is other natural resources.[57] In all his lectures in and on India, Hill dwelt upon the need to control population. Several times he repeated, "You can not keep cats without drowning the kittens," which, put in terms of *Homo sapiens* instead of *Felis cattus*, simply meant, "You can not have a higher standard of life without lim-

---

[49] S. S. Bhatnagar, "Indian Scientists and the Present War," *Sci. and Cult.*, 1940, *VI:*195.

[50] Krishna, *Organisation of Industrial Research* (cit. n. 43), p. 315.

[51] K. Motwani, *Science and Society in India* (Bombay: Hind Kitab, 1945), p. vii.

[52] A. V. Hill, *Scientific Research in India* (London: Royal Society, 1944).

[53] A list of Hill's correspondents in India is given in Hill Papers, AVHL II, 5/115, Churchill College, Cambridge (hereafter cited as Hill Papers, Cambridge).

[54] Hill to F. W. Oliver, 5 June 1945, Hill Papers, London, MDA, A7.9.

[55] A. V. Hill, speech before the East India Association, 4 July 1944, *Asiatic Review*, Oct. 1944, *XI:*351–6.

[56] A. V. Hill, Messel Lecture before the Society of Chemical Industry, 1944, later published in A. V. Hill, *The Ethical Dilemma of Science and Other Essays* (New York: Rockefeller Institute Press, 1960), p. 375.

[57] Hill to John Mathai, 5 Feb. 1954, Hill Papers, Cambridge, AVHL II, 4/79.

iting reproduction."[58] Hill was not the first to stress population control.[59] But the political and industrial leadership of India remained complacent. To quote Hill:

> When I asked one of the authors of the Bombay Plan why population was practically not mentioned in their report, he replied that his colleagues could not agree and so had decided to leave the population problem to God. I asked him why they did not leave industry and housing to God, too . . . If public health and food are to be planned, so inevitably must population be, or all our efforts will be brought to nought.[60]

Hill's diagnosis was correct, but the remedies that he suggested were mixed in nature and value. He wanted government to recognize that science was not just a hand-maiden but "an equal partner in statecraft." In England, the Medical Research Council (1914), the DSIR (1916), and the Agricultural Research Council (1931) were not under respective user ministries, but under the lord president of the Council. In India, meteorology was under the Postal Department, the Geographical Survey was under Labour, the Survey of India was under the Education, Health and Land Department, and the CSIR was under the Department of Commerce. If scientific research in India was to make a concerted and coordinated contribution to national development, it had to be brought under "some more systematic plan" or "under one umbrella." So Hill recommended the creation of a Central Organization for Scientific Research with six research boards, covering medicine, agriculture, industry, engineering, war research, and geological and botanical surveys. He felt that, under a central organization, the different subjects would be treated with "some degree of uniform encouragement." The user-departments or ministries could have separate development or improvement councils that would be entrusted with the task of translating pure research into production.[61]

Hill made two additional points. Health and population problems were to be taken seriously, and agriculture and industry were to be brought together. As a biologist, he emphasized the research component in medical education, and asked for an all-India medical center.[62] He regretted that biophysics was completely neglected in the country, except at the Bose Institute in Calcutta. When Bhabha was planning for the founding of the Tata Institute for Fundamental Research in 1944, Hill advised him to "take biophysics under its wing."[63] Agricultural progress, he argued, would depend very largely on mechanization, land utilization, fertilizers, irrigation, transport, roads, food processing, and a great variety of other technoscientific factors. Probably he wanted to reduce India's traditionally heavy dependence on agriculture, but would

---

[58] *Ibid.*, 6/4/548.

[59] Eugenic ideas had already found some acceptance in India. N. S. Phadke pleaded for it in his *Sex Problem in India* (Bombay: Tarporvala & Sons, 1927), pp. 328–41. See also R. P. Paranjpye, *Rationalism in Practice* (Calcutta: Calcutta Univ. Press, 1935), pp. 20–3.

[60] Hill Papers, Cambridge, AVHL II 4/42.

[61] Hill, *Scientific Research in India* (cit. n. 52), pp. 40–4. Hill would not prescribe similar centralization for England "because already there is so much of scientific work going on in different departments that to centralise would do harm rather than good." Hill to Bhatnagar, 21 June 1948, Hill Papers, Cambridge, AVHL II.

[62] The All India Institute of Medical Sciences is now known more as a referral hospital than a research center.

[63] " . . . many of the most important application[s] of physics will be in biology." Hill to Bhabha, 22 June 1944, Hill Papers, London, MDA, A4.6.

not say so openly.[64] He preferred a safer, middle course. As he put it, "the factor of safety in India is far too low for luxuries like bloody revolutions or for monkeying about with machinery already groaning under a heavy overload."[65] This was a subtle denunciation of both the Communists and the Gandhians. Hill had a sense of mission, probably à la Kipling. To a British metallurgist, he wrote:

> India is at the parting of the ways, and that either she may go ahead to an efficient and prosperous economy or may sink back into inefficiency, disorder and disaster. We can of course, like Pilate, wash our hands of it and clear out, but that scarcely relieves us of the responsibility for having got them so far and then having left before the job is finished.[66]

Another interesting facet of Hill's interest in India is his understanding of the Indian scientists. Of C. V. Raman, India's best-known scientist, Hill wrote, "he is queer fish . . . Nobel Prize more or less turned his head! Meghnad Saha is also a rather an odd fellow but much more reasonable than Raman."[67] "Birbal Sahni has done good work in Palaeo-botany. Unfortunately he too is a politician and is unreasonable and stiff-necked."[68] Bhatnagar was his best bet, "a fighting Punjabi, extremely energetic, dashing hither and thither like mercury but always to good purpose, a most loyal and friendly fellow."[69]

Other scientists with whom Hill developed lasting friendships were H. J. Bhabha, S. S. Sokhey (director, Haffkine Institute in Bombay), S. L. Bhatia (Indian Medical Service), J. C. Ghosh (director, Indian Institute of Science), and J. N. Mukherjee (director, Indian Agricultural Research Institute). To Bhabha he advised, "I very much hope that you will get Chandrasekhar (a renowned astrophysicist) . . . There are not many of his quality in India that you can afford to give him away to the Americans."[70] How prophetic! Chandrasekhar later decided to settle in the United States.

Another influential British scientist who took a keen interest in India was J. D. Bernal. Bernal firmly believed that "science does not exist in a social and economic vaccum," and he debunked the idea of pure science as a "convenient fiction" that enabled "the wealthy to subsidize science without fearing to endanger their interests and scientists to avoid having to ask awkward questions as to the effects of their work in building up the black hell of industrial Britain."[71] Bernal wanted India to make the most scientific use of its resources. For example, he believed water was "even more important than power in the economy of India." So he advocated sand reservoirs (as in Soviet Asia) rather than open tanks and enclosed agriculture (as in the United States) for northern parts of India. Control of soil erosion was vital, as

---

[64] In contrast, more than five decades before Hill, a perceptive judge had asked, "Why are we to suppose that the inhabitants of India, ingenious, quick, receptive, skil[l]ful of hand, and authentic in taste, are to confine themselves, as in the past, almost exclusively to agriculture, and will not some day be in a position to take a hint from England, which for its small population, imports more than two millions' worth of food per week?" Justice Cunningham, "The Public Health in India," *Journal of the Society of Arts*, 1888, *XXXVI*:241–65.

[65] Hill Papers, London, MDA, A5.12.

[66] Hill to A. MacCance, 16 May 1944, Hill Papers, London, MDA, A5, 12.

[67] Hill to the Vice Chancellor, Cambridge University, 26 March 1946, Hill Papers, London, MDA, A5.5.

[68] Hill to F. W. Oliver, FRS, 22 Dec. 1944, Hill Papers, London, MDA, A.7.

[69] Hill to Ralph Glyn, M. P., 18 Oct. 1944, Hill Papers, London, MDA, A5.14.

[70] Hill to Bhabha, 22 June 1944, Hill Papers, London, MDA, A4.6

[71] J. D. Bernal, "Science and Liberty," *Science and Society*, 1938, *I*, 6:348–50.

was microbiology for tapping biological resources. Bernal was also one of the earliest to stress the value of solar energy.[72]

Like Hill, Bernal was a keen observer of people. He was present at the opening of both the National Chemical Laboratory in Poona and the National Physical Laboratory in Delhi. Of the Delhi ceremony he recounted later:

> It was astonishing the respect which the authorities in India give to science, it is difficult to imagine an opening of any building in England which would be attended by the King, the Prime Minister and two other ministers but that was the case there, because the retiring Governor-General (Mr. Rajgopalachari) was making his last public appearance on this occasion . . . As the speeches went on in their inanity they gradually got me down, finally I had to listen to Patel (Home Minister) himself talking about the restoration of law and order which seemed hardly worthwhile. As a result I threw away the speech I had prepared and even proceeded to give quite a different one, telling them that they would never get anywhere unless they did the job with popular support and *put some more money into the universities and the teachings* and saw that the plans which were made were actually carried out instead of remaining on paper.[73]

REFLECTIONS: DISUNITY IN THE DEVELOPMENT DISCOURSE

The twentieth century, like the eighteenth century, was an era of transition for India. Swadeshi and *swaraj* reverberated in the air. These were more than political slogans; they symbolized an intense yearning for change.[74] The direction of change, however, remained unclear. Many critiques appeared, and many options were debated. A renowned art critic, Ananda Coomaraswamy, criticized his contemporary Swadeshi protagonists for having ignored the skilled artisans and village craftsmen.[75] Officials like Alfred Chatterton (director of industries, Madras) differed and asked for industrial development under government patronage.[76] Towards this end, the Holland Commission attempted some sort of "a national planning in a bureaucratic guise." But nothing could impress the British government, which conveniently and regularly took refuge in the situation created by war and depression.

By the 1930s, nationalism had gathered strength, and thanks to the movement's demand for total indigenization, the British government could conveniently leave the core sectors of agriculture, health, and education in Indian hands. Indian leaders

[72] "Notes on the Utilisation of Research on Short and Long-term Developments in India," J. D. Bernal Papers, MSS Add. 8287, box 52.4.55, Cambridge Univ. Library, Cambridge (hereafter cited as Bernal Papers).

[73] *Ibid.*, box 91.L1.33. The inauguration in Poona was no less interesting. To quote Bernal again, "Bhatnagar was running the whole show. He had even provided complete speeches for everyone to make. I saw the one that had been given to Irene [Curie]. It was mostly an account of the wonderful energy, enterprise, foresight and public spirit of Sir Bhatnagar himself. However, Irene was sufficiently high-spirited to alter it all and hardly to mention the admirable gentleman in her speech. After about two hours [of] speeches the doors were opened with some difficulty with a golden key by the beloved leader (Pt. Nehru) and we all trooped in. The effect however was somewhat disconcerting because behind the façade there was very little indeed; most of the building had not been built at all. There was only one storey and that contained no apparatus that actually worked." *Ibid.*

[74] As a recent work argues, "It would be erroneous to conceive Swadeshi's nativism as an atavistic upsurge of a reified tradition in the face of modernization. Rather, nationalism's nativist particularism must be situated within a broader understanding of the perceived decentering dynamic of capitalist expansion." Manu Goswami, "From Swadeshi to Swaraj: Nation, Economy, Territory in Colonial South Asia, 1870–1907," *Comparative Studies in Society and History*, 1998, *40*, 4:609–37.

[75] Ananda Coomaraswamy, *Art and Swadeshi* (Madras: Ganesh, 1910).

[76] For details see Nasir Tyabji, *Colonialism and Chemical Technology* (Delhi: Oxford Univ. Press, 1995).

and government now vied to raise development issues. Dissenters were gradually marginalized. To the government, Gandhi was important politically, but not otherwise. His vision of a new India was not fully shared even by his own followers!

The period 1937–1947 deserves more attention. Besides NPC publications, the Bombay Plan, and Hill's report, other significant studies emerged. Health and industrial research, for example, were explored in committees under the chairmanship of Joseph Bhore and Shanmukham Chetty, respectively.[77] All these committees had at least one thing in common: they asked for greater institutionalization. Everyone recognized the importance of coordination, and even for this they asked for a coordinating institution![78] But the proliferation of institutions did not necessarily result in quality research, nor could it bring together academic science and industrial needs. Indians were encountering a situation similar to Britain's, where the work of the Board of Education, the University Grants Commission, and the DSIR reflected widely varying pressures.[79] All of India's major political leaders, and even the colonial government, had joined in the development discourse chorus.[80] The new watchwords were planning, industrialization, institutionalization, and coordination.

While there seemed to be unity of purpose among the sectors taking part in the debate, in reality this was not so. The discourse was run by "experts": middle-class professionals who stood for the state and, through the state, for the nation. Politicians and bureaucrats added their own flavor. And all this was done in the name of the "masses" who "entered the picture only as the somewhat abstract ultimate beneficiary."[81] In the name of the masses, the authors and defenders of the Bombay Plan asked for a shift from an "over-agriculturalised" economy to an industrial economy.[82] At the plan's core was the notion that industrial production was more capitalistic than agricultural production. Even those who argued for industrialization felt the pressure of class interests:

> The real bottleneck is the attitude of capitalists, and civil servants in league with them, who oppose all industrial production in excess of the proved absorptive capacity of the market and threaten that such production is unprofitable. They forget that planned development ultimately aims at giving plenty to all and not profits to few.[83]

A more viable proposition was planned development using the socialist (Soviet) pattern. Nehru's heart lay here, but the pragmatist in him led to compromises at

---

[77] *Report of the Health Survey and Development Committee*, 4 vols. (Delhi: Manager of Publications, 1946); *Report of the Industrial Research Planning Committee* (Delhi: Council of Scientific and Industrial Research Publications, 1945).

[78] The Chetty Committee recommended a National Research Council (consisting of representatives of science, industry, labor, and administration) to coordinate all research activities and to function as a National Trust for Patents. *Ibid.*

[79] Roy MacLeod and E. K. Andrews, "The Origins of DSIR: Reflections on Ideas and Men, 1915–1916," *Public Administration*, Spring 1970, pp. 23–48.

[80] Bidyut Chakraborty, "Jawaharlal Nehru and Planning, 1938–41: India at Crossroads," *Modern Asian Studies*, 1992, *16*, 2:275–87; I. Talbot, "Planning for Pakistan: The Planning Committee of the All India Muslim League 1943–46," *Mod. Asian Stud.*, 1994, *28*, 4:877–86; Vinod Vyasulu, "Nehru and the Visvesvaraya Legacy," *Economic & Political Weekly*, 29 July 1989, pp. 1700–4.

[81] Benjamin Zachariah, "The Development of Professor Mahalanobis," *Economy and Society*, 1997, *26*, 3:434–44. I am grateful to Sanjoy Bhattacharya for this subtle review article.

[82] An interesting debate for and against the Bombay Plan can be seen in P. C. Malhotra and A. N. Agarwala, "Agriculture in the Industrialists' Plan," *Indian Journal of Economics*, 1945, *XXV*, 99:502–10.

[83] J. C. Ghosh, "Industrial Planning for India," *Sci. and Cult.*, 1947, *XII*:345–7.

every step. Lest right-wing leaders feel alienated, Nehru spoke of independence and a democratic structure first, to be followed by socialism and planning.[84] After independence, he perfected the art of mixed economy and mixed politics.[85] He tried to combine Gandhi and Visvesvaraya, and finally could do justice to neither.

Another casualty of the era of mixed policies was the traditional distinction between pure and applied science. Bhatnagar, Bhabha, and many others looked for a composite structure that combined the two. But to purists like Meghnad Saha, pure science was the seed of applied science, and "to neglect pure science would be like spending a large amount on manuring and ploughing the land and then omit the sowing of any kind."[86] Satyen Bose and C. V. Raman had similar views. Academic science still held a greater appeal to the Indian mind.

However, the notion that science and technology were two sides of the same coin had two interesting results for an undeveloped country. First, it meant assigning a far greater authority and responsibility to the state.[87] Second, in the name of coordinating the two, the tendency in practice was to centralize. Hill himself argued for centralization (which he would not prescribe for Britain), and this suited Bhatnagar. Centralization and concentration of power was to become the hallmark of the scientific establishment in postindependent India. Bhatnagar acquired the reputation of being an "empire-builder."[88] He built a chain of eleven national laboratories from 1947–1954 and twenty more would follow. Was this done at the cost of the university system? Even Hill was alarmed to see "the great developments in Government research laboratories in India," and warned Bhatnagar "lest by getting all the best people away from the universities you may dry up the source of scientific talent, or at least training, for the next crop of scientists."[89] How right he was!

The development discourse in India was thus both intense and instructive. Its reconstruction pleas and plans were neither inane nor mere pious hopes: they were sincere attempts to attain a well-deliberated goal. In recent years, the concept of planning for development has been discredited, and the state now appears more "predatory" than developmental.[90] In today's India, at the threshold of another century, one sorely misses the quality and intensity of the early to mid-twentieth century development debates.

[84] S. Gopal, ed., *Selected Works of Jawaharlal Nehru* (New Delhi: Orient Longman, 1976), vol. 9, pp. 377–99.

[85] In a chance meeting in Beijing in 1954, Nehru confided to J. D. Bernal, "Most of my Ministers are reactionary and scoundrels but as long as they are my Ministers I can keep some check on them. If I were to resign they would be the Government and they would unloose the forces that I have tried ever since I came to power to hold in check . . . I have to work with the people who are actually influential with the country. They may not be the kind of people I like but it is the best I can do." To this Bernal added, "He treated the rule of a country of hundreds of millions as if it were the management of a college in Cambridge." Bernal Papers, MSS Add. 8287, box 48, B.3.349.

[86] *Sci. and Cult.*, 1943, *IX:*571.

[87] This was not something new. The colonial requirements had introduced the concept of state scientist. The postcolonial state strengthened it further.

[88] G. F. Heany Papers, Memories (TS), box VII, f. 340, Centre for South Asian Studies, Cambridge.

[89] Hill to Bhatnagar, 11 May 1951. To this Bhatnagar replied, "The universities in this country have not suffered for want of government help but the public interest in the universities has declined largely because the universities are having vice-chancellors not on the consideration of their attainments but of their political affinity." Bhatnagar to Hill, 18 May 1951, Hill Papers, Cambridge, AVHL II.

[90] S. Chakravarti, "Predatory State: The Black Hole of Social Science," *The Times of India*, 22 Sept. 1999.

Development as Experiment:
Science and State Building in Late Colonial and Postcolonial Africa, 1930–1970

Christophe Bonneuil[*]

ABSTRACT

This paper explores the continuous role that science has played in the establishment of a colonial and post-colonial "development regime" in Africa. Examining development schemes that flourished between 1930 and 1970, the paper shows how African agrarian societies became objects of both state intervention and expert knowledge. In pursuing large scale social engineering and social experiments, these schemes constituted a particular—colonial?—way of managing the African environment and of crafting knowledge on African societies. In constructing development ideologies and practices in the late colonial and post independence periods, they also played an important part in the construction of the African state. Their approaches shaped the future of tropical medicine, agriculture, and development studies. Ironically, they also created the preconditions for later interest in the values of indigenous knowledge.

INTRODUCTION

IN 1958, THE BRITISH BIOLOGIST EDGAR B. WORTHINGTON, general secretary of the intercolonial Scientific Council for Africa South of the Sahara from 1950 to 1955, reflected on the problems of African development in these terms:

> Can anyone to-day forecast just how the numerous factors involved in the progressive development of backward people will react on each other? . . . The pilot scheme, owing to its limited liability, offers great advantage in testing the theories in advance by intensifying the effort locally. This is a normal procedure when a new discovery or technique is introduced into an enterprise or into agricultural practice. It is less simple when a human society is the object of the study, because mistakes are less easy to rectify.[1]

In his first book, *Science in Africa* (1938), which resulted from his work on the African Research Survey, an unprecedented continentwide study funded by the Car-

* Centre National de la Recherche Scientifique, Centre Koyré d'Histoire des Sciences et des Techniques, Paris, France; christophe.bonneuil@wanadoo.fr.

[1] Edgar B. Worthington, *Science in the Development of Africa* (Hertfort, U.K.: Stephen, Austin & Sons, 1958), p. 26.

negie Corporation in 1931, Worthington had already pictured Africa as a "fruitful field in history for experiment concerning the place of expert scientific knowledge."[2] But what he meant by this was that more scientific knowledge of African environments and societies was needed for a sounder and more progressive colonial policy, and that Africa was a continent full of research opportunities. Twenty years later, he was more precise. By then, African development had become an equationlike problem that could be solved by experiment. Planned pilot schemes constituted the laboratories where development could be experimented with, using Africans as subjects. He viewed these schemes as laboratory experiments in "acceleration of progress" that would provide a forward-looking perspective and models that could be used elsewhere to monitor the development process.[3] From "science for development" to "development as experimental science," Worthington's changing views reveal the evolving discourses and practices of science and development. Exploring these discourses and practices at work in planned-development schemes, this essay considers the role that science played in the building of the developmentalist state in tropical Africa in the colonial and postcolonial period.[4]

By the term "developmentalist state," I mean a specific stage in the history of African societies, situated between the early colonial state and the post-1980 crisis of the state in Africa. One can locate the birth of the developmentalist state in tropical Africa in the 1930s, when colonial governments confronted the disorders and the threats of the Great Depression, adopted a more *dirigiste* agenda, intervened more directly in the economy, and took steps towards planning and state regulation. Major welfare and development policies also emerged in the 1930s and were key milestones in state building in Africa. Despite political changes during and after the Second World War and decolonization, strong continuities persisted between the 1930s and 1970s.[5] Postindependence (mostly urban) African elites sought to end

[2] *Idem, Science in Africa: A Review of Scientific Research Relating to Tropical and Southern Africa* (Oxford: Oxford Univ. Press, 1938), p. 17. See Helen Denham (now Tilley), "Africa as a 'Living Laboratory': The African Research Survey and *Science in Africa*, 1920–1945," communication to the Science in Africa conference (Oxford, 7 Mar. 1998), kindly communicated by the author. She is preparing a doctoral dissertation at Oxford on the survey, entitled "Africa as a 'Living Laboratory'—The African Research Survey and the British Colonial Empire: Consolidating and Applying Environmental, Medical, and Anthropological Ideas, 1920–1945."

[3] Worthington, *Science in the Development of Africa* (cit. n. 1) p. 26. Note that, writing only a few years after the complete failure of the Groundnut Scheme (a mammoth mechanized peanut-growing program) in Tanganyika, his mention of "mistakes" nevertheless reveals a consciousness of the dangers of this experimentation and a caution concerning unprepared large-scale schemes.

[4] I am aware that my synthetic perspective neglects important contrasts between different colonies and countries, especially between cash-crop and non-cash-crop areas; between settler and nonsettler colonies; among French, Belgian, and British governance styles; and between Marxist and non-Marxist postindependence regimes. But I believe that despite these important differences, there are common trends that singularize the developmentalist era and the role that science played.

[5] This periodization is suggested by works such as Claudine Cotte, *La Politique économique de la France en Afrique Noire, 1936–1946* (Paris: Thèse de l'Univ. Paris, 7, 1981); Stephen Constantine, *The Making of the British Colonial Development Policy, 1914–1940* (London: Franck Cass, 1984); Jacques Marseille, *Empire colonial et capitalisme français: Histoire d'un divorce* (Paris: Albin Michel, 1984). It differs substantially from Young's periodization: early colonial state, institutionalized colonial state in the interwar years, post-1945 contested colonial state, independent African state, post-1980 predatory state. (See Crawford Young, *The African Colonial State in Comparative Perspective* [New Haven: Yale Univ. Press, 1994].) This difference can be explained, first, by Young's mainly political approach, which leaves some important economic, institutional, and social continuities aside. Second, whereas Young sees the welfare and development policy only as a belated legitimation discourse, I consider that this policy was a keystone in the construction of the state itself,

both the economic dependence inherited from colonialism and traditions viewed as delaying their country's entrance into modernity. They therefore largely followed paths opened in the late colonial period and opted for policies and forms of "authoritarian social engineering" first experimented with by the colonial rulers.[6] One can speak of the developmentalist state emerging when the colonial state gave priority to a form of power concerned with changing ("improving") living conditions, so as to disable old forms of life and subjectivity and to turn African societies into objects of its cognitive apparatus and rationalizing interventions.[7] Exploring these new forms of power and knowledge, this essay contributes to an archaeology of "development," the word being taken as a regime of practices of intervention and knowledge whose objects were the agrarian societies of Africa.[8]

Science played a central role in the making of this development regime and its maintenance after decolonization. The study of the multilayered role that science played in the colonial enterprise has become an active field.[9] Conversely, it is now acknowledged that science in the colonies, far from representing a mere transfer of European science, was shaped by the colonial context. For example, specific natural, institutional, and social environments framed not only local practices of the same disciplines, but also the rise and development of new objects and disciplines—the shift from colonial to tropical medicine being the best-documented case.[10] Moreover, encounters with indigenous knowledge and practices brought a level of hybridity that only now is becoming widely acknowledged. More deeply, recent scholarship has viewed "colonial science" not merely as a science practised in the colonies (the descriptive sense in which George Basalla used the term[11]) but as a kind of knowledge specifically colonial both in the way it was crafted and in that it represented a discourse that conceptualized European domination and shaped the subjectivity of the colonized people.[12] Such reciprocal ties between science and colonialism have

---

as have scholars like Gerd Spittler, *Verwaltung in einem afrikanischen Bauernstaat: Das koloniale Französisch-Westafrika, 1919–1939* (Freiburg: Atlantis, 1981); David Ludden, "India's Development Regime," in *Colonialism and Culture*, ed. Nicholas Dirks (Ann Arbor: Univ. of Michigan Press, 1992), pp. 247–87; and James Scott, *Seeing Like a State: How Certain Schemes to Improve the Human Condition have Failed* (New Haven: Yale Univ. Press, 1998).

[6] Significantly, Young refers to the postindependence state in Africa as the "integral state." Young, *The African Colonial State* (cit. n. 5), pp. 287–90.

[7] David Scott, "Colonial Governmentality," *Social Text*, 1995, *43:*191–220.

[8] On the concept of "development regime," see Ludden, "India's Development Regime" (cit. n. 5).

[9] For recent overviews of this field, see Michael Adas, "A Field Matures: Technology, Science, and Western Colonialism," *Technology and Culture*, 1997, *38:*478–87; Roy MacLeod, "Passages in Imperial Science: From Empire to Commonwealth," *J. Interdis. Hist.*, 1993, *4:*117–50; Paolo Palladino and Michael Worboys, "Science and Imperialism," *Isis*, 1993, *84:*91–102; Richard H. Drayton, "Knowledge and Empire," in *The Oxford History of the British Empire: The Eighteenth Century*, ed. P. J. Marshall (Oxford: Oxford Univ. Press, 1998), pp. 231–52; Patrick Petitjean, Catherine Jami, and Anne Marie Moulin, eds., *Science and Empires: Historical Studies about Scientific Development and European Expansion* (London: Kluwer, 1992); Marie-Noëlle Bourguet and Christophe Bonneuil, "De l'inventaire du globe à la 'mise en valeur' du monde: Botanique et colonisation (fin XVIIIe siècle– début XXe siècle). Présentation," *Revue Française d'Histoire d'Outre-Mer*, 1999, *322–23:*9–38.

[10] Michael Worboys, "The Emergence of Tropical Medicine," in *Perspectives on the Emergence of Scientific Disciplines,* eds. G. Lemaine, R. MacLeod, M. Mulkay *et al.* (The Hague: Mouton, 1976), pp. 75–98.

[11] George Basalla, "The Spread of Western Science," *Science*, 1967, *156:* 611–22.

[12] Megan Vaughan, "Health and Hegemony: Representation of Disease and the Creation of the Colonial Subject in Nyasaland," in *Contesting Colonial Hegemony: State and Society in Africa and India*, eds. Dagmar Engels and Shula Marks (London: British Academic Press, 1994), pp. 173–201, on p. 201; Shula Marks, "What is Colonial about Colonial Medicine? And What Happened to Imperi-

an inertia. No wonder that science and technology, shaped in colonial contexts, remained major factors of the colonial legacy.

Large-scale, prepackaged development schemes offer a particularly promising field—but one largely unexplored by historians of science and technology—for research into the relations among science, the state, and society from the colonial to the postcolonial periods. As Worthington's quoted remarks rightly suggest, these schemes were at the center of development discourses and practices. By 1960, they involved more than one million Africans. The first section of this chapter analyzes these attempts to redesign African lives and modes of production from above as emblematic of the growing power of scientists in the culture of development. The second section considers how new power relationships, and new knowledge of African environments and societies, were coproduced in the age of the developmentalist state.

**DEVELOPMENT CULTURES: EXPERTS, PLANNING, AND THE
MODERNIZATION OF AFRICA FROM ABOVE**

An Era of Settlement Schemes

Prepackaged settlement schemes flourished in Africa between the 1930s and the 1970s (Table 1 and Figure 1).[13] Some of them were driven by public health concerns. As early as 1906, the British in Uganda evacuated African populations from areas inhabited by the tsetse fly. During the interwar years, resettlements were undertaken for similar reasons in British East and West Africa, as well as in the Belgian Congo.[14] Other schemes aimed at the settlement of unpopulated but potentially fertile regions. Moving the Peanut Belt eastward was the aim of the Terres Neuves program in Senegal. Between 1934 and 1937, 4 thousand Sereer people were relocated (often by force) under this scheme in the Terres Neuves.[15]

Other schemes were driven by a concern with irrigation—for example, the Gezira Scheme, run by the British in the Sudan, and the Office du Niger, created in 1932 in French Sudan (now Mali).[16] These mammoth projects aimed to bring millions of acres under irrigation and to engineer whole societies around the production of cot-

---

alism and Health?," *Social History of Medicine*, 1997, *10*:205–19. See also Jean Comaroff and John Comaroff, *Of Revelation and Revolution: Christianity, Colonialism and Consciousness in South Africa* (Chicago: Univ. of Chicago Press, 1991), chap. 3, and Valentin Yves Mudimbe, *The Invention of Africa: Gnosis, Philosophy and the Order of Knowledge* (Bloomington: Indiana Univ. Press, 1988).

[13] There were, during the same period, many similar schemes in North Africa and South and Southeast Asia that I cannot discuss here.

[14] See Robert Chambers, *Settlement Schemes in Tropical Africa* (London: Routledge, 1969), pp. 18–22; Michael Worboys, "The Comparative History of Sleeping Sickness in East and Central Africa, 1900–1914," *Hist. Sci.*, 1994, *32*:89–102; Kirk A. Hoppe, "Lords of the Flies: British Sleeping Sickness Policies as Environmental Engineering in the Lake Victoria Region, 1900–1950," Working Papers in African Studies, no. 203 (Boston: Boston Univ. African Studies Center, 1995); Maryinez Lyons, *The Colonial Disease: A Social History of Sleeping Sickness in Northern Zaire* (Cambridge: Cambridge Univ. Press, 1992).

[15] Christophe Bonneuil, "Penetrating the Natives: Peanut Breeding, Peasants and the Colonial State in Senegal (1900–1950)," *Sci. Technol. Soc.*, 1999, *4*:273–302.

[16] On the Gezira Scheme see Victoria Bernal, "Cotton and Colonial Order in Sudan: A Social History with Emphasis on the Gezira Scheme," in *Cotton, Colonialism and Social History of Sub-Saharan Africa,* eds. Allen Isaacman and Richard Roberts (Portsmouth, New Hampshire: Heinemann, 1995), pp. 96–118.

Table 1. Settlement Schemes in Tropical Africa

Planned-resettlement schemes (driven by either productivist or sanitary concerns, or due to dam construction)

Anchau Rural Development and Settlement Scheme, Nigeria: 5,000 people relocated out of sleeping sickness areas (1930s)

Terres Neuves scheme, Oriental Saloum (Senegal): 4,000 Sereer people resettled (1934–1939)

Numerous "planned-resettlement" schemes in British East and West Africa; examples: resettlement of Mwea in Kenya (included irrigation) and Volta River resettlements in Ghana

Several resettlements due to dam construction

Irrigation settlement schemes (two mammoth undertakings)

Gezira scheme, Anglo-Egyptian Sudan (begun 1906): 420,000 hectares (1939) and 25,000 tenant households (by 1957)

Office du Niger, French Sudan (first scheme begun 1926, Office established 1932): 34,700 inhabitants on 35,673 hectares (1959)

Mechanization settlement schemes (tenancy system)

Farming settlement schemes, Tanganyika (1952–), managed by Tanganyika Agricultural Corporation in Kongwa, Nachingwea, and Urambo (after failure of East African Groundnut Scheme): a few hundred tenant households farming a few thousand hectares by 1961

Boulel (Terres Neuves) and Sefa (Compagnie Générale des Oléagineaux Tropicaux, Casamance), Senegal: 10,660 people on 4,460 hectares (1959)

Niger agricultural project (1949–1954) and Sokoto Mechanized Rice Scheme, Nigeria

Volta Dam project, Ghana (1964–1967); relocation into 52 new towns, with mechanized agriculture: 80,000 people (1965)

More integrated technology transfer and welfare settlement schemes (involved changes in land use and tenure, housing and village political organization, agricultural techniques, soil conservation, and often educational and health programs)

"Paysannats," Belgian Congo (1936–1960): 140,000 inhabitants (1955)

Zande scheme, Sudan (1943–): 60,000 families resettled (1946–1950)

Swinnerton Plan, Kenya (1954–)

Ujaama Villages, Tanzania (1967–1977): *first phase* (1973), 2 million inhabitants in 5,628 Ujamaa villages; *second phase* (forced villagization, 1977), 13 million inhabitants in 7,500 Ujamaa villages

---

ton to fulfill the needs of industry at home.[17] In these schemes, the land use, farming system, work, and life of the thousands of African tenants, brought from other regions, were strictly constrained by a disciplinary order. Each village was overseen by an African officer, under the authority of European inspectors who were themselves

---

[17] In fact, only 36,000 were irrigated in 1960 because of various technical and human problems. See Jean-Michel Bordage, *De la terre, de l'eau et des hommes: Colons et techniciens de l'Office du Niger, 1932–1985* (Tours: Thèse de l'Univ. of Tours, 1991); Emil Schreyger, *L'Office du Niger au Mali de 1932 à 1982: La Problématique d'une grande entreprise agricole dans la zone du Sahel* (Wiesbaden: Steiner, 1984); Monica Van Beusekom, "Colonisation indigène: French Rural Development Ideology at the Office du Niger, 1920–1940," *Int. J. African Hist. Stud.*, 1997, *30*:299–323.

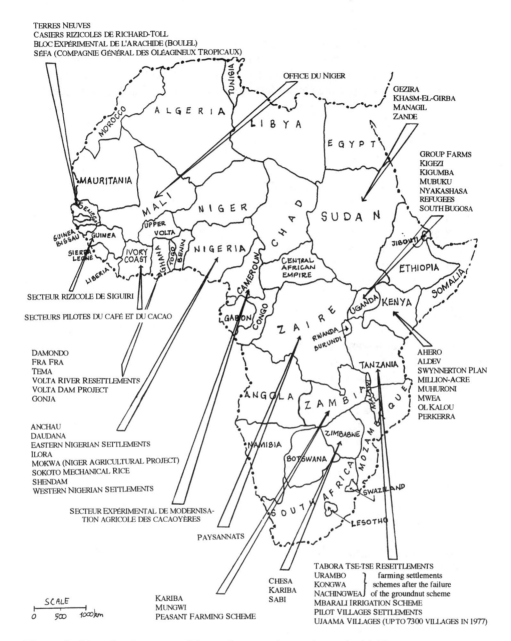

TERRES NEUVES
CASIERS RIZICOLES DE RICHARD-TOLL
BLOC EXPÉRIMENTAL DE L'ARACHIDE (BOULEL)
SÉFA (COMPAGNIE GÉNÉRAL DES OLÉAGINEUX TROPICAUX)

OFFICE DU NIGER

GEZIRA
KHASM-EL-GIRBA
MANAGIL
ZANDE

GROUP FARMS
KIGEZI
KIGUMBA
MUBUKU
NYAKASHASA
REFUGEES
SOUTH BUGOSA

SECTEUR RIZICOLE DE SIGUIRI

SECTEURS PILOTES DU CAFÉ ET DU CACAO

DAMONDO
FRA FRA
TEMA
VOLTA RIVER RESETTLEMENTS
VOLTA DAM PROJECT
GONJA

AHERO
ALDEV
SWYNNERTON PLAN
MILLION-ACRE
MUHURONI
MWEA
OL KALOU
PERKERRA

ANCHAU
DAUDANA
EASTERN NIGERIAN SETTLEMENTS
ILORA
MOKWA (NIGER AGRICULTURAL PROJECT)
SOKOTO MECHANICAL RICE
SHENDAM
WESTERN NIGERIAN SETTLEMENTS

SECTEUR EXPÉRIMENTAL DE MODERNISA-
TION AGRICOLE DES CACAOYÈRES

PAYSANNATS

SCALE
0   500   1000 km

KARIBA
MUNGWI
PEASANT FARMING SCHEME

CHESA
KARIBA
SABI

TABORA TSE-TSE RESETTLEMENTS
URAMBO   farming settlements
KONGWA   schemes after the failure
NACHINGWEA   of the groundnut scheme
MBARALI IRRIGATION SCHEME
PILOT VILLAGES SETTLEMENTS
UJAAMA VILLAGES (UP TO 7300 VILLAGES IN 1977)

Figure 1. *Map showing some of the settlement schemes in tropical Africa.*

supervised by a unit manager. Villages were configured so as to be functional—of standardized size (300 inhabitants, in the Office du Niger), housing, and cost. Built on a checkerboard plan, the villages had a central green so people could easily be gathered. African tenant farmers were prisoners of an organization of work in which they were mere factors of production.

After the Second World War, mechanization of agriculture became the order of the day and motivated many prepackaged schemes. Impressed by the American and Soviet agricultural models, France and Britain sought to increase African agricultural production with the help of mechanical modes of production to cope with the postwar foreign-exchange crisis. After the failure of projects based on wage labor (the most notorious being the Groundnut Scheme in British Tanganyika), mechanization schemes flourished in almost all the colonies in the 1950s and assumed an organization based on tenant farming.[18]

Rather than being resolutely focused on a single theme (public health, irrigation, colonization of marginal lands, agricultural production, mechanization), many schemes simultaneously pursued several goals. The *paysannats* in the Belgian Congo are examples. In public health policy as well as in agricultural development, the authoritarian and interventionist character of the Belgian colonial regime prepared the authorities in Zaire to engage in social engineering on a larger scale than their French and British counterparts. The *paysannats,* established from 1936 onward, were therefore bold and huge settlement and land-use schemes that by 1955 included no less than 140,000 Africans.[19]

Following decolonization, the *paysannats* faded away, but most other large-scale development projects that started in the late colonial period continued after independence. Whereas agricultural development programs in tropical Africa during the last years of colonial rule saw a tendency to move away from bold settlement and mechanization schemes towards piecemeal improvements, a spectacular renewal of large technocratic schemes marked the postindependence years at the initiative of the new ruling elites, who expected to turn their modernist aspirations into reality.[20] The Office du Niger, for instance, survived decolonization quite well: after 1960, farmers became—in an almost unchanged Office—the guinea pigs for the collectivization policies of the new regime. In President Kwame Nkrumah's Ghana, despite the failure of the Gonja Development Scheme in the late colonial period, another huge mechanization scheme was started in 1964, in fifty-two new towns to which 80,000 people were relocated after the construction of the Volta Dam. Independent Tanganyika also followed the path opened by British East Africa's settlement schemes. President Julius Nyerere sought to modernize rural societies through the collection of people in villages providing education, health facilities, and newly rationalized farming methods. Not only did the Rural Settlement Commission take over nine schemes formerly managed by the colonial Tanganyika Agricultural Corporation—including three holdovers from the Groundnut Scheme—but it also established eight

[18] On the Groundnut Scheme, see J. S. Hogendorn and K. M. Scott, "The East-African Groundnut Scheme: Lessons of a Large Scale Agricultural Failure," *African Economic History,* 1981, *10*:81–115. For similar but smaller experiences in Senegal, see Marina Diallo Cô-Trung, *La Compagnie générale des oléagineux tropicaux en Casamance de 1948 à 1962: Autopsie d'une opération de mise en valeur coloniale* (Paris: Karthala, 1998); for postindependence government mechanized farms in several African countries see Hamid Aït Amara and Bernard Founou-Tchuigoua, eds., *L'Agriculture africaine en crise dans ses rapports avec l'etat, l'industrialisation et la paysannerie* (Paris: L'Harmattan, 1989).

[19] Maryinez Lyons, *The Colonial Disease: A Social History of Sleeping Sickness in Northern Zaire* (Cambridge: Cambridge Univ. Press, 1992); on the *paysannats,* see Bogumil Jewsiewicki, *Modernisation ou destruction du village africain: L'Economie politique de la "modernisation agricole" au Congo Belge* (Brussels: Centre d'Etude et de Documentation Africaines, 1983); and Wemo Mengué, *Le Transfert de savoir d'une métropole vers une colonie: Le Cas de l'Institut National pour l'Etude Agronomique du Congo Belge (INEAC)* (Paris: Thèse de l'Univ. Paris 7, 1998).

[20] Chambers, *Settlement Schemes* (cit. n. 14), p. 31.

Pilot Village Settlements.[21] This "villagization" policy was made more vigorous between 1973 and 1977, when five million people were relocated.[22]

Designing Development

The similarities and parallels between late colonial European and postindependence rural modernization policies in Africa that attempted to "redesign rural life and production from above" tell us "something generic about the project of the modern developmentalist state," whether colonial or postcolonial.[23] Beyond the diversity of the concerns that motivated them (productivist, sanitary, soil conservation, etc.) and of the technical recipes (irrigation, motorization, improved seeds, etc.) that they implemented, all of these large-scale schemes had indeed much in common. First, all put experts in power. The schemes were designed and often headed by scientists or technical officers, commanding a hierarchy of bureaucrats, inspectors, and overseers. The creation and management of settlement schemes stimulated the growth of technical and research services.[24] Scientists and technical officers (in such fields as agriculture, soil and forestry, education, public health, and public works) gained greater status and power with the emergence and affirmation of the developmentalist state. As a former colonial officer wrote, "[I]n the later stages of colonial rule they considerably outnumbered the Administration in the field and progressively exerted more influence at all levels of government."[25]

These experts were mobilized by the colonial state to help appropriate and master African environments, pathologies, and societies. In British Africa, the 1930s saw increasing concern with soil erosion and deforestation as well as with malnutrition and public health.[26] These brought to power a flood of experts. The gospel of soil conservation legitimized scientific measures (including confinement of people in settlement schemes, where access to land and grazing were restrained) against "irresponsible" Africans who had to be prevented from destroying their environment.[27]

---

[21] D. J. Morgan, *The Official History of Colonial Development*, 5 vols. (London: Macmillan, 1980), vol. 4, *Changes in British Policy, 1951–1970*, pp. 83–86.

[22] Scott, *Seeing like a State* (cit. n. 5), pp. 223–34. See also Goran Hyden, *Beyond Ujamaa in Tanzania: Underdevelopment and an Uncaptured Peasantry* (Berkeley and Los Angeles: Univ. of California Press, 1980), and Henry Mapolu, "Impérialisme, Etat et Paysannerie en Tanzanie," in Amara and Founou-Tchuigoua, *L'Agriculture africaine en crise* (cit. n. 18), pp. 71–88.

[23] Scott, *Seeing Like a State* (cit. n. 5), pp. 184 and 224.

[24] Significantly, the Belgian Congo, where the *paysannats* had become tropical Africa's largest settlement schemes in the 1950s, also ranked first in agricultural research.

[25] Chambers, *Settlement Schemes* (cit. n. 14), p. 17.

[26] On concerns about soil erosion, see William Beinart, "Soil Erosion, Conservationism and Ideas about Development: A Southern African Exploration, 1900–1960," *J. Southern African Stud.*, 1984, *11*:52–83. For Lord Hailey, the most urgent problem in all British territories was the introduction of methods of maintaining soil fertility without recourse to shifting cultivation. See Lord Hailey, *An African Survey* (Oxford: Oxford Univ. Press, 1938), p. 969. In French Africa, the soil question only became prominent some years later with the creation of a Comité des Sols at the Office de la Recherche Scientifique Coloniale in 1943, and then when a survey was ordered by the Ministère des Colonies in 1945 that led to the creation of a Bureau des Sols in each colony. See Fonds AOF, 3R58, Archives du Senegal, Dakar. On nutrition, see, Michael Worboys, "The Discovery of Colonial Malnutrition between the Wars," in *Imperial Medicine and Indigenous Societies*, ed. David Arnold (Manchester: Manchester Univ. Press, 1988), pp. 208–25.

[27] William Beinart, "Agricultural Planning and the Late Colonial Technical Imagination: The Lower Shire Valley in Malawi, 1940–1960," in *Malawi: An Alternative Pattern of Development*, proceedings of seminar at Centre of African Studies, Univ. of Edinburgh, May 1985 (Edinburgh: Univ. of Edinburgh, Centre of African Studies, 1985), pp. 95–148.

Social changes within African societies, the rise of new African elites, and the will to govern in a more scientific way also led to a call for social scientists and attempts to integrate science into policy making. In British Africa, a growing research school in social anthropology viewed "contact," "colonial situations," and "development"—rather than "tradition"—as its central objects of investigation.[28]

These new concerns, together with the will to systematize all branches of knowledge on Africa and improve cooperation among colonial governments, engendered numerous new scientific institutions, conferences, and surveys. The first pan-African Agricultural Conference was organized in 1929, followed in 1935 by a Pan-African Health Conference sponsored by the League of Nations. While in British Africa the African Research Survey emphasized the relationship between scientific research, economic development, and good governance, in France two Congrès de la Recherche Scientifique Coloniale were held in Paris in 1931 and 1937. These led in 1942 to the creation of the Office de la Recherche Scientifique Coloniale, a research agency that established branches in all French colonies after the Second World War.[29] In the social sciences, the International Institute of African Language and Cultures was established in 1926 in London. In 1936, the Institut Français d'Afrique Noire was founded in Dakar, and the Rhodes-Livingstone Institute was established in Lusaka (Northern Rhodesia, then Zambia) in 1937. After the war, anticolonialism encouraged cooperation among the colonial powers, which established the Scientific Council for Africa South of the Sahara in 1949.

This call for expertise spurred the emergence and growth of scientific communities. From fewer than one thousand researchers—still mostly European—in the late 1930s, by 1950 the number had risen to several thousand in Belgian, French, and British Africa.[30] As the number and authority of scientists and technical officers increased, so the problems of development, such as soil erosion, deforestation, and malnutrition, which had sometimes been called upon as evidence of colonialism's extractiveness and unsustainability,[31] were reconceptualized as mere technical problems to be solved with appropriate expertise. The deployment of development expertise and discourse often functioned as a depoliticizing machine.[32]

[28] See C. Rosseti, "B. Malinowski, the Sociology of 'Modern Problems' in Africa and the 'Colonial Situation,'" *Cahiers d'Etudes Africaines*, 1985, *25,* 4:477–504; Benoît de L'Estoile, "The 'Natural Preserve of Anthropologists: Social Anthropology, Scientific Planning and Development," *Social Science Information*, 1997, *36,* 2:343–76; Lynette L. Schumaker, "Fieldwork and Culture in the History of the Rhodes–Livingstone Institute, 1937–1964" (Ph.D. diss., Univ. of Pennsylvania, 1994); Henrika Kuklick, *The Savage Within: The Social History of British Anthropology, 1885–1945* (Cambridge: Cambridge Univ. Press, 1991).

[29] Christophe Bonneuil and Patrick Petitjean, "Science and French Colonial Policy: Creation of the ORSTOM: From the Popular Front to the Liberation via Vichy, 1936–1947," in *Science and Technology in a Developing World,* eds. T. Shinn, J. Spaapen, and V. V. Krishna, Sociology of the Sciences Yearbook 1995 (Dordrecht: Kluwer, 1997), pp. 129–78.

[30] Thomas O. Eisemon, Charles H. Davis, and Eva-Marie Rathberger, "Colonial Legacies: Transplantation of Science to Anglophone and Francophone Africa," *Science and Public Policy*, 1985, *12:*191–202. The training of numerous African graduates came only after the Second World War. (The British began training them much earlier than the French.)

[31] Examples of such "green anticolonialism" are Jean-Paul Harroy, *Afrique, terre qui meurt: La Dégradation des sols africains sous l'influence de la colonisation* (Brussels: Marcel Hayez, 1944), and Pierre Boiteau, "Biologie et colonialisme," *La Nouvelle Critique: Revue du Marxisme Militant*, (Nov. 1952):76–88.

[32] See James Ferguson, *The Anti-Politics Machine: "Development," Depoliticization, and Bureaucratic Power in Lesotho* (Cambridge: Cambridge Univ. Press, 1990). Technocratic ideology in Africa has been well analyzed by an African follower of Herbert Marcuse: Sidiki Diakite, *Violence techno-*

Another feature common to the various schemes of the developmentalist era is that their physical and social space was designed in accordance with "plans" produced by the scientific bureaucracy. Agricultural and social activities were construed in terms of uniform fields and villages, with rigid schedules. Farmers were told what to plant and what cropping systems to use. The timing of each farming operation was centrally controlled. In the Office du Niger scheme, for example, work started at the sound of a bell, and anybody caught in the village during field time risked having his food ration cut. This rigid order reflected the ambition of the developmentalist state to reorganize agricultural production and to hasten African society into modernity. Though they were not given large financial means before the 1940s, premises of planning emerged in the 1930s in the French and British Empires and developed strongly in the following decades.[33] This did not end with decolonization. With the assistance of foreign experts, independent states, especially Marxist regimes, looked also to planning with the aim of hastening growth and modernization. The faith in large, integrated, planned projects rather than grass-roots initiatives and piecemeal improvement was typical of the high modernist ideology of the developmentalist state.

Large development schemes were products of this culture of planning that gave the state responsibility for organizing (colonial, then national) economic development and assumed that society was a complex machine that only experts could operate optimally.[34] In these schemes a developmentalist discourse of experimentation flourished. From the 1930s onward, development narratives are filled with the deliberate use of words like "experiment," "experimentation," and "test"—on the part not only of experts and scientists, but also of colonial officers and journalists.[35] Significantly, the first step towards the creation of the Office du Niger, conducted in 1926 by relocating nine families near the agricultural station of Nienebalé (80 km downstream from Bamako) to cultivate cotton under the supervision of the station, which provided irrigation, was called a "settlement experiment" (*expérience de colonisation*). Showing "that the yield for indigenous farmers working for their own was much higher than their yield the previous year when they were employed as

---

logique et développement (Paris: L'Harmattan, 1985). The 1980s saw the rise of civil society in Africa and a repoliticization of development issues.

[33] Among the initiatives of the 1930s are the Colonial Development Fund in 1929, the Conférence Impériale in 1934, and early planning attempts in British East Africa in the late 1930s. For the 1940s, the main achievements were the ten-year colonial plan of the Vichy government (1942), the Fonds d'Investissement pour le Developpement Economique et Social in 1946, and the Colonial Development and Welfare Acts in 1940 and 1945. See Claudine Cotte, *La Politique économique de la France en Afrique Noire, 1936–1946* (Paris: Thèse de l'Univ. Paris 7, 1981); Constantine, *British Colonial Development Policy* (cit. n. 5); Michael Worboys, "Science and British Colonial Imperialism, 1895–1940" (Ph.D. diss., Univ. of Sussex, 1979); Catherine Coquery-Vidrovitch, "L'Impérialisme français en Afrique noire: Idéologie impériale et politique d'investissement, 1924–1975," *Relations Internationales,* 1976, 7:261–82.

[34] A. F. Robertson, *People and the State: An Anthropology of Planned Development* (Cambridge: Cambridge Univ. Press, 1984).

[35] See, for instance, book titles like these: K. D. S. Baldwin, *The Niger Agricultural Project: An Experiment in African Development* (Oxford: Blackwell, 1957); E. O. W. Hunt, *An Experiment in Resettlement* (Kaduna, Nigeria: Government Printer, 1957); Elsbeth Huxley, *A New Earth: An Experiment in Colonialism* (London: Chatto & Windus, 1960); John C. de Wilde, *Expériences de développement agricole en Afrique Tropicale,* 3 vols. (Paris: Maisonneuve & Larose, 1967–1968). See also Herbert Frankel's articles "The Kongwa Experiment" in the *London Times,* 2 and 5 Oct., 1950, and Terrasson de Fougère, "Expériences de colonisation indigène au Soudan Nigérien," *Compte-Rendus de l'Académie des Sciences Coloniales,* 1928–1929, 12:293–307.

waged labourers," the experiment confirmed the officials' vision of the progressive
individualist African farmer.[36] "The communism existing in the black society does
not resist the lure of profit," concluded the governor with satisfaction.[37] As a result
of this "experiment," which scientifically falsified the idea of an African "communism," the Office du Niger would be based on a tenancy system. But when the nine
African families expressed their desire to go back home, they were told by officials
"that the experiment was not finished, and that they had to stay" in the program.[38]
The experimentalist gospel indeed often helped make authoritarian and productivist
obsessions look like the pursuit of knowledge. Does not a good experiment require
control of all parameters? Sometimes, the rhetoric of experimentation (as in the case
of the Groundnut Scheme in Tanganyika) also helped to justify the huge amounts of
money lost in such schemes and to excuse in advance all errors.

In French Africa, the idea that the practice of colonial domination was a permanent process of experimentation on African societies was conceptualized as early as
1935 by Robert Delavignette, a district commissioner in West Africa, who in 1937
became director of the French school of colonial administration.[39]

TAMING AGRARIAN SOCIETIES INTO OBJECTS OF DEVELOPMENT

More than mere showcases of a development ideology that advocated the transformation of Africa from above and brought scientists into power, settlement schemes
were key attempts to shift the balance of power between agrarian communities and
the state. Challenging both colonialists' self-representation and the anticolonialist
historiography that overestimated the success of European control over African lives,
many historians have recently underlined the epistemic and political weakness of
the colonial state.[40] In Africa, the early colonial state confronted, in the late nineteenth century, environments, knowledge, and social relations that had evolved prior
to and independently of its plans. Most of rural life and production remained out
of the reach of the early colonial state. The difficulties faced by administrators in

[36] Archives CIRAD-CA, Colonie du Soudan Français, "Rapport agricole 1927," p. 39.

[37] De Fougère, "Expériences" (cit. n. 35), on p. 295.

[38] *Ibid.*, p. 294.

[39] The first mention of the "experimental method" is found in Robert Delavignette, "Pour le paysannat noir, pour l'esprit africain," *Esprit* 1 Dec. 1935, 367–90: "Mais l'exploration, la découverte,
l'invention n'est pas finie; il faut la poursuivre en profondeur, dans le repli des coutumes du pays.
Sous la colonie, voir les pays pour mieux régler l'invention et d'autre part éprouver les pays par la
colonie. C'est dans cette attitude expérimentale que réside la plus sûre garantie d'humanité de l'action
colonisatrice." (But exploration, discovery, and invention are not over; we have to continue this task
in the thickness of local customs. Under the "colony" we have to see the country, so as to better
adjust invention and put the countries to the test of the colony. In this experimental behavior lies the
better guarantee of the humanity of colonization) (p. 389). See also *idem, Les Vrais Chefs de l'empire*
(Paris: Gallimard, 1939), p. 21, 30, 210, and 213. It is ironic—and significant of the spreading of
the discourse of experiment beyond settlement schemes—that Delavignette, who contributed greatly
to making experiment a central theme of colonial discourse, opposed the Office du Niger and other
such technocratic schemes.

[40] On the political and epistemic weakness of the (colonial and postcolonial) state in Africa, see
Gerd Spittler, *Verwaltung* (cit. n. 5); *idem*, "Administration in a Peasant State," *Sociologia Ruralis*,
1983, *23:*130–44; Robert Debusmann, "Bureaucratie contre paysans: Un Modèle sociologique du
pouvoir colonial," in *La Recherche en histoire et l'enseignement de l'histoire en Afrique centrale
francophone, colloque international* (Aix-en-Provence: Univ. of Provence, 1997), pp. 105–16; Henri
Brunschwig, *Noirs et blancs dans l'Afrique noire française* (Paris: Flammarion, 1983), especially pp.
105–33; Jean-François Bayart, "La Politique par le bas en Afrique noire: Questions de méthode,"
Politique Africaine, 1981, *1:*53–82.

understanding and controlling rural Africa were the result of several factors, including low population density; the huge diversity of cultures and ecologies; the variety of work techniques, languages, units of measure, and family structures; the weak connection between household economies and the market—the peasant being better known by the state cognitive apparatus when integrated into the market economy—and strategies of passive resistance. To bring agrarian societies under its epistemic (and hence political) grasp, the colonial state undertook to transform materially the social and environmental conditions of life in rural areas. This strategy—which characterizes the emerging developmentalist state—worked through transportation, irrigation, and agricultural "modernization," education, standardization of units, and integration of producers into the market. Prepackaged settlement schemes were major building blocks of this enterprise. They were the laboratories where the developmentalist state attempted to shape agrarian societies and environments so as to render them compliant to "development": more productive, more commensurable to expert knowledge, and more amenable to state intervention.

Legible Villages

Whereas the African village remained a social hieroglyph for the early colonial state, the planned-development schemes of the developmentalist era were major attempts to capture the peasantry into stable, legible, and more productive units that would make taxation, conscription, and "enlightened" intervention easier.[41] For example, new villages were often located along main roads and sometimes used for regrouping and stabilizing itinerant or scattered populations. Village layout and housing as well as social life were also designed from above, so as to turn villages into functional units of command and control: not organic historical and cultural units but units of supervision and experimentation (Figure 2).[42]

Land tenure and land ownership are fundamental elements of social control, and experts and bureaucrats attempted to reorganize and standardize land tenure and use. In Kenya and Southern Rhodesia, where European settlement had reduced the amount of land available for African use, several schemes were designed to confine and control African farmers in their own areas (for example, by limiting areas of cultivation, introducing irrigation, or limiting animal stocks and grazing rights).[43] Even in colonies without European settlers, growing concern about overpopulation and soil erosion led to schemes designed to confine populations within smaller areas so as to favor more intensive cultivation: the Zande Scheme in British Sudan, for instance, as well as some Congolese *paysannats*, were attempts to transform slash-and-burn cultivation into standardized sedentary land use. Though limited, this control

[41] Colonial and postcolonial statecraft presupposed the rationalization and standardization of the village into a more productive, a "legible and administratively more convenient format." Scott, *Seeing Like a State* (cit. n. 5), p. 3.

[42] But on the other hand, elements of what was believed to be "traditional" were grafted (reinvented or maintained) onto this unit so as to make it more acceptable and to ensure social stability. One vivid example is the fact that Mossi settlers brought from Yatenga (upper Volta region) to the Office du Niger were grouped in villages that remained under the authority of the descendant of the king of Yatenga (nominated *chef de province* by the French rulers). Negotiation with or reinforcement of local structures of power (especially chiefship) is a common feature in the 'blocs' of development.

[43] On irrigation schemes driven by these concerns in the Kikuyu region in Kenya and in Southern Rhodesia in the 1950s, see Chambers (cit. n. 14), *Settlement Schemes*, p. 25.

Figure 2. *Geometrizing nature and society: a view of the Office du Niger. (From Office du Niger,* Le Delta ressussité *[Ségou: Mali, 1960]).*

of land represented a major step in the history of the African state in the twentieth century. Except in the few areas where land was appropriated by European settlers, the early colonial state had failed to incorporate decisions about land tenure into its realm and never succeeded in establishing village institutions strong enough to challenge the authority of kinship organization. For these reasons, the African state remained (and still often is) a "state without territory," unable to transform the basic relations of production in the countryside.[44] Settlement schemes therefore represent a shift in governmentality from ruling "non-state spaces" to administering "state spaces," that is, places where state power was reinforced by a new mode of production.[45]

Geometrization, simplification, standardization, and discipline ensured not only the social order and legibility sought by the state but also the experimental order necessary to produce expert knowledge. The geometrization of land use and the replacement of polycropping by monocropping were also simplifying strategies that aimed at transforming the richness and complexity of local practices into more uniform and controllable systems, more amenable to expert modes of knowing and intervening.[46] In many ways, settlement schemes were up-scaled agricultural-station programs. Often established near a preexisting research station, they were generally set up to experiment with technical recipes proposed by scientists in such areas as new varieties, new crop rotations, and new tools: the schemes were trials of strength (*épreuves*, as Bruno Latour would put it) in which a possible extension to the outside world of the validity of the results and artifact of the station was tested. The Terres Neuves scheme in Senegal provides a good example. This authoritarian scheme served as a laboratory for the testing and multiplication of "improved" peanut varieties developed by researchers at Bambey experiment station. Because the first trials of the new varieties entrusted to African chiefs were disappointing and their value was much debated, scientists at Bambey gained clearance to organize and supervise a test themselves in two villages in the Terres Neuves in 1935. Totally new, these villages were perfect sites for the experiment: the administration had cleared the land with penal labor and distributed it, built roads and wells, recruited and transported the Sereer settlers, and fed them during the hungry season. No area was better known to the administration. In no village could the social structures and agricultural activities be more easily bent to the imperatives of control. Only in this vast social and agricultural laboratory, under tight supervision, were the yields of "improved" varieties measurable—and they proved remarkable. This first controlled large-scale trial cleared the way for one of the most successful seed distribution schemes in Africa.[47] The Paysannat Turumbu, established in 1942 in the neighborhood of the agricultural station of Yangambi (Belgian Congo), the headquarters of the Institut National pour l'Etude Agronomique du Congo Belge, also worked as an extension of the research station. Involving no less than 5,300 people in 1953, it served as a center for the large-scale testing of varieties selected in the station and for seed multiplication.

In this scaling-up process, "improved" techniques and seeds were not the only

[44] Goran Hyden, "State without Territory: Africa in Comparative Perspective," address presented at The Relationships between State and Civil Society in Africa and Eastern Europe conference (Bellagio, Italy, 5–11 Feb. 1990), p. 8.

[45] I borrow the phrase "state space" from Scott, *Seeing Like a State* (cit. n. 5), p. 187.

[46] "[S]cientific agricultural research has an elective affinity with agricultural techniques that lie within reach of its powerful methods. Maximizing the yields of pure-stand crops is one technique where its power can be used at best advantage . . . [A]gricultural agencies . . . have tended to simplify their environments in ways that make them more amenable to their system of knowledge." *Ibid.*, p. 291.

[47] By 1951, half of the 800,000 hectares under peanut cultivation in Senegal were planted with "improved" varieties. This was a major achievement at a time when the green revolution was still in the air. For a detailed study, see Christophe Bonneuil, "Penetrating the Natives: Peanut Breeding, Peasants and the Colonial State in Senegal (1900–1950), *Sci. Technol. Soc.*, 1999, *4*:273–302.

things disseminated from the research station. Some important elements of the station's experimental order—working methods, including rules for keeping varieties pure, thorough weeding, practices of precision (such as mapping and measuring plots and yields, and tactics for observing of farmer's activities)—were also imposed on the villages that were involved in the settlement schemes.[48] In the Office du Niger scheme, thousands of farmers, caught within a standardized and rigid organization of space and work, were turned into objects of experimentation. For instance, as the researchers of the Office were seeking a solution to the problem of maintaining soil fertility, the standard crop rotation scheme imposed upon all tenants changed four times between 1937 and 1947.[49]

The Paysannat Turumbu also illustrates how agriculturalists' attempted to simplify and "rationalize" African farming systems, and how the capture of farmers as subjects of experimentation, led to a better knowledge of the conditions of agriculture in the rain forest of the Congo Basin. Until the 1930s, Belgian agronomists and foresters harshly condemned shifting cultivation, viewing it as responsible for deforestation. Agricultural scientists worked hard to find alternatives to the "Bantu primitive system" of shifting cultivation and instead promoted ploughing, monocropping, and short grass fallowing with leguminous plants. But because they never managed to find a system capable of maintaining the fertility of the (rather poor and fragile) soils of the Congo Basin, scientists were led to acknowledge the ecological efficiency of multicropping (which better protects the soil against sun and rain) and of long forest fallowing. In their view, however, the indigenous system had still to be scientifically "improved." The Paysannat Turumbu hence became in the 1940s a laboratory for testing a "rationalized" land-use and cropping system under equatorial-forest conditions. Access to land was obtained from the customary chief, but farmers' land use had to conform to a standard "corridor system": one band (rectangular and standardized in size) of cleared land alternating with one band of forest, so as to accelerate the recolonization of cleared lands by forest during the fallow period (Figures 3a and 3b). Furthermore, crop rotation included a shorter fallow period (twelve years, instead of twenty in Turumbu farming practices). The experiment was not successful. After some time, it was found that the shorter fallow period could not maintain soil fertility, so managers of the scheme had to instruct farmers to lengthen it—returning to Turumbu practice![50]

An important point made by recent social studies of science has been to view the production of scientific knowledge as a local and situated activity and to show that scientific knowledge and artifacts travel only with their ecologies—that is, if their new social and natural environments can be reengineered in a way similar to that of the place where they were first elaborated.[51] Similarly, one can view the settlement

[48] For a detailed study following how elements of the experimental order of the station were imposed on the farmers (through successive steps: indigenous farm within the station, Terres Neuves settlement scheme, and hiring farmers to multiply seeds), see *ibid.*

[49] Bordage, *De la terre, de l'eau et des hommes* (cit. n. 17), p. 141.

[50] J. Henry, "Les Bases théoriques des essais de paysannat indigène, entrepris par l'Ineac au Congo Belge," in *Contribution à l'étude du problème de l'économie rurale indigène au Congo Belge, Bulletin Agricole du Congo Belge*, special number, 1952, *43*:159–192; J. Muller and F. Vervier, "Paysannat et coopérative Turumbu," *Bulletin d'Information de l'Institut National de l'Etude Agronomique du Congo,* 1953, 2:115–22.

[51] Harry M. Collins, *Changing Order: Replication and Induction in Scientific Practice* (London: Sage, 1985); Joseph Rouse, *Knowledge and Power; Towards a Political Philosophy of Science* (Ith-

schemes as crucial sites for aligning rural societies with the conditions and practices of the station. The schemes were designed to answer under controlled conditions the questions agricultural scientists asked.[52] They were hence "experimental systems," in Hans-Jörg Rheinberger's sense, insofar as they constituted an arrangement of objects and people designed to produce experimental data.[53] As an experimental system ought to do, a settlement scheme was also a web designed to record unexpected phenomena, since it helped to capture within the experimental realm, through the experimental manipulation of farmers' lives and practices, previously unseen aspects of indigenous farming knowledge.

The experimental design of the settlement schemes was a hybrid construct with features of both village and station. This hybridity was essential. On the one hand the schemes were supposed to tell something about how new techniques would work in "indigenous farming conditions"; on the other hand, they created state-space conditions in order to operate under controlled conditions similar to that of the station.[54] From the scientists' perspective, the making of valid knowledge required that African farming practices be integrated into and manipulated within a controlled experimental design. It was only after having experimented themselves in the Yangambi station that Belgian agricultural scientists acknowledged that shifting cultivation was a rational practice. Similarly, only after having experimented with short forest fallowing on a large scale in the *paysannat* did they acknowledge that a long fallow period (as practised by Turumbu farmers) was necessary. It is therefore precisely because they were hybrid experimental systems that the settlement schemes established certain lines of commensurability between expert knowledge and farmers' knowledge.[55] In settlement schemes, bringing (and refashioning) the "indigenous" into the experimental realm was a strategy for domesticating the opacity of village

---

aca: Cornell Univ. Press, 1987), on pp. 209–47, Bruno Latour, *Science in Action: How to Follow Scientists and Engineers through Society* (Cambridge, Mass.: Harvard Univ. Press, 1987); Simon Schaffer, "Glass Works: Newton's Prisms and the Uses of Experiment," in *The Use of Experiment,* eds. D. Gooding, T. Pinch, and S. Schaffer (Cambridge: Cambridge Univ. Press, 1989), pp. 67–104.

[52] In the examples that I have discussed, these were some of the questions: Are Bambey-improved peanut varieties really suitable for release to Senegalese farmers? Is the Sudanese farmer really communist? Which cropping system best maintains the fertility of the soil in irrigated cotton farming in the delta of the Niger or in the equatorial forest? Which farming operation can profitably be mechanized and which not?

[53] For an elaboration of the concept of "experimental system," see Hans-Jörg Rheinberger, *Towards a History of Epistemic Things: Synthesizing Proteins in the Test Tube* (Stanford: Stanford Univ. Press, 1997). Unlike Rheinberger, who considers experimental systems only within the space of the pure-science laboratory and uses this concept to draw a demarcation between science and technology (p. 32), I tend to see a strong similarity between Michel Foucault's *dispositifs* and Rheinberger's "experimental systems."

[54] In reality these "indigenous conditions" were state-space conditions that differed strongly from indigenous non–state-space conditions. In the state space the practices were designed from above, the farmers (sometimes) came from distant regions with differing farming conditions, household plots were (often) chosen and assigned by the experts, and farmers were working and living under permanent scrutiny and control.

[55] I am indebted to the historian of anthropology Michael Bravo for showing the hard work required to create bridges (commensurability) between knowledge traditions. See Michael Bravo, "The Accuracy of Ethnoscience: A Study of Inuit Cartography and Cross-cultural Commensurability," communication to the Nature's History conference on History of Science and Environmental History, Max-Planck Institute for the History of Science (Berlin, Aug. 1997).

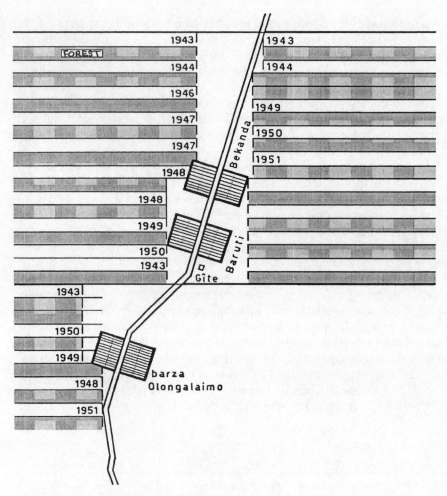

Figure 3a. *The "corridor system" at the Paysannat Turumbu (Belgian Congo).*

communities and the otherness of farmers' practices. This allowed a cognitive penetration of African agrarian societies by an analytic/experimental expert-knowledge system. In this way (and at these costs), the farmers' world was made amenable to an experimental world view, and indigenous knowledge was translated into a format that allowed its circulation and accumulation within the academic community.

Legible Households

The subject of the new system of domination that was at work in the settlement schemes was the individual household head rather than the village community. His name and the composition of his family (the model of the nuclear family was usually favored) were recorded when he was granted a tenancy, and he gained access to land not through a chief but through the managers of the scheme, to whom he paid charges. The household head also had the legal responsibility for conducting the

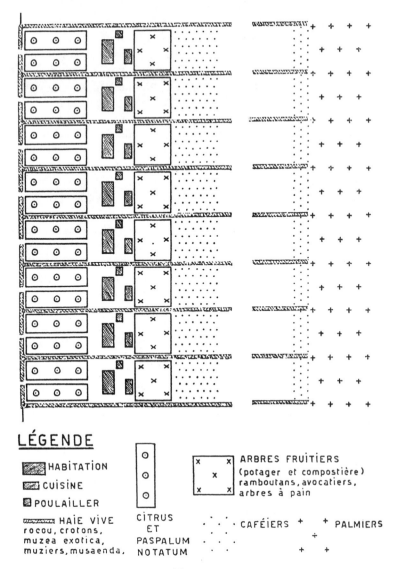

LÉGENDE

HABITATION

CUISINE

POULAILLER

HAIE VIVE
rocou, crotons,
muzea exotica,
muziers, musaenda,

CITRUS
ET
PASPALUM
NOTATUM

ARBRES FRUITIERS
(potager et compostière)
ramboutans, avocatiers,
arbres à pain

CAFÉIERS   PALMIERS

Figure 3b. *Village layout in Yalibwa.*

agricultural operations on his plot according to the experts' and inspectors' prescriptions. An individual relationship between the individual household head and the state was therefore established.[56]

One cannot underestimate the novelty of this direct relationship. Basically, the early colonial state in twentieth-century Africa had no grasp on individual inhabitants of rural areas. Very few possessions had a genuine census until late in the

---

[56] This was also not without consequences for the balance of power at the household level itself. Victoria Bernal notes that the exclusion of women from tenancy reinforced or instituted a position of dependence of women within the family in the Gezira Scheme (British Sudan) and promoted the model of a patriarchal peasant household. See Bernal, "Cotton and Colonial Order" (cit. n. 16), pp. 104 and 109.

colonial period (or even after independence), and the estimates of population were unreliable. The budget of most African colonies supposedly rested on income from either an individual head tax or a hut tax. But in fact tax collection looked in many areas rather like the taking of a collective tribute from village communities. For gathering information and ruling at the village level, colonial administrators depended heavily on local African intermediaries (interpreters, village headmen) who had their own agendas.[57] At the opposite extreme from this mysterious collective entity represented by more or less reliable spokesmen, the village in development "blocs" was a grouping of well-known families, each member of the household being recorded and receiving food rations or regular medical assistance. In tenancy systems (in use in most irrigation and mechanization schemes), it was possible for the scheme managers, who controlled the marketing of crops, to document in detail the incomes and expenditures of farmers. This cleared the way for farm budget studies, which became a key object for rural economics.[58] Time spent in each agricultural operation by each member of the household (a key object in the rise of economic anthropology as a field of research in the 1960s) was also more easily recorded in such schemes, where households were intensely scrutinized by researchers attached to the scheme.[59] In planned-development schemes, the individual household had in this way become at once the subject of the colonial (and then postcolonial) state and a legible object of its expert knowledge (Figures 4a and 4b).

Repression and Invention of "Indigenous Knowledge"

Driven by high modernist, top-down diffusionist, and narrow experimentalist concerns, prepackaged development schemes acted as powerful mechanisms to repress indigenous knowledge and initiatives. Farmers living in settlement schemes saw their interactions with the environment oversimplified. Their subsistence relied on loans, stable prices, technical assistance, and expensive inputs, and they were usually deprived of the local and flexible strategies that allowed farmers to cope with risk and prevent food deficit (diversification of crops, lowland cultivation, complementing farm income with wage labor, and private gardens).[60] In these state spaces, the peasantry was deskilled, and the autonomy of the village community was reduced. The local capacity for innovation was diminished. A division of labor between innovation and execution was reinforced. "Developers" now had the power

---

[57] In remote regions, a district officer on a tour of inspection in a village could not even be absolutely sure that the man who welcomed him was the genuine village head! Delavignette reports such a situation when he was district officer in the French Sudan. Cf. *Les Vrais Chefs de l'empire*, pp. 124–25. See also the novel by Amadou Hampaté Bâ, *L'Etrange Destin de Wangrin ou les roueries d'un interprête africain* (Paris: Union Générale d'Edition, 1973).

[58] See, for instance, W. P. Cocking and R. F. Lord, "The Tanganyika Agricultural Corporation's Farming Settlement Scheme," *Tropical Agriculture*, 1958, *35*, 2:85–101, which contains a study of the income and expenditures of the tenants of the Nachingwea scheme.

[59] In Senegal, thanks to tight administrative control, it was in the Terres Neuves scheme that agricultural statistics developed in the 1930s: the first statistical data on average per-crop acreage cultivated by an adult male were collected there, and the first land-use maps were drawn there.

[60] See Henrietta L. Moore and Megan Vaughan, *Cutting Down Trees: Gender, Nutrition, and Agricultural Change in the Northern Province of Zambia, 1890–1990* (Portsmouth, N.H.: Heineman, 1994), pp. 138–9 (on the prohibition of *citemene* gardens in successive schemes in northern Rhodesia); Beinart, "Agricultural Planning" (cit. n. 27), p. 124 (prohibition of lowland cultivation at the Nyamphota Village Land Improvement Scheme in late colonial Malawi); Bernal, "Cotton and Colonial Order" (cit. n. 16), p. 114 (prohibition on selling sorghum).

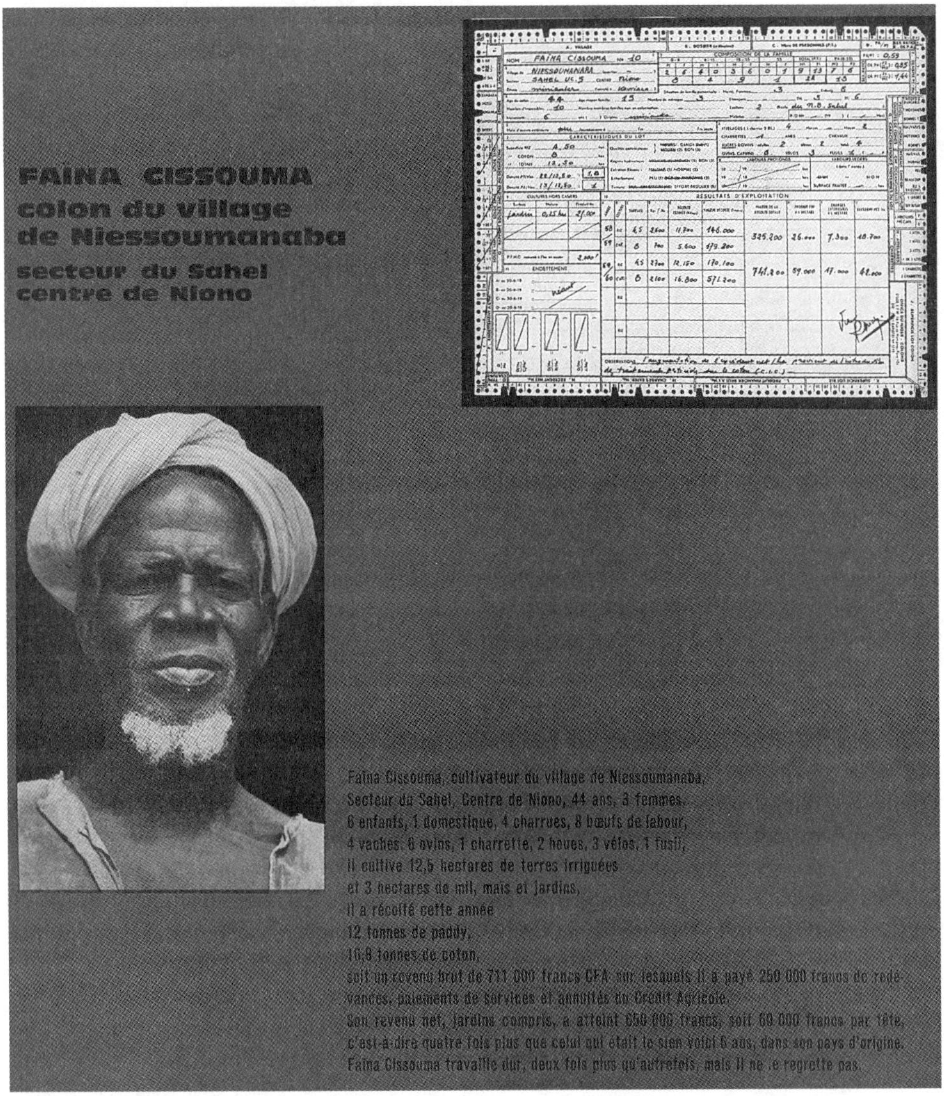

Figure 4a. *The farmer as an object of knowledge and control. (a) A tenant farmer as described in a leaflet from the Office du Niger just after independence. [Translation of the text: "Faina Cissouna, farmer of the village of Niessoumanaba Sahel sector and Nioro Center (of the Office du Niger), 44 years old, 3 wives, 6 children, 1 servant, 4 plows, 8 oxes, 4 cows, 6 ovins, 1 cart, 2 hoes, 3 bikes, 1 gun. He cultivates 12.5 hectares (31 acres) of irrigated land and 3 hectares (7.5 acres) of millet, corn, and vegetables. This year he has harvested 12 metric tons of paddy and 16.8 metric tons of cotton, which he sold for 711,000 francs CFA and out of which he paid 250,000 francs CFA in taxes, dues, and loan repayments. His net income amounted to 650,000 francs CFA, which comes to 60,000 francs CFA per person. This is six times more than the income from his land six years ago. Faina Cissouna works hard, twice as hard as he used to, but he does not regret it."] (From the Office du Niger,* Le Delta ressussité *[Ségou: Mali, 1960]).*

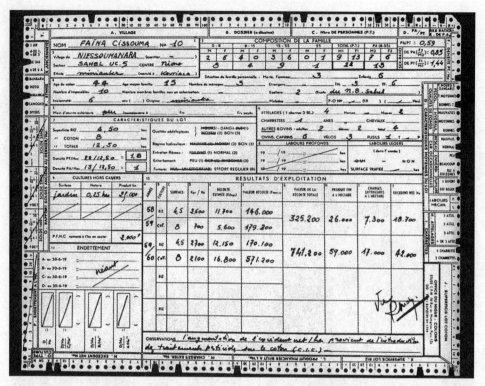

Figure 4b. *Closeup of the information card shown in (4a), which documents the structure of Faina Cissouna's family, his belongings, and his farming budget.*

to experiment with development on the "developed." Read from a political perspective, the repression of indigenous knowledge was therefore not merely a consequence of ignorance, blindness caused by the scientist faith, the colonial bias, or the professional ethos of experts: the deskilling of African farmers was also an intrinsic element in the affirmation of the state, its institutions, and its agents (whether European or African). Read from an epistemic perspective, the repression of indigenous knowledge helped to subordinate the farmers' sphere of knowledge and practices to the realm of experimental design (whose keystone was the isolation and control of individual variables). Farmers' practices were considered invalid until their scientific validity was proven by controlled experiment.[61]

On the other hand, however, through the experimentation on African women and men, some limits and blind spots of Western analytic-experimental approaches to African farming could be made visible to experts. Such binary tensions as uniformity-diversity, maximization of yield-minimization of risk, monocropping-polycropping, permanent fields-shifting cultivation, isolation of experimental variables-monitoring of a complex and variable ecosystem, could appear as problems and be progressively acknowledged as new directions for research. The "dis-

---

[61] "From a narrow scientific view, *nothing* is *known* until and unless it is proven in a highly controlled experiment." Scott, *Seeing Like a State* (cit. n. 5), p. 305.

covery" of the rationality of shifting cultivation by Belgian agronomists in the Congo in the 1940s and 1950s exemplifies how settlement schemes worked as (heavy and oppressive but) evolutive experimental systems that helped capture elements of farmers' knowledge.

Many important data and pioneering scholarship on indigenous agricultural knowledge emerged in settlement schemes. Before anthropologist Harold Conklin's influential monograph on shifting cultivation in the Philippines, an agricultural scientist who worked in the Zande Scheme in the British Sudan (60,000 households in 1950), Pierre de Schlippé, published in 1954 the most comprehensive study of shifting cultivation and a major contribution toward its acceptance as a rational and sustainable system.[62] One of the first agriculturalists to employ anthropology in his research, De Schlippé trained six African surveyors, each of whom studied some twenty households. (The daily activities of each adult had to be carefully reported.) Such scrutiny faced resistance from farmers, afraid that the data collected might be used against their interests. Village headmen, who feared that their authority might be diminished, also opposed this approach. But such opposition, which could have made the study impossible in a non-state space, could be overcome in the authoritarian context of the Zande Scheme. De Schlippé's influential notions of "field type" and his case for the rationality of an indigenous "system of agriculture" were therefore made possible by this particular logistic of scrutinizing farmers.

In addition to De Schlippé, several other influential scholars in development studies, including David Brokensha,[63] René Tourte,[64] and Robert Chambers,[65] whose works around 1980 were critical influences in the turn towards the "farmers first," "indigenous farming knowledge," and "farming systems" approaches, began their careers and gained access to the field in settlement schemes. More generally, the work of dozens of agricultural scientists, demographers, rural economists, geographers, anthropologists, medical and nutrition scientists, and sociologists attached to settlement schemes proved decisive in the development of agronomy, rural economics, and economic anthropology. In the 1950s and 1960s, when quantitative analysis was fashionable and data collection painstaking, fieldwork research presupposed compliant Africans, ready to answer long series of questions and to accept intrusions

---

[62] Pierre de Schlippé, "The Zande System of Agriculture," *Sols Africains/African Soils*, 1954, *3*, 1:52–63; *idem*, *Shifting Cultivation in Africa: The Zande System of Agriculture* (London: Routledge & Kegan Paul, 1956). De Schlippé's work influenced Conklin. Harold C. Conklin, *Hanunoo Agriculture: A Report on an Integral System of Shifting Cultivation in the Philippines* (Rome: Food and Agriculture Organization, 1957).

[63] David Brokensha, coeditor in 1980 with O. Warren and O. Werner of *Indigenous Knowledge Systems and Development* (Lanham, Md.: Univ. Press of America) started his work as an anthropologist for the Volta Dam resettlement project in Ghana. See his "Volta Resettlement and Anthropological Research," *Human Organization*, 1963, *22*:286–90.

[64] Tourte initiated the Unités Expérimentales in Senegal in the late 1960s. Although not a settlement scheme, this project was a tightly controlled, tightly scrutinized on-farm experiment. A decade later she founded the Department for Farming System Research (DSA) in the Centre International de Recherche Agronomique pour le Développement (the French overseas agricultural research agency). Social science research in Senegal (especially economic anthropology) owes much to the access to the field provided by the Unités Expérimentales scheme.

[65] Chambers started his career as a colonial district officer in Kenya and had directed several settlement schemes before he undertook a comparative study of settlement schemes as his doctoral dissertation at the University of Manchester in 1967. See Chambers, *Settlement Schemes* (cit. n. 14), and *idem, Rural Development: Putting the Last First* (London: Longman, 1983).

into their lives. Africans amenable to data extraction were more easy to find in planned-development schemes, where they had already been tamed as objects of intervention and experimentation.[66]

Thus the promotion of indigenous farming knowledge in the 1970s and 1980s did not emerge simply from more "open" or more grass-roots-driven scholars. As paradoxical and ironical as it may seem, decades of vertical technocratic intervention and cognitive penetration of agrarian societies through prepackaged development schemes have certainly been preconditions for the emergence of the present vast scholarship on indigenous knowledge and African farming systems. Indigenous knowledge may therefore well have attained its recent intellectual significance from its appropriation by the state and experts.

CONCLUSION

The colonizers of Africa acted like and represented themselves as discoverers, proud to throw light on dark regions, to conquer and unveil rebel nature and societies. By contrast, the colonialists of the 1930s viewed themselves as experimenters. Colonizing then meant completing conquest through modernization, and the transformation of African societies and environments with the assistance of the "experimental method." By 1900, the dominant metaphor of development was that of a continuous process (growth from childhood to adulthood) that would be guided by knowledge of laws, presented as scientific. By contrast, after the 1930s, development came to be seen as an experiment, and Africa as a laboratory. This shift from "governing, thanks to the light of science" to "governing as an experimental activity" is an essential feature of the emergence of a development regime in Africa.

The cornerstones of the experimental culture of development, planned settlement schemes, were state spaces where administrators and experts attempted to shape the natural and social environment after their own image, in ways that made them more amenable to their—Western—modes of intervention and systems of knowledge. These attempts did in fact seldom fully succeed. Farmers' responses made the schemes far more complex systems than those envisioned by officials. And simplified standard uses of nature proved ineffective under the fragile and variable environmental conditions in Africa. Planned-development schemes of the golden age of the developmentalist era are now seen as monsters, as mammoth projects that resulted in economic, environmental, and social failures. They are used by today's develop-

---

[66] David Norman, a pioneer in the farming systems approach, wrote interestingly about his early fieldwork in Nigeria: "The commitment required to do such studies and ensure complete farm records meant that spontaneous interaction with farmers often was sacrificed. Thus, the farmer tended to become an object from which data were extracted rather than a colleague from whom one can learn in an interactive mode. The major limitation of so much dependency on extracting data from farmers via enumerators was brought home to me when, after years of painstaking data collection and analysis, I concluded that farmers were rational in growing crops in mixtures. . . . I then thought, I [will simply] ask farmers why they grow crops in mixtures. After one week, I obtained answers . . . similar to those from the detailed surveys." David W. Norman, "The Farming Systems Approach: A Personal Evolution," in *On the History, Status, and Future Direction of the Farming Systems Approach*, ed. M. Collinson (in press), kindly communicated by the author. See also *idem*, "Rationalizing Mixed Cropping under Indigenous Conditions: The Example of Northern Nigeria," *Journal of Development Studies*, 1974, *11*:3–21; David W. Norman, E. B. Simmons, and H. M. Hays, *Farming Systems in the Nigerian Savanna: Research and Strategies for Development* (Boulder: Westview, 1982).

ers as a foil to the virtues of the new trends in development policies, which favor schemes that are small, participative, not managed by the state, and adapted to local ecologies.[67] But these retrospective judgments may miss a crucial point for the historian: that these schemes were key elements in the building of the state and the making of expert knowledge in Africa. The "experimentalization" of agrarian communities—which may be seen as a colonial mode of knowing—played a central role in gaining a better knowledge of the conditions of farming in tropical Africa, of agrarian societies, and of the way that development experts should intervene. African farmers were turned into both the subject of the state and the object of development studies. Prepackaged development schemes were laboratories where a new governmentality, whose subject is the individual household head rather than the village community, and new scholarship and academic knowledge on African societies and environments were co-constructed.

[67] See Paul Richards, *Indigenous Agricultural Revolution: Ecology and Food Production in West Africa* (London: Hutchinson, 1985); Chambers, *Rural Development* (cit. n. 66); Georges Dupré, ed., *Savoirs paysans et développement* (Paris: Karthala–Orstom, 1991).

Bio-prospecting or Bio-piracy: Intellectual Property Rights and Biodiversity in a Colonial and Postcolonial Context

John Merson[*]

ABSTRACT

Despite the rhetoric of decolonization following World War II, developing countries are, if anything, more dependent now on the science and technology of the developed world than they were in colonial times. This has led some critics to describe their situation as "neo-colonial." This paper will explore the issue in relation to the biotechnology industry, and to the 1993 United Nations Convention on Biodiversity. This convention challenged the assumption that the earth's biological and genetic resources are part of the "global commons" by giving property rights over these resources to the nation-states. While the objective of encouraging states to conserve biodiversity is universally endorsed, the strategy of using property law to do so is not. The search for new genetic and biological resources has become a major priority for the agrichemical and pharmaceutical industries, and despite continuation of the colonial tradition of appropriating indigenous knowledge and resources, new and more equitable models are being explored and developed within the convention's framework. These strategies, while controversial, offer the hope of a new and more just "International Genetic Order."

Lands long given up to imperial enterprise suffer by imperial withdrawal for two related reasons. Unless their economies remain neo-colonial, they are ill equipped to bear the loss of habitual modes of employment. Nor can they rapidly overcome habits of dependence ingrained by long subservience to the will, as to the interests, of others.
—David Lowenthal, "Empires and Ecologies: Reflections on Imperial History"

I N DISCUSSING THE SHIFT FROM COLONIAL TO POSTCOLONIAL science, it has been customary to make a distinction between settler countries, such as Canada, North America, Australia, Argentina, and New Zealand, and those where the indigenous populations were colonized, including India, Malaysia, Indonesia, Korea, and Africa. Countries such as China, Japan, and Thailand, although

---

[*] Science and Technology Studies, University of New South Wales, Sydney, Australia.

independent, were nonetheless economically shaped by the policies of imperial powers throughout the nineteenth and early twentieth centuries.[1]

The economic development of settler cultures, particularly the United States, provided new market structures and mechanisms for funding innovative research. However, this was rarely the case in nonsettler colonies, where the indigenous population and environment were primarily exploited for cheap labor and natural resources. Even following independence, the ability to link local scientific research with the economy remained limited, owing to cultural ambivalence, a low level of industrialization, and lack of private-sector support.[2]

In the heady days of decolonization following the end of the Second World War, India, Indonesia, Korea, Malaysia, and many African states placed under national control the fledgling scientific and technological research facilities that had once served their colonial masters. However, with the exception of South Korea and Taiwan, the shift from colonial to postcolonial science has meant very little in terms of the capacity to use the tools of research to shape economic development. Most newly decolonized states invested heavily in education, and especially in the training of scientists and technicians. However, their reliance on import substitution schemes for the transfer of industrial technology provided few opportunities for local technicians to innovate in the application of science. Exceptions to this general rule can be found, notably in agriculture. Yet even where local plant breeding and agricultural extension services were maintained, by the 1960s they were subsumed by the imperatives of industrialized agriculture that came with the green revolution. Although it increased yields, the green revolution also reinforced dependency on foreign technical advisers and industrial inputs such as high-yielding plant varieties, chemical fertilizers, insecticides, and herbicides, along with farm machinery and irrigation systems.

As countries gradually opened their economies to the global trading system in the second half of the twentieth century, an international division of labor emerged. The poorer ex-colonial countries of Africa, Asia, and Latin America continued to play the traditional role of suppliers of cheap primary resources in exchange for advanced industrial goods and military technology. Despite the postwar establishment of the United Nations, the World Bank, and the International Monetary Fund (IMF), whose charters were ostensibly to help emerging nation-states find an equal place in the new international economic order, the global division of labor has reduced many of these states to a neo-colonial economic status.

This situation was reinforced in the 1970s due to declining commodity prices, population growth, and increased costs of imported industrial goods.[3] As Third World governments confronted a rising level of public debt, they were forced by the World Bank and the IMF to emphasize the production of cash crops for the export

[1] Alfred W. Cosby, *Ecological Imperialism: The Biological Expansion of Europe, 900–1900* (Cambridge: Cambridge Univ. Press, 1986); and Ian Inkster, *Science and Technology in History* (London: Macmillan, 1991).

[2] For an account of the problem in Indonesia, see Steven Hill, Anthony March, John Merson, and Falatehan Siregar, "Science and Technology: Partnership in Development," in *Expanding Horizons: Australia and Indonesia into the 21ˢᵗ Century* (Australia: East Asia Analytical Unit, Department of Foreign Affairs and Trade, 1994).

[3] Susan George, *A Fate Worse than Debt* (Harmondsworth, U.K.: Penguin, 1990).

market, and to adopt structural adjustment programs that led to cutbacks in public-sector expenditure. This was disastrous for local scientific research and indigenous technological development. It also meant that with ever-burgeoning populations and often corrupt ruling elites, natural resources were exploited with ruthless disregard for the environmental consequences. Tropical forest ecosystems were destroyed at an unprecedented rate to make way for plantation agriculture: beef in the Brazilian Amazon, wood pulp in Thailand, and palm oil in Indonesia.

In this sense, globalization has created a situation in which, despite the rhetoric of national sovereignty, most developing countries remain in a condition of dependency. This paper will review some of the difficulties faced by developing nations as they attempt to move beyond a neo-colonial relationship within the global trading system towards a legitimate place in the corporate research networks that dominate production in the emerging biotechnology industries. This is particularly critical in the case of the economically poor but biologically rich nations of the tropical and subtropical regions.

BIOTECHNOLOGY AND THE CONVENTION ON BIODIVERSITY

The tropical regions of the world, which occupy only six percent of the earth's surface and are economically the poorest and least developed, contain the greatest diversity of the world's fauna and flora. The biologist Edward O. Wilson has estimated that at the beginning of the 1990s, human activity had already eliminated 55 percent of original forest cover in the wet tropical regions. Since then we have been reducing the remaining 45 percent at a rate of around 1 percent per annum, or an acre per second.[4]

In 1992, amid growing international concern that the economic pressures on developing countries in the tropical regions were leading to destruction of a large part of the earth's biological heritage, the United Nations mounted its historic Rio Earth Summit on Environment and Development. The passage at Rio of an International Convention on Biological Diversity, and its ratification by most countries within the UN system, was a remarkable achievement. In the colonial context, biological resources had been regarded as being part of the global commons and were not subject to property rights, except where specific plant breeding occurred. The drafters of the convention believed that the best strategy to protect biological and genetic resources was to give states explicit property rights.

The Convention on Biodiversity was therefore an attempt to encourage Third World governments to conserve their existing forests or to harvest them sustainably. The argument conservationists put forward was that the long-term commercial value of these biological resources to the world's chemical, pharmaceutical, and biotechnology industries was much greater than their value for extracted timber and agriculture, especially as tropical soils can have short-term use for agricultural production. In support, botanists pointed out that at least seven thousand of the most commonly used drugs in Western medicine are derived from plants. Much of this comes from the biologically rich environments of the tropics, and is worth U.S. $32 billion a

---

[4] Edward O. Wilson, "Biodiversity, Prosperity and Value," in *Ecology, Economy, Ethics: The Broken Circle*, eds. F. Herbert Bormann and Stephen R. Kellert (New Haven / London: Yale Univ. Press, 1991), pp. 3–10.

year in sales worldwide. However, while Third World countries supply and maintain the bulk of these resources, they receive only U.S. $551 million in return.[5]

To appreciate the significance of this landmark agreement, it is necessary first to review the use and development of biological resources during the colonial and postcolonial periods. For even though the convention is a well-meaning attempt to address some of the global inequalities that are a legacy of colonialism, defining genetic and biological resources as state property has opened a Pandora's box of controversy and complexity.

SCIENCE AND THE COLONIZATION OF THE NATURAL WORLD

The collection and trade of plants for use as foods, drugs, or insecticides dates back to the earliest hunter-gatherer communities. The knowledge and use of local plants was important in the development of medical practices. One of the earliest documented compendiums of medicinal plants was the *Pen Ts'ao* by the Chinese herbalist Shen Nung. Written in 2500 B.C., it listed some 366 plant drugs, some of which, like ephedra, were also used in the West. A thousand years later, the *Ebers Papyrus* listed opium, aloes, and henbane among the drugs in use in Egypt in 1500 B.C. By A.D. 78, Dioscorides, in his *De Materia Medica*, described 600 plants including specific extracts such as aloe, ergot, and opium. While many of these materials were traded throughout the ancient world, it was not until the development of European colonial empires that moving plants from one side of the globe to another took on real economic significance.[6]

Within fifty years of such American crops as the potato, maize, and tobacco arriving in Europe, they were also being cultivated in China. Cotton and cane sugar transferred from India to the Caribbean and the Americas formed the basis of the plantation system in the New World. By the late seventeenth century, the Dutch East India Company plantations in Java had given the Netherlands an international monopoly on spices such as pepper, cloves, and cinnamon, and also political control of Java itself. However, the growth of the administrative and trading port of Batavia brought new diseases to the Dutch, such as malaria, dengue fever, and a range of bacterial infections unknown in Europe. As Leonard Blusse has documented, the old city of Batavia was abandoned in 1730 because of the inability to control the disease being spread by severe pollution of the city's canal system.[7] Pollution from sugar manufacture and the system's failure to flush out sewage turned the canals into breeding grounds for tropical diseases against which the Europeans had neither immunity nor drugs.

New medicines were thus sought not only to find cures for diseases that racked the growing cities of Europe, but also to sustain colonial communities in remote parts of the world. By the time of James Cook's voyages to the South Pacific, the

---

[5] The Crucible Group, *People, Plants and Patents: The Impact of Intellectual Property on Trade, Plant Biodiversity, and Rural Society* (Ottawa: International Development Research Centre, 1994).

[6] Varro E. Tyler, "Natural Products and Medicine: An Overview," in *Medicinal Resources of the Tropical Forest: Biodiversity and its Importance to Human Health*, eds. Michael J. Balick, Elaine Elisabetsky, and Sarah A. Laird (New York: Columbia Univ. Press, 1996), pp. 3–10.

[7] Leonard Blusse, *Strange Company: Chinese Settlers, Mestizo Women and the Dutch in VOC Batavia* (Amsterdam: Foris Publications, 1986), pp. 15–34.

British Admiralty saw the inclusion of botanical research under Joseph Banks and his Swedish colleague Carl Solander as a legitimate part of a voyage of exploration.

In the late eighteenth and early nineteenth centuries, the establishment of botanical gardens in Europe and in the colonies was of growing scientific, medical, and agricultural importance. Botanical gardens at Kew and Leiden became major centers for adapting economic and medicinal plants from around the world for cultivation. Botanical gardens established in the colonies were to become part of a sophisticated international network. In Java, for instance, the famous botanical gardens at Bogor and the scientific institutes in Bandung became major centers of research, at times eclipsing metropolitan centers in Holland. In Australia, botanical gardens were primarily concerned with the adaptation of European food crops to alien environments, and only later became centers for the collection and scientific exploration of native plants. As Linden Gillbank has observed, "imperial powers sought to control the cultivation of useful plants, with colonial botanical gardens providing crucial testing grounds for the suitability of plants to new climates. The Australian botanical gardens were part of a well controlled network of British colonial gardens which manipulated global botanical resources for the economic interests of Britain."[8]

The transfer of plant and animal species from one colonial region to another led to both enormous profit and environmental disaster. Establishing plantations for tea in Ceylon and India, and for South American rubber trees in Malaysia, provided a lasting economic foundation for Britain's imperial aspirations, especially after the valuable tropical timber resources had been exploited. Despite Brazil's efforts to stop the export of rubber-tree seeds and seedlings in the late nineteenth century, within twenty years of the first rubber trees being planted in Malaysia, Brazil's share of the rubber trade had dropped from 98 percent to virtually nothing.[9] The growing imperial and international market for rubber, tea, coffee, sugar, cotton, and wool meant that colonial administrations were actively encouraging the clearing of forested regions to make way for these lucrative export crops. It also meant that biological research was largely focused on crossbreeding to develop plants suited to specific environmental conditions.

By the late nineteenth century, the destruction of forest ecosystems in Australia had become so extensive that Frederick von Mueller, director of the Melbourne Botanical Gardens, argued for the need to preserve biodiversity. "Floral Commons . . . should be reserved in every great country for some maintenance of the original vegetation, and therewith for the preservation of animal life concomitant to peculiar plants."[10] It is perhaps ironic that colonists and scientists could talk of the principle of a "floral commons" when, in reality, the expansion of European property laws to the colonies meant the appropriation of lands and resources commonly held and used by aboriginal communities.

---

[8] Linden Gillbank, "The Life Sciences: Collections to Conservation," in *The Commonwealth of Science: ANZAAS and the Scientific Enterprise in Australasia 1888–1988*, ed. Roy MacLeod (Melbourne: Oxford Univ. Press, 1988), p. 100.

[9] L. H. Brockway, "Plant Science and Colonial Expansion: The Botanical Chess Game," in *Seeds and Sovereignty*, ed. Jack R. Kloppenburg (Durham, N.C.: Duke Univ. Press, 1988), pp. 49–66.

[10] Frederick von Mueller, "Inaugural Address," *Report of the American Association for the Advancement of Science* (Melbourne, 1890), vol. 2. As quoted in Gillbank, "The Life Sciences" (cit. n. 8), p. 117.

THE GROWTH OF BIOCHEMICAL KNOWLEDGE

The nineteenth-century application of the experimental methods of science to the extraction of active agents from biological material played a crucial role in establishing the chemical industry, particularly in Germany. Extracting and identifying the active pharmacological agents from many of the better-known plants yielded major advances in medical treatment. In 1803, the German pharmacist Freidrich Serturner experimented with the newly discovered techniques of isolating organic acids. He tried the technique on opium and ended up not with an acid but the first alkaloid. This led not only to the development and use of morphine in the control of pain, but also to the discovery of the active pharmaceutical agents in a number of other important medicinal plants, many derived directly from colonial sources. Quinine was first extracted in 1819, atropine in 1831, cocaine in 1860, ergotamine in 1918, and tubocurarine in 1935.[11] The medical importance of these new alkaloids, especially the use of quinine in combating the debilitating effects of malaria in the colonies, cannot be underestimated. As with rubber seedlings from Brazil, many South American states attempted to prevent the export of cinchona bark or seedlings (from which quinine is derived), but with little success. The dominance of European powers, and the disregard for intellectual property rights other than those possessed by European industry, made such efforts fruitless.[12]

Institutes for the study of tropical diseases and medicine began to emerge towards the end of the century. The impact of European diseases on native populations was also a major concern for colonial administrations. The documentation of local medical practices, and the use of native plant material, soon became part of the process of fighting disease.[13] The value of new drugs and the internationalization of the chemical and pharmaceutical industries led to a far more systematic exploration of the plants used by native peoples under colonial rule.

Medicine in the colonies was not concerned only with the plight of colonists and natives. The introduction of sheep, cattle, horses, pigs, and poultry was a risky business given their transfer to very different environments and ecosystems. Again, diseases unknown in Europe meant that crossbreeding for resistance was critical. Overcoming new animal diseases gave veterinary medicine a unique and critical role in the economic survival of colonial agriculture.[14]

INTELLECTUAL PROPERTY RIGHTS AND BIOLOGICAL RESOURCES

The colonial assumption that all plant and genetic resources were part of a "common biological heritage" was at least tacitly accepted until 1930, when the United States

---

[11] Tyler, "Natural Products and Medicine" (cit. n. 6), p. 4.

[12] Calestous Juma, *The Gene Hunters: Biotechnology and the Scramble for Seeds* (Princeton, N.J.: Princeton Univ. Press, 1989).

[13] The 1898 Cambridge Anthropological Expedition to the Torres Strait is an example of one of the first systematic anthropological explorations. See A. Herle and J. Philp, *Torres Strait Islanders* (Cambridge: Univ. of Cambridge Museum of Archaeology and Anthropology, 1998).

[14] For an interesting account of colonial veterinary policy and practices, see William Beinart, "Vets, Viruses and Environmentalism at the Cape," in *Ecology and Empire: Environmental History of Settler Societies*, eds. Tom Griffiths and Libby Robin (Melbourne: Melbourne Univ. Press, 1997), pp. 87–101.

passed the Plant Patent Act. This legislation, which allowed for the patenting of asexually reproduced plants, was passed mainly under pressure from plant breeders in the ornamental garden market. However, it opened a Pandora's box on plant breeder's rights. In the 1940s, Europeans passed laws allowing for the protection of sexually reproduced plants. By 1961, international trade in hybrid species had grown to the point where an international convention on plant breeder's rights, known as the UPOV Convention, was established by the Union for New Varieties of Plants.[15] With the development of recombinant DNA techniques, the genetic engineering of new varieties of agricultural plants became a reality, leading to the patenting of genetically engineered strains of common agricultural crops such as cotton, soybeans, and corn. These new varieties, with built-in resistance to common pests, offered economic advantages in terms of reduced reliance on chemical inputs. As a result, they tended to undermine the value of traditional varieties held within farming communities as common property. This clash between property law and customary rights also arose as pharmaceutical and agrichemical corporations began to explore the medicinal plants used by traditional communities around the world.

The environmental problems caused by the overuse of synthetic fertilizers, insecticides, and herbicides meant that there was a constant search for new and less-damaging chemical compounds. By the 1970s the search for new biodegradable insecticides was a major goal, especially given adverse publicity and consumer concerns about residues of the organochlorine and phosphate insecticides in common use. The development of one of the best known of these new bio-insecticides, Bio-Neem, illustrates the problems of equity that arise when communally held intellectual property is used to create products for the international market.

In 1971 an American timber importer, Robert Larsen, became interested in the widespread use of the berries of the neem tree (*Azadirachta indica*) by villagers throughout India. For over two thousand years, the oil from the neem has been used in India as an insecticide, a fungicide, a contraceptive, and an antibacterial agent. Over a number of years at his company headquarters in Wisconsin, Larsen carried out experiments on the extraction of azadirachtin from neem oil, which he found to be a powerful insect growth inhibitor. In 1985 he took out a patent on his process of extracting azadirachtin, which he then sold to W. R. Grace & Co. in 1988. In 1993, after considerable R & D, Grace released a new bio-insecticide called Bio-Neem or Margosan-O. Its active ingredient, azadirachtin, has unique characteristics as a bio-insecticide in that it is lethal to at least two hundred types of insects, as well as to species of mites and nematodes, yet it is completely harmless to birds, mammals, and beneficial insects such as bees. Given the huge demand for the product, especially for the control of greenhouse pests throughout America, Grace entered into a deal with a private Indian firm, P. J. Margo, to start a neem processing plant in Karnataka, in South India. The plant was soon processing twenty tons of neem seeds a day and the new insecticide was taken up in greenhouses across the United States, from California to Florida.

Dr. Martin Sherwin, president of Grace's commercial development division in Florida, argued that Indian industry was set to benefit from further development of this product. The W. R. Grace patent, he argued, created a more valuable resource from neem. Before 1988, neem berries were processed primarily for oil, which was

[15] The Crucible Group, *People, Plants and Patents* (cit. n. 5), pp. 55–65.

used as a surfactant in medicinal soaps and other domestic products, and the berries' waste material was then sold as a fertilizer. The Grace patent involved a three-stage extraction process that produced azadirachtin along with both the oil and the fertilizer.[16] This, by any account, was a considerable improvement. As a consequence, six Indian companies followed suit and set up operations based on separate patents taken out in India for azadirachtin extraction.

With the bio-insecticide industry estimated to be worth around $1 billion, the demand rose for neem berries collected from over fourteen million trees throughout India. But the small-scale, local neem-oil industry was in no position to compete with the technological resources and industrial power available to W. R. Grace. Following 1993's General Agreement on Tariffs and Trade–Trade-Based Intellectual Property (GATT–TRIPS) agreement, India and other developing countries were given until 1998 to develop an intellectual property system compatible with other trading nations, in order to protect their interests in international patents such as Grace's. While patenting options and intellectual property rights had been available to Indian manufacturers, no legal mechanism recognized the collective intellectual property interests of traditional users. On these grounds, critics such as Vandana Shiva argued that the Grace patent fell little short of "bio-piracy."[17] W. R. Grace has not, as yet, even tried to extend its United States patent to India because of the commercial risks and political sensitivity involved. The company's industrial and global market dominance provides it with sufficient security.

European intellectual property law was not designed to represent knowledge held collectively, as is the case with neem. "Indigenous heritage," to use Dr. Erica-Irene Daes's term, tends to fall outside the normal patenting and other commercial/legal mechanisms.[18] However, international concern about protecting intellectual property held collectively within traditional indigenous cultures has led a number of Third World countries, such as the Philippines, to develop sui generis legislation. There is no doubt that the innovations involved in the W. R. Grace patent are a legitimate improvement on the methods of extracting azadiractin from neem oil. But in the view of Tony Simpson and Vanessa Jackson, lawyers specializing in intellectual property rights, the World Trade Organization TRIPS agreement, which aims to provide international protection for such patents, provides no recognition of the intellectual property interests of the community in India that first discovered and used neem-based products. In their view, such disregard for cultural knowledge

> will deepen the North/South rift, with ensuing unfair and unequal exchange; the agreement will facilitate increased occurrence of bio-piracy of biological and genetic resources from indigenous peoples; and communities and cultures may be irreversibly damaged by the forced introduction of foreign concepts of intellectual property law (such as the concepts of exclusive ownership and alienability), and the further erosion of their means of self determination.[19]

[16] As stated in a 1994 interview with the author and broadcast on an Australian Broadcasting Corporation Radio's National Science Show program, "Gene Prospecting," 18 Feb. 1995.

[17] Vandana Shiva, "Biodiversity, Biotechnology and Profit: The Need for a People's Plan to Protect Biological Diversity," *The Ecologist*, 1990, *20*, 2:44–7.

[18] Dr. Erica-Irene Daes is the United Nations' special rapporteur for the Subcommittee on Prevention of Discrimination and Protection of Minorities, and is chairperson of the Working Group on Indigenous Populations. As quoted in Tony Simpson and Vanessa Jackson, "Effective Protection of Indigenous Cultural Knowledge: A Challenge for the Next Millennium," *Indigenous Affairs*, 1998, *3*:45.

[19] *Ibid.*

THE MERCK/INBIO DEAL

The case of neem illustrates the conflict that commonly arises when traditional industries confront the commercial and technological power of a global corporation like W. R. Grace. However, there have been cases in which global corporations entered into more equitable relationships with Third World countries in the development of new chemical and pharmaceutical products.

In 1991, the Central American nation of Costa Rica had a GNP of about $5.2 billion, only slightly more than half the $8.6 billion in annual sales of pharmaceutical giant Merck Corporation. Yet in the same year, Costa Rica's National Biodiversity Institute, INBio, signed a bio-prospecting agreement with Merck worth $1.135 million over two years. This strategy was designed to help maintain the rich biodiversity locked in Costa Rica's national parks and forestry reserves, which represent 27 percent of the country. Despite its tiny size, Costa Rica holds around 4 percent of the world's diverse range of plants and animals.

The Merck/INBio joint venture was the first in what has now become a growing trend of agreements involving biologically rich but economically poor countries. From INBio's point of view, this deal gave them not only much-needed finance but also access to advanced screening technology and training for their researchers. It has also allowed local villagers in forest areas to be trained as "parataxonomists" and to participate in the scientific identification and collection of species, building on extensive traditional knowledge of the fauna and flora. Dr. Anna Siddenfeld, who was in charge of research at INBio in 1991, argued that the great value of the deal was Merck's transfer of assaying technology to INBio, and the training of staff in its use. "Data from the work of the first group indicates that 15 parataxonomists generate well in excess of 50,000 prepared specimens a month, and the inventory collection at INBio at present contains more than two and a half million specimens."[20] It has not only helped Costa Rica to justify the preservation of its forest ecosystems, but also to begin the enormous task of accurately documenting its diverse range of plants and animals. The value-added element in this sort of *in situ* screening is that the royalty returns to INBio and Costa Rica are trebled in the case of a successful project. "It is common to receive royalties of 1–6% of net sales for unscreened chemical samples, 5–10% for material backed by preclinical information on its medical activity, and 10–15% for factional and identified material with effective data."[21]

Through this experience, INBio has been put in a position to explore and develop products with other international groups, but from a position of relative strength. For example, INBio has fostered the development of Costa Rica's first bio-insecticide industry in conjunction with the British Technology Group (BTG). It is a nematocide produced from the seeds of a leguminous tree found in Costa Rica's dry tropical forests. This was developed in conjunction with Dr. Dan Jonstone after the discovery that the seeds of this tree were not eaten by either birds or animals. After eight years of testing at the Royal Botanical Gardens at Kew, chemists isolated a pyrolidine alkaloid called DMDP. This alkaloid is both nontoxic to humans and

---

[20] Ana Sittenfeld, "Tropical Medicinal Plant Conservation and Development Projects: The Case of the Costa Rican National Institute of Biodiversity (INBio)," in Balick, Elisabetsky, and Laird, *Medicinal Resources of the Tropical Forest* (cit. n. 6), p. 336.

[21] Walter Reid *et al.*, "Biodiversity Prospecting," in *ibid.*, p. 161.

biodegradable, so it can be sprayed on plants at very low doses to protect the roots from attack by nematodes. (The only other chemical protection suitable for tropical conditions requires extensive application and is highly toxic.) BTG has taken out a patent on the product in conjunction with INBio, and Costa Rica will be producing the new bio-insecticide locally. If successful, this product could give Costa Rica a new industry in the cultivation and harvesting of raw materials, and in the processing and production of the final product. Trials are being carried out in Costa Rica on tropical plants, and in Britain on temperate crops such as tomatoes and potatoes.

Costa Rica's experience demonstrates the logical connection between the maintenance of biodiversity in forest ecosystems and the development of profitable new industries. This example is often used to vindicate the arguments put forward by those drafting the Convention on Biodiversity. It also demonstrates what can be achieved where there is genuine North/South collaboration.

Since Costa Rica's success, a number of bilateral deals have been struck between large pharmaceutical/chemical companies and state bio-prospecting agencies. Mexico's National Biodiversity Commission was set up in 1992, and similar institutions have been established in Peru and Brazil. The Indonesian government and the Asian Development Bank have agreed to establish a Biodiversity Marketing and Commercialisation Board. In Australia, the AMRAD Corporation established similar bio-prospecting agreements with the Tiwi people of the Northern Territory in 1994, and formed the Northern Lands Council in 1995 for collecting and assaying specimens on aboriginal land. More recently, these agreements have extended to the Malaysian state of Sarawak (in 1996 and 1998).[22] Pharmaceutical companies around the world are developing similar agreements. Some, like the Sharman Corporation of San Francisco, have targeted remote tribal communities and are using their traditional plant medicines as the basis for developing new drugs. Others, like the U.S. National Cancer Institute and the British company Biotics Ltd., are acting as brokers between research groups in developing countries that are independently exploring their indigenous pharmacopoeia, such as in China, and the international drug companies looking for new chemical compounds or genetic materials.

THE BIODIVERSITY CONVENTION AND INDIGENOUS RIGHTS

The Biodiversity Convention agreed to at Rio, and subsequently ratified by most of the 160 countries present, represents an important step in trying to overcome the colonial heritage that is inherent in the economic and technological power relations around the globe. It could be argued that the enthusiasm with which so many Third World countries embraced the convention reflects their very real concern over their loss of control. Critics of the GATT–TRIPS agreement, such as Shiva and Narji, argue that the economic forces driving globalization are leading to a form of neo-colonialism.[23] They cite the fact that global corporations, such as W. R. Grace and Merck, have the capital to dominate and shape international markets. However, Dr. Lyn Capporal, the Merck representative involved in setting up the deal with Costa

---

[22] From AMRAD Prospectus, 1998.

[23] Vandana Shiva and G. S. Nijar, a lawyer with the Third World Network, view enforced compliance with the GATT–TRIPS agreement, as well as the Organization for Economic Cooperation and Development's proposed Multilateral Agreement on Investment, as reinforcing neo-colonial economic relations between the North and South.

Rica, points out that while Merck might be the largest pharmaceutical company in the world, it controls only 5 percent of the world market, and has far less power then its critics imagine.

It is still unclear whether the countries that were signatories to the convention, will find the principle of property rights over the biological and genetic resources within their territories meaningful enough to encourage the conservation of biodiversity. For many economic, legal, and scientific factors may weaken and undermine the anticipated benefits. Some of these factors are reflected in the pronounced inequality, in both scientific and industrial resources, that exists between the countries holding the bulk of the world's biodiversity and the global agrichemical and pharmaceutical corporations most capable of making use of it. Also, while the convention focuses on giving property rights to states, it also recognizes that in many cases the maintenance of biodiversity rests in the hands of indigenous communities that have been part of the ecological balance for many thousands of years. The convention refers to the need to protect the interests of these indigenous communities as part of any conservation strategy.[24] However, the state ownership of biological and genetic resources may simply reinforce past patterns of appropriation and dispossession. Consider, for example, the situation of the Australian aborigines.

The Mabo High Court decision of June 1992, which recognized aboriginal land ownership prior to European settlement, destroyed the illusion that Australia had been *terra nullius*, or an empty land, as colonists had conveniently believed. Their contention that aborigines had had no land use or traditions of ownership meant that the country was available for settlement without the colonial government having to enter into treaties or purchase agreements. In reality, archaeological research by Rhys Jones has shown that aboriginal systems of land management ("fire stick farming") had transformed the Australian environment from the time of their arrival around sixty thousand years ago.[25] Evidence of this practice was first reported by James Cook as early as the 1770s, and by the explorer Ernest Giles over a century later (in 1889):

> The natives were about, burning, burning, everywhere burning; one would think they were the fabled salamander race, and lived on fire instead of water.[26]

The regime of burning the land in a mosaic pattern in cycles of up to thirteen years was responsible for preventing raging fires from destroying valuable fruit-bearing trees, as well as useful plants and animals.[27] In other words, the traditional biodiversity of Australia, particularly its unique fire-dependent or -tolerant species, could well be considered a byproduct of aboriginal culture. This fact was observed as early as 1838 when Sir Thomas Mitchell, surveyor general of New South Wales, noted the change in vegetation that had occurred as a result of the aboriginals being forced off the land areas around Sydney.

---

[24] Articles 8j and 10c.

[25] Rhys Jones, "Fire Stick Farming," *Australian Natural History*, Sept. 1968, *16*, 3:224–8.

[26] Tim Flannery, *The Future Eaters* (Sydney: Reed, 1994), p. 217.

[27] This was clearly demonstrated after aboriginal tribes were forced off the pasture lands west of the Great Dividing Range. Forest scrub returned, creating conditions in the early part of the nineteenth century allowing for some of the most destructive bush fires in the colonies' history.

Kangaroos are no longer to be seen there; the grass is choked by underwood; neither are there natives to burn the grass . . . the omission of the annual periodical burning by natives, of the grass and young saplings, has already produced in the open forests nearest Sydney, thick forests of young trees, where, formerly, a man might gallop without impediment, and see whole miles before him.[28]

The significance of the Mabo decision is that, in recognizing aboriginal traditions of land ownership, and that their land management practices shaped the biodiversity across large areas of the country, it can therefore be argued that this biodiversity is, in part, an artifact of aboriginal culture. Clear evidence of this is to be found in the continued use of a system of mosaic burning to maintain biodiversity in the famous World Heritage Kakadu National Park in Arnhem Land. For aborigines still living close to their land, this burning practice is often referred to as "looking after the country."

As a consequence of the Convention on Biodiversity, a number of Australian states have passed legislation claiming ownership of all native botanical and genetic resources within their jurisdiction. However, while articles of the convention specifically state that indigenous interests and rights are to be taken into account, these have yet to be recognized in state legislation. Similar conflicts of interest will be found in many countries as stakeholders identified by the convention begin to assert their rights. There are other factors associated with the past international movement of plant species that could make the convention's notion of property rights difficult to implement. For example, many ornamental and economic plant species are held either in gene banks as germ plasm, or in botanical gardens and arboretums around the world. Often these have been in such collections for hundreds of years and the location of the original specimen is no longer known. Biotechnology companies, concerned about the increased complexities of obtaining bio-prospecting agreements in many Third World countries, are turning to these vast collections. For example, in 1996 the United States biotechnology company Phytera signed agreements with seven European botanical gardens to obtain seeds and tissues from tropical plants in their collections. While this contravenes the spirit of the 1993 Convention on Biodiversity, it has not been legally challenged. The deal involves Phytera paying the gardens $15 per plant specimen delivered and 0.25 percent of the profit from products derived from the specimens. The gardens involved will get 2.5 percent if products are developed under license from Phytera.[29]

In the case of food plants, the free international flow of germ plasm is essential for scientists and plant breeders in both rich and poor countries alike, as they develop new species to keep ahead of resistance to disease and climate change. But as Balick and Kloppenburg argue, "Third world plant breeders can have access to the USDA's gene bank—[but] in practice, 'free exchange' is not the even-handed opportunity its proponents have made it out to be, and the benefits of collecting biochemical and genetic information have accrued to the North in extremely disproportionate fashion."[30]

Of equal importance has been the operation of gene banks under an agreement

[28] Sir Thomas Mitchell, cited in Flannery, *The Future Eaters* (cit. n. 26), p. 220.

[29] *New Scientist*, 29 June 1996.

[30] Jack R. Kloppenburg and Michael J. Balick, "Property Rights and Genetic Resources: A Framework for Analysis," in Balick, Elisabetsky, and Laird, *Medicinal Resources of the Tropical Forest* (cit. n. 6.), p. 181.

between the Consultative Group on International Agricultural Research and the UN Food and Agriculture Organisation. This global network of international agricultural research centers represents the world's largest collection of germ plasm for crops and forestry. Six hundred thousand accessions are made available to researchers free of charge each year, with many of these going to developing countries.[31] This system is now governed by trustees from sixty countries, with funding flowing from forty countries. With the transformation of agriculture and the increasing loss of global biodiversity, this international network of gene banks is an essential asset, but the value of free access is clearly a matter of an individual country's knowledge and resources.

The efforts of overzealous governments to use claims of ownership to stem the free flow of germ plasm and genetic materials may backfire on their own scientists and plant breeding programs, creating what has been called a "seed war." However, some Third World groups argue that these gene banks should operate on a fee system. Like the copyright laws that govern the international use of music or art, the banks could provide some return to the country whence the germ plasm originated. While this would be in the spirit of the Convention on Biodiversity, it could prove to be a commercial nightmare for both plant breeders and the emerging biotechnology industry.

This complex environment of competing interest groups has led Balick and Kloppenburg to observe that a "New International Genetic Order is clearly in the offing. The degree to which this altered order is truly 'new,' or whether it is simply a kinder or gentler version of the old exploitative relationship, remains to be seen."[32]

CONCLUSION

In considering the legacy of colonialism, there is an obvious tendency to blame the colonial powers for the scientific, technological, and economic underdevelopment plaguing many Third World countries. Despite the rhetoric, decolonization has not resulted in any fundamental changes in the locality of scientific and technological innovation. The countries' dependence on the research and development of the advanced economies of Europe, the United States, and Japan is, if anything, greater now than during the colonial era. This is largely due to the integration of most of the world's economies following the Second World War and, more recently, to the ending of the cold war. This dependency is particularly true of agriculture and biotechnology following the genetic engineering revolution of the 1970s and 1980s. This transformation led to large-scale capital investment in the biosciences, which have followed high-energy physics into becoming "Big Science."[33] The financial returns from the global marketing of new agrichemicals and pharmaceuticals are so great that corporations such as Novartis are prepared to invest $250 million in an Institute for Functional Genomics in La Jolla, California, and the Wellcome Trust and other pharmaceutical companies in Britain are putting $45 million into a major

---

[31] The Crucible Group, *People, Plants and Patents* (cit. n. 5), p. 92.

[32] Kloppenburg and Balick, "Property Rights and Genetic Resources" (cit. n. 30), p. 183.

[33] The emergence of biological "Big Science" is similar to the transformation of physics in the 1960s as outlined in John Ziman, *The Force of Knowledge* (Cambridge: Cambridge Univ. Press, 1976).

research project addressing diagnostic nucleotide polymorphism. The scale of these investments means that highly centralized international research teams will dominate the applied market in many fields of biotechnology, leaving behind small groups of scientific researchers in ex-colonial countries. Even in Australia, New Zealand, Canada, and India, where there have been strong traditions of biological research, these changes could reduce applied research to niche or peripheral fields.

The era of globalization has seen the emergence of corporations having international dominance in key markets, such as W. R. Grace, Monsanto, Merck, and Microsoft. Their global networks of production and distribution have led to increasing standardization and the accentuation of inequalities.[34] In the case of the cotton industry, countries that once had independent agricultural and research traditions now face a situation in which all inputs—from seeds and fertilizers to insecticides, herbicides, and machinery—are imported from international agribusiness firms. These same corporations also buy back the crop for processing in the industrialized centers of the North. The only stake that "peripheral" nation-states have in the final product is in providing labor, land, and water to the "core" industry group, all of which represent low value-added returns.[35] While critics have described this situation as neo-colonial in that it perpetuates the traditional colonial condition of technological and economic dependency, there is one obvious difference. We are no longer discussing colonial empires, but rather a global market system that the world community accepts as being in everyone's long-term interest, even though it has engendered transnational corporations of unprecedented economic power. This being the case, multilateral agreements through the United Nations system, such as the Convention on Biodiversity and the WTO/TRIPS protocols, may be the only realistic strategy to redress the economic inequalities that are the root cause of the high levels of environmental destruction in Third World countries.[36] Reshaping or redefining rights and interests within the global trading system may eventually lead to more constructive and beneficial relationships, along the lines of the Merck/INBio deal.

For the scientific community, the shift towards viewing the world's biological and genetic resources as the property of nation-states, rather than as part of the global commons, is only just beginning to sink in. Some have argued that the Convention on Biodiversity is unworkable and unenforceable, given the geographical distribution of species across political boundaries and the vast dispersion of plants and animals around the world. Others contend that the loss of biodiversity in the tropical and subtropical regions of the world is so great that the convention must be made to work, or at least serve as the starting point for alternative strategies that can halt the destruction.

Allying mechanisms to protect our biodiversity with the legitimate interests of the world community in using its genetic riches is something that will be battled out in courts and legislative assemblies well into the next millennium. What many call bio-

[34] Robert Reich, *Work of Nations: Preparing Ourselves for the 21st Century* (New York: Knopf, 1991).

[35] This use of the concept of "core and peripheral states" is drawn from the work of C. Hamelink, "Information Imbalance: Core and Periphery," in *Questioning the Media*, eds. John Downing *et al.* (London: Sage Publications, 1990).

[36] For a useful discussion of this issue, see Larry Hempel, *Environmental Governance: The Global Challenge* (Washington, D.C.: Island Press, 1996).

prospecting, others in the Third World call bio-piracy, in that they see the real benefits of any discovery ultimately flowing into the coffers of global chemical and pharmaceutical companies, and not to the knowledge and resource providers.

Ethnobotanists and biotechnology enthusiasts argue that the future returns from
forest and wetland ecosystems could be enormous. The successful development of
a new drug that could halt the scourge of AIDS, or of a valuable new agrichemical
compound, could be worth hundreds of millions in royalties each year. At the same
time, the WTO/TRIPS system of intellectual property rights is coming up against
very real problems. These flow from trying to make antiquated patent and copyright
systems apply to everything from genetic information in living organisms to the
collective knowledge and property rights of indigenous peoples. While the 1993
Convention on Biological Diversity is flawed by internal contradiction and legal
loopholes, it is nonetheless an important start in trying to halt the annihilation of the
world's ever-diminishing range of plant and animal life. One thing is clear: unless
the convention becomes part of a much broader process of overcoming the colonial
and neo-colonial legacy of entrenched global inequality, there is little likelihood of
its objectives ever being realized.

Bibliography

This bibliography is intended as a point of entry to the rapidly expanding study of colonial science, beginning at the intersection of the history of science and imperial history. It includes key works in Spanish and Portuguese as well as other titles, some well known and others less visible but deservedly recommended by the editor and authors of this volume.

Aceves Pastrana, Patricia. "La Difusión de la química de Lavoisier en el Real Jardín Botánico de México y en el Real Seminario de Minería (1788–1810)." *Quipu,* 1990, *7*:5–35.
———. *La Química en Europa y América (siglos XVIII y XIX).* Mexico City: UNAM, 1994.
———. *Química, botánica y farmacia en la Nueva España a finales del siglo XVIII.* Mexico City: UAM, 1993.
Adas, Michael. *Machines as the Measure of Men: Science, Technology, and Ideologies of Western Dominance.* Ithaca: Cornell Univ. Press, 1989.
———. *Technology and European Overseas Enterprise.* Aldershot: Variorum, 1996.
Allchin, Douglas, and Robert DeKosky. *An Introduction to the History of Science in Non-Western Traditions.* Seattle: History of Science Society, 1999.
Alvares, C. *Homo Faber: Technology and Culture in India, China and the West from 1500 to the Present Day.* The Hague: Martinus Nijhoff Publishers, 1980.
Anderson, Benedict. *Imagined Communities: Reflections on the Origin and Spread of Nationalism.* London: Verso, 1983.
Anderson, Warwick. "Climates of Opinion: Acclimatization in Nineteenth Century France and England." *Victorian Studies,* 1992, *35*:135–57.
———. "Disease, Race, and Empire." *Bulletin of the History of Medicine,* 1996, *70*:62–7.
———. "Excremental Colonialism: Public Health and the Poetics of Pollution." *Critical Inquiry,* 1995, *21*:640–69.
———. "Immunities of Empire: Race, Disease, and the New Tropical Medicine, 1900–1920." *Bull. Hist. Med.,* 1996, *70*:94–118.
———. "Where Every Prospect Pleases and Only Man is Vile": Laboratory Medicine as Colonial Discourse." *Crit. Inq.,* 1992, *18*:506–29.
Anderson, Warwick, and Mark Harrison. "Race and Acclimatization in Colonial Medicine." *Bull. Hist. Med.,* 1996, *70*:62–118.
Andrade, Lencastre P. de. "Explorações científicas na India durante os reinados de D. Manuel, D. João III e D. Sebastião—Os primeiros naturalistas portugueses do século." *O Instituto,* 1926, *74*:118–301.
Annino, Antonio, Luis Castro Leiva, and François-Xavier Guerra, eds. *De los imperios a las naciones: Iberoamérica.* Zaragoza: IberCaja, 1994.
Arasaratnam, S. "Recent Trends in the Historiography of the Indian Ocean, 1500–1800." *Journal of World History,* 1990, *2*:225–48.
Arboleda, Luis Carlos. "Science and Nationalism in New Granada on the Eve of the Revolution of Independence." In *Science and Empires: Historical Studies about Scientific Development and European Expansion,* eds. Patrick Petitjean, Catherine Jami, and Anne-Marie Moulin. Dordrecht: Kluwer Academic Publishers, 1992, pp. 247–58.
Archila, Ricardo. *Historia de la medicina en Venezuela: Epoca colonial.* Caracas: Tipografía Vargas, 1961.
Arnold, David. *Colonizing the Body: State Medicine and Epidemic Disease in Nineteenth-Century India.* Berkeley: Univ. of California Press, 1993.
———. *The Problem of Nature: Environment, Culture and European Expansion.* Oxford: Blackwell, 1996.
Arnold, David, ed. *Imperial Medicine and Indigenous Societies.* Manchester: Manchester Univ. Press, 1988.

Attwood, B., and J. Arnold, eds. *Power, Knowledge and Aborigines*. Melbourne: LaTrobe Univ. Press, 1992.

Austen, Ralph A., ed. *Modern Imperialism, Western Overseas Expansion, and Its Aftermath, 1776–1965*. Lexington, Mass.: Heath, 1969.

Azevedo, Fernando de. *As ciências no Brasil*, 2 vols. São Paulo: Edições Melhoramentos, 1955.

Baark, Erik, Aant Elzinga, and Bengt-Erik Borgström. *Technological Change and Cultural Impact in Asia and Europe: A Critical Review of the Western Theoretical Heritage*. Lund and Stockholm: Committee for Future Oriented Research, 1980.

Baber, Zaheer. *The Science of Empire*. New York: State Univ. of New York, 1996.

Babini, J. *La Evolución del pensamiento científico en la Argentina*. Buenos Aires: Ediciones La Fragua, 1954.

Baines, G., and N. Williams. "Partnerships in Tradition and Science." In *Traditional Ecological Knowledge: Wisdom for Sustainable Development*, eds. N. M. Williams and G. Baines. Canberra: Centre for Resource and Environmental Studies: ANU, 1993, pp. 1–6.

Bairoch, Paul. "Le Bilan économique du colonialisme: Mythes et réalités." *Itinerario*, 1980, *1*:29–41.

Banes, Daniel. "The Portuguese Voyages of Discovery and the Emergence of Modern Science." *Journal of the Washington Academy of Sciences*, 1988, *78*, 1:47–58.

Barley, Nigel, ed. *The Golden Sword: Stamford Raffles and the East*. London: British Museum Press, 1999.

Basalla, George. "The Spread of Western Science." *Science*, 1967, *156*:611–22.

———. "The Spread of Western Science Revisited." In *Mundialización de la ciencia y cultura nacional*, eds. Antonio Lafuente, Alberto Elena, and María Luisa Ortega. Madrid: Ediciones Doce Calles, 1993, pp. 599–603.

Baumgart, W. *Imperialism: The Idea and Reality of British and French Colonial Expansion, 1880–1914*. New York: Oxford Univ. Press, 1982.

Bayly, C. A. *Empire and Information: Intelligence Gathering and Social Communication in India, 1780–1870*. Cambridge: Cambridge Univ. Press, 1997.

———. *Imperial Meridian: The British Empire and the World, 1780–1830*. London: Longman, 1989.

Bazaco, Evergisto. *History of Education in the Philippines: Spanish Period, 1565–1898*. Manila: Univ. of Santo Tomas Press, 1953.

Beinart, William. "Vets, Viruses and Environmentalism at the Cape." In *Ecology and Empire*, eds. Tom Griffiths and Libby Robin. Melbourne: Melbourne Univ. Press, 1997, pp. 87–101.

Bell, Heather. *Frontiers of Medicine in the Anglo-Egyptian Sudan, 1899–1940*. Oxford: Oxford Univ. Press, 1999.

Beltrán, E. "Fuentes mexicanas de la historia de la ciencia." *Anales de las Sociedad Mexicana de Historia de la Ciencia y de la Tecnología*, 1970, *2*:57–112.

Benians, E. A., James Butler, and C. E. Carrington, eds. *The Cambridge History of the British Empire*. Vol. III, *The Empire-Commonwealth, 1870–1919*. Cambridge: Cambridge Univ. Press, 1959.

Berlin, Brent. *Ethnobiological Classification: Principles of Categorization of Plants and Animals in Traditional Societies*. Princeton: Princeton Univ. Press, 1992.

Bethell, Leslie, ed. *Colonial Brazil*. New York: Cambridge Univ. Press, 1987.

———. *Colonial Spanish America*. New York: Cambridge Univ. Press, 1987.

———. *Ideas and Ideologies in Twentieth Century Latin America*. Cambridge: Cambridge Univ. Press, 1996.

———. *Mexico Since Independence*. New York: Cambridge Univ. Press, 1991.

Betts, Raymond F. *Assimilation and Association in French Colonial Theory 1890–1914*. New York: Columbia Univ. Press, 1961.

Bhabha, H. K. *The Location of Culture*. New York: Routledge, 1994.

Blanckaert, Claude, R. Chartier, C. Cohen, and P. Corsi, eds. *Le Muséum au premier siècle de son histoire*. Paris: Editions du Muséum National d'Histoire Naturelle, 1997.

Blaut, J. M. *1492: The Debate on Colonialism, Eurocentrism, and History*. Trenton, N.J.: Africa World Press, 1992.

———. *The Colonizer's Model of the World*. New York: The Guilford Press, 1993.

Bonneuil, Christophe. "Crafting and Disciplining the Tropics: Plant Science in the French

Colonies." In *Science in the Twentieth Century,* eds. J. Krige and D. Pestre. Amsterdam: Harwood, 1997, pp. 77–96.

———. "Des Savants pour l'empire." *Cahiers pour l'Histoire du CNRS, 1939–1989,* 1990, *10:*83–102.

———. *Des Savants pour l'empire: La Structure des recherches scientifiques coloniales au temps de 'la mise en valeur des colonies française' 1917–1945.* Paris: Editions de l'ORSTOM, 1991.

———. *Mettre en ordre et discipliner les tropiques: Les Sciences du végétal dans l'empire français, 1870–1940.* Paris: Editions des Archives Contemporaines, forthcoming.

———. "Mise en valeur: De l'Empire colonial et naissance de L'agronomie tropicale." In *Du Jardin d'essai à la station experimentals,* eds. Christophe Bonneuil and Mina Kleiche. Paris: CIRAD, 1993, pp. 17–65.

———. "'Penetrating the Natives': Peanut Breeding, Peasants and the Colonial State in Senegal (1900–1950)." *Science, Technology, and Society,* 1999, *4,* 2:273–302.

Bonneuil, Christophe, and Mina Kleiche. *Du Jardin d'essais colonial à la station expérimentale, 1880–1930.* Paris: CIRAD, 1993.

Bonneuil, Christophe, and Patrick Petitjean. "Science and French Colonial Policy: Creation of the ORSTOM: From Popular Front to the Liberation via Vichy, 1936–1945." In *Science and Technology in a Developing World: Sociology of Sciences, Yearbook 1995,* eds. Terry Shinn, Jack Spaapen, and Venni Krishna. Dordrecht: Kluwer Academic Publishers, 1997, pp. 129–78.

Bourguet, Marie-Noëlle. "La Collecte du monde: Voyage et histoire naturelle, fin XVIIe–début XIXe siècles." In *Le Muséum au premier siècle de son histoire,* eds. C. Blanckaert *et al.* Paris: Editions du Muséum, 1997, pp. 163–96.

Bourguet, Marie-Noëlle, and Christophe Bonneuil. "De l'Inventaire du globe à la 'mise en valeur' du monde: Botanique colonisation (fin XVIIIe siècle – début XIXe siècle). Présentation." *Revue Française d'Histoire d'Outre-Mer,* 1999, *322–323:*9–38.

Bourguet, Marie-Noëlle, Bernard Lepetit, Daniel Nordman, and Maroula Sinarellis. *L'Invention scientifique de la Méditerranée: Egypte, Morée, Algérie.* Paris: Ecole des Hautes Etudes en Sciences Sociales, 1998.

Bourguet, Marie-Noëlle, and Christian Licoppe. "Voyages, mesures et instruments: Une Nouvelle expérience du monde au siècle des Lumières." *Annales HSS,* 1997, *5:*1115–51.

Bowen, Margarita. *Empiricism and Geographical Thought: From Francis Bacon to Alexander von Humboldt.* London: Cambridge Univ. Press, 1981.

Boxer, C. R. *The Dutch Seaborne Empire 1600–1800.* London: Knopf, 1965.

———. *The Portuguese Seaborne Empire 1415–1825.* London: Hutchinson, 1969.

Bravo, Michael T. "Precision and Curiosity in Scientific Travel: James Rennell and the Orientalist Geography of the New Imperial Age." In *Voyages and Visions: Towards a Cultural History of Travel,* eds. Jas Elsner and Joan-Pau Rubiés. London: Reaktion, 1999, pp. 162–83.

Bret, Patrice. *La Géographie des philosophes: Géographes et voyageurs français au XVIIIe siècle.* Paris: Ophrys, 1975.

———. *L'Égypte au temps de l'expédition de Bonaparte, 1798–1801.* Paris: Hachette, 1998.

Brockway, L. H. "Plant Science and Colonial Expansion: The Botanical Chess Game." In *Seeds and Sovereignty,* ed. Jack R. Kloppenburg. Durham: Duke Univ. Press, 1988, pp. 49–66.

Browne, Janet. "A Science of Empire: British Biogeography before Darwin." *Revue d'Histoire des Sciences,* 1992, *45,* 4:453–75.

Burkholder, Mark A., and Lyman L. Johnson. *Colonial Latin America.* New York: Oxford Univ. Press, 1990.

Cain, P. J., and A. G. Hopkins. *British Imperialism: Innovation and Expansion, 1688–1914.* London: Longman, 1993.

Cañizares, Jorge. "Entre Maquiavelo y la jurisprudencia natural: William Robertson y la disputa del Nuevo Mundo." *Quipu,* 1991, *8,* 1:279–91.

———. "New World, New Stars: Patriotic Astrology and the Invention of Indian and Creole Bodies in Colonial Spanish America, 1600–1650." *The American Historical Review,* Feb. 1999, *104,* 1:33–68.

———. "Spanish America in Eighteenth-Century European Travel Compilations: A New

'Art of Reading' and the Transition to Modernity." *Journal of Early Modern History,* 1998, *2,* 4:329–49.

Carter, Harold B. *Sir Joseph Banks, 1743–1820.* London: British Museum, 1988.

———. *Sir Joseph Banks 1743–1820: A Guide to Biographical and Bibliographical Sources.* Winchester: St. Paul's Bibliographies, 1987.

Carter, Paul. *The Road to Botany Bay: An Exploration of Landscape and History.* New York: Alfred A. Knopf, 1988.

Cayola, Lourenço. *Sciencia de colonização,* 2 vols. Lisbon: Typographia da Cooperativa Militar, 1912.

Cell, John. "Anglo-Indian Medical Theory and the Origins of Segregation in West Africa." *Amer. Hist. Rev.,* 1986, *91,* 2:307–35.

Chakrabarty, Dipesh. "Postcoloniality and the Artifice of History: Who Speaks for 'Indian' Pasts?" *Representations,* 1992, *32:*1–26.

Chambers, David Wade. "Centre Looks at Periphery: Alexander von Humboldt's Account of Mexican Science and Technology." *Journal of Iberian and Latin American Studies,* 1996, *2,* 1:94–113.

———. "Does Distance Tyrannize Science?" In *International Science and National Scientific Identity,* eds. R. W. Home and Sally Gregory Kohlstedt. Dordrecht: Kluwer, 1991, pp. 19–39.

———. "Locality and Science: Myths of Centre and Periphery." In *Mundialización de la ciencia y cultura nacional,* eds. Antonio Lafuente, Alberto Elena, and María Luisa Ortega. Madrid: Ediciones Doce Calles, 1993, pp. 605–18.

———. "Period and Process in Colonial and National Science." In *Scientific Colonialism: A Cross-Cultural Comparison,* eds. Nathan Reingold and Marc Rothenberg. Washington, D.C.: Smithsonian Institution Press, 1987, pp. 297–321.

———. Preface to *Maps Are Territories: Science is an Atlas,* ed. David Turnbull. Chicago: Univ. of Chicago Press, 1993.

Chambers, David Wade, James E. McClellan, and H. Zogbaum. "Science/Nation/Culture in the Caribbean Basin." In *The Cambridge History of Science,* ed. Ron Numbers, vol. 8. Cambridge: Cambridge Univ. Press, forthcoming.

Chartrand, Luc. *Histoire des sciences au Quebec.* Montréal: Les Editions du Boreal, 1987.

Chatterjee, Partha. *Nationalist Thought and the Colonial World: A Derivative Discourse?* London: Zed Books, 1986.

———. *The Nation and Its Fragments: Colonial and Postcolonial Histories.* Princeton: Princeton Univ. Press, 1993.

Chayut, Michael. "The Hybridisation of Scientific Roles and Ideas in the Context of Centres and Peripheries." *Minerva,* 1994, *32,* 3:297–308.

Cittadino, Eugene. *Nature as the Laboratory: Darwinian Plant Ecology in the German Empire 1880–1900.* Cambridge: Cambridge Univ. Press, 1990.

Cohen, I. B. "The New World as a Source of Science for Europe." *Actes du IX Congrés International d'Histoire des Sciences.* Madrid, 1960, pp. 96–130.

Coombes, Annie E. *Reinventing Africa: Museums, Material Culture, and Popular Imagination in Late Victorian and Edwardian England.* New Haven: Yale Univ. Press, 1994.

Cooper, Frederick, and Ann Laura Stoler, eds. *Tensions of Empire: Colonial Cultures in a Bourgeois World.* Berkeley: Univ. of California Press, 1997.

Coquery-Vidrovitch, Catherine. "L'Impérialisme français en Afrique noire: Idéologie impériale et politique d'investissement, 1924–1975." *Relations Internationales,* 1976, *7:*261–82.

Crawford, Elisabeth, Terry Shinn, and Sverker Sörlin, eds. *Denationalizing Science: The Contexts of International Scientific Practice.* Dordrecht: Kluwer Academic Publishers, 1993.

Crosby, Alfred W. *Ecological Imperialism: The Biological Expansion of Europe 900–1900.* Cambridge: Cambridge Univ. Press, 1986.

———. *The Columbian Exchange: Biological and Cultural Consequences of 1492.* Westport, Conn.: Greenwood Press, 1972.

Crowder, Michael. *Senegal, a Study of French Assimilation Policy.* Oxford: Oxford Univ. Press, 1962.

Cueto, Marcos. *Excelencia científica en la periferia.* Lima: CONCYTEC, 1989.

Cueto, Marcos, ed. *Missionaries of Science: The Rockefeller Foundation in Latin America.* Bloomington: Indiana Univ. Press, 1994.

Cunningham, A., and P. Williams. "De-centring the 'Big Picture': *The Origins of Modern Science* and the Modern Origins of Science." *British Journal of the History of Science,* 1993, *26:*407–32.

Cunningham, A., and B. Andrews, eds. *Western Medicine as Contested Knowledge: Studies in Imperialism.* Manchester: Manchester Univ. Press, 1997.

Curtin, Philip. *Death by Migration: Europe's Encounter With the Tropical World in the Nineteenth Century.* Cambridge: Cambridge Univ. Press, 1989.

———. "Medical Knowledge and Urban Planning in Tropical Africa." *Am. Hist. Rev.,* 1985, *90,* 3:594–613.

———. "The Environment beyond Europe and the European Theory of Empire." *Journal of World History,* 1990, *1,* 2:131–50.

Daston, Lorraine. "The Ideal and Reality of the Republic of Letters in the Enlightenment." *Science in Context,* 1991, *4:*367–86.

Davis, Ralph. *The Rise of the Atlantic Economies.* Ithaca: Cornell Univ. Press, 1973.

de Jesus, C., ed. *The Tobacco Monopoly in the Philippines: Bureaucratic Enterprise and Social Change, 1766–1880.* Quezon City: Ateneo de Manila Univ. Press, 1980.

De Lannoy, Charles, and Herman Vander Linden. *Histoire de L'expansion coloniale des peuples européens: Portugal et Espagne.* Brussels: Henri Lamertin, 1907.

de Martonne, Edouard. *Le Savant Colonial.* Paris: Editions Larose, 1931.

Deacon, Harriet. "Cape Town and Country Doctors in the Cape Colony during the First Half of the Nineteenth Century." *Social History of Medicine,* 1997, *10,* 1:25–52.

Denoon, Donald. *Settler Capitalism: The Dynamics of Dependent Development in the Southern Hemisphere.* Oxford: Oxford Univ. Press, 1983.

Desmond, Ray. *Kew: The History of the Royal Botanic Gardens.* London: Harvill, 1995.

Diamond, Jared. *Guns, Germs and Steel: The Fates of Human Societies.* London: Jonathon Cape, 1997.

Díaz, Elena, Yolanda Texera, and Hebe Vessuri. *La Ciencia periferica.* Caracas: Monte Avila Editores, 1983.

Díaz-Piedrahita, Santiago. "Caldas y la historia natural." In *Francisco José de Caldas.* Bogotá: Molinos Velásquez Editores, 1994, pp. 111–123.

Díez Torre, Alejandro R. *et al. De la ciencia ilustrada a la ciencia romática.* Aranjuez: Doce Calles, 1995.

Díez Torre, Alejandro R. *et al. La ciencia española en Ultramar.* Aranjuez: Doce Calles, 1991.

Dirks, Nicholas B. *Colonialism and Culture.* Ann Arbor: University of Michigan Press, 1992.

Domingues, Angela. *Viagens de exploração geográfica na Amazónia em finais do século XVIII: Política, ciência e aventura.* Lisbon: Centro de Estudos de História do Atlántico, 1991.

Domingues, F. C. "Science and Nationalism: Portugal in the Late 18th Century." *History of European Ideas,* 1993, *16,* 1–3:91–6.

Domingues, Maria Heloísa B. "A Idéia de progresso no processo de institucionalização nacional das ciências no Brasil: A Sociedade Auxiliadora da Indústria Nacional." *Asclepio,* 1996, *48,* 2:149–62.

Dorn, Walter L. *Competition for Empire, 1740–1763.* New York: Harper and Row, 1940.

Drayton, Richard H. "Imperial Science and a Scientific Empire: Kew Gardens and the Uses of Nature, 1772–1903." Ph.D. dissertation, Yale Univ., 1993.

———. "Knowledge and Empire." In *The Oxford History of the British Empire,* vol. II, *The Eighteenth Century,* ed. P. J. Marshall. Oxford: Oxford Univ. Press, 1998, pp. 231–52.

———. *Nature's Government: Science, Imperial Britain and the Modern World.* New Haven and London: Yale Univ. Press, 2000.

———. "Science, Medicine and the British Empire." In *The Oxford History of the British Empire,* vol. V, *Historiography,* ed. Robin W. Winks. Oxford: Oxford Univ. Press, 1999, pp. 264–75.

Drouin, Jean-Marc. "De Linné à Darwin: Les Voyageurs naturalistes." In *Eléments d'histoire des sciences,* ed. Michel Serres. Paris: Bordas, 1989, pp. 321–35.

Drummond, Ian M. *British Economic Policy and the Empire, 1919–1939.* London: George Allen and Unwin, 1972.

Dubow, Saul. *Scientific Racism in Modern South Africa.* Cambridge: Cambridge Univ. Press, 1995.

Dunmore, John. *French Explorers in the Pacific,* 2 vols. Oxford: Clarendon Press, 1965–69.

———. *Visions and Realities: France in the Pacific, 1695–1995.* Waikane, N.Z.: Heritage, 1997.

Edney, Matthew H. *Mapping an Empire: The Geographical Construction of British India, 1765–1843.* Chicago: Univ. of Chicago Press, 1997.

———. "The Patronage of Science and the Creation of Imperial Space: The British Mapping of India, 1799–1843." *Cartographica,* 1993, *30,* 1:61–7.

Elena, Alberto. "La Configuración de las periferias cientificas: Latinoamérica y el mundo islámico." In *Mundialización de la ciencia y cultura nacional,* eds. Antonio Lafuente, Alberto Elena, and María Luisa Ortega. Madrid: Ediciones Doce Calles, 1993, pp. 139–46.

———. "Models of European Scientific Expansion: the Ottoman Empire as a Source of Evidence." In *Science and Empires: Historical Studies about Scientific Development and European Expansion,* eds. Patrick Petitjean, Catherine Jami, and Anne-Marie Moulin. Dordrecht: Kluwer Academic Publishers, 1992, pp. 259–67.

Elliott, John H. *Spain and its World 1500–1700.* New Haven: Yale Univ. Press, 1989.

———. *The Old World and the New 1492–1650.* Cambridge: Cambridge Univ. Press, 1970.

Emerson, Rupert. *From Empire to Nation: The Rise to Self-Assertion of Asian and African Peoples.* Cambridge, Mass.: Harvard Univ. Press, 1960.

Engstrand, Iris H. W. *Spanish Scientists in the New World: The Eighteenth-Century Expeditions.* Seattle: Univ. of Washington Press, 1981.

Etherington, Norman. *Theories of Imperialism: War, Conquest and Capital.* London and Canberra: Croom Helm, 1984.

Fieldhouse, D. K. *Colonialism 1870–1945: An Introduction.* London: Macmillan, 1988.

———. *Economics and Empire, 1830–1914.* London: Weidenfeld and Nicolson, 1973.

———. *The Colonial Empires: A Comparative Survey from the Eighteenth Century.* London: Weidenfeld and Nicolson, 1965.

Figueirôa, Silvia F. de M. *As Ciências geológicas no Brasil: Uma História social e institucional, 1875–1934.* São Paulo: Hucitec, 1997.

———. "German-Brazilian Relations in the Field of Geological Sciences during the 19th Century." *Earth Sciences History,* 1990, *9,* 2:132–37.

Figueirôa, Silvia F. de M., and Maria Margaret Lopes, eds. *Geological Sciences in Latin America: Scientific Relations and Exchanges.* Campinas: Instituto de Geociências/UNICAMP, 1995.

Findlen, Paula. *Possessing Nature: Museums, Collecting, and Scientific Culture in Early Modern Italy.* Berkeley: Univ. of California Press, 1994.

Fleming, C. A. "Science, Settlers and Scholars: The Centennial History of the Royal Society of New Zealand." *Bulletin of the Royal Society of New Zealand,* 1987, *25:*1–353.

Fleming, Donald. "Science in Australia, Canada and the United States: Some Comparative Remarks." *Proceedings of the Xth International Congress of the History of Science* (Ithaca, 1962), *1:*179–96.

Fonseca, Maria R. F. da. "A Única ciência é a pátria: O Discurso científico na construção do Brasil e do México, 1770–1815." Ph.D. dissertation, Faculdade de Filosofía, Letras e Ciências Humanas, University São Paulo, 1996.

Foucault, Michel. "Governmentality." In *The Foucault Effect: Studies in Governmentality,* eds. Graham Burchell *et al.* Chicago: Univ. of Chicago Press, 1991, pp. 87–104.

Fournier, P. *Voyages et découvertes scientifiques des missionnaires naturalistes français travers le monde, pendant cinq siécles.* Paris: Paul Lechevalier, 1932.

Fox, Robert. "The Savant Confronts his Peers: Scientific Societies in France, 1815–1914." In *The Organization of Science and Technology in France, 1810–1914,* eds. G. Weisz and R. Fox. Cambridge and Paris: Cambridge Univ. Press and MSH, 1980, pp. 241–82.

Fradera, Josep. "La Experiencia colonial Europea del siglo XIX (Una Aproximación al debate sobre los costes y beneficios del colonialismo europeo)." In *Europa en su historia,* ed. Pedro Ruiz Torres. Valencia: Universidad de Valencia, 1993.

Francis, Mark. "Anthropology and Social Darwinism in the British Empire, 1870–1900." *Australian Journal of Politics and History,* 1994, *40:*203–15.

Frank, Andre. *ReORIENT: Global Economy in the Asian Age.* Berkeley: Univ. of California Press, 1998.

Franklin, Sarah. "Science as Culture, Cultures of Science." *Annual Review of Anthropology,* 1995: 163–84.

Frost, Alan. *Arthur Phillip, 1738–1814: His Voyaging.* Melbourne: Oxford Univ. Press, 1987.

―――. "The Antipodean Exchange: European Horticulture and Imperial Designs." In *Visions of Empire: Voyages, Botany, and Representations of Nature,* eds. David Philip Miller and Peter Hanns Reill. Cambridge: Cambridge Univ. Press, 1996, pp. 58–79.

Gallagher, John, and Ronald Robinson. "The Imperialism of Free Trade." *The Economic History Review,* 1953, *6:*1–15.

Gardener, William. "Botany and the Americas." *History Today,* 1966, *16:*849–55.

Gascoigne, John. *Joseph Banks and the English Enlightenment: Useful Knowledge and Polite Culture.* Cambridge: Cambridge Univ. Press, 1994.

―――. *Science in the Service of Empire: Joseph Banks, The British State and the Uses of Science in the Age of Revolution.* Cambridge: Cambridge Univ. Press, 1998.

Geertz, Clifford. *Local Knowledge: Further Essays in Interpretive Anthropology.* New York: Basic Books, 1983.

Geertz, Clifford, ed. *Old Societies and New States: The Quest for Modernity in Asia and Africa.* London: The Free Press of Glencoe, 1963.

Gellner, Ernest. *Nations and Nationalism.* Ithaca: Cornell Univ. Press, 1983.

Gibson, C. *Spain in America.* New York: Harper, 1966.

Gilli, S. "The Relationship between Science, Technology and the Economy in Lesser Developed Countries." *Social Studies of Science,* 1993, *23:*201–15.

Gilman, Sander. *Difference and Pathology: Stereotypes of Sexuality, Race and Madness.* Ithaca: Cornell Univ. Press, 1985.

Gilmartin, David. "Scientific Empire and Imperial Science: Colonialism and Irrigation Technology in the Indus Basin." *Journal of Asian Studies,* 1994, *53,* 4:1127–49.

Glick, Thomas. "Establishing Scientific Disciplines in Latin America: Genetics in Brazil, 1943–1960." In *Mundialización de la ciencia y cultura nacional,* eds. Antonio Lafuente, Alberto Elena and María Luisa Ortega. Madrid: Ediciones Doce Calles, 1993, pp. 363–76.

―――. "La Ciencia latinoamericana en el siglo XX." *Arbor,* 1992, *142:*233–52.

―――. "Science and Independence in Latin America (with Special Reference to New Granada)." *Hispanic American Historical Review,* 1991, *71:*306–34.

Gollwitzer, Heinz. *Europe in the Age of Imperialism, 1880–1914,* trans. David Adam and Stanley Baron. New York: Harcourt, Brace and World, 1969.

Gonçalves, M. *Comunidade científica e poder.* Lisbon: Federação Portuguesa das Associações e Sociedades Científicas, 1993.

Graham, Richard. *Britain and the Onset of Modernization in Brazil, 1850–1914.* Cambridge: Cambridge Univ. Press, 1968.

Graham, Richard, ed. *The Idea of Race in Latin America, 1870–1940.* Austin: Univ. of Texas Press, 1990.

Greenblatt, Stephen. *Marvellous Possessions: The Wonders of the New World.* Chicago: The Univ. of Chicago Press, 1991.

Greenway, John. *The Last Frontier: A Study of Cultural Imperatives in the Last Frontiers of America and Australia.* Melbourne: Lothian, 1972.

Griffiths, Tom, and Libby Robin, eds. *Ecology and Empire: Environmental History of Settler Societies.* Melbourne: Melbourne Univ. Press, 1997.

Grote, Andreas, ed. *Macrocosmo in Microcosmo: Die Welt in der Stube. Zur Geschichte des Sammelns, 1450 bis 1800.* Opladen: Leske and Budrich, 1994.

Grove, Richard H. *Green Imperialism: Colonial Expansion, Tropical Island Edens and the Origins of Environmentalism, 1600–1860.* Cambridge: Cambridge Univ. Press, 1995.

―――. "The Transfer of Botanical Knowledge between Asia and Europe 1498–1800." *Journal of the Japan-Netherlands Institute,* 1991, *3:*160–72.

Guha, Ranajit, ed. *Subaltern Studies I: Writings on South Asian History and Society.* Delhi: Oxford Univ. Press, 1982.

Hannaway, Caroline. "Distinctive or Derivative? The French Colonial Medical Experience, 1740–1790." In *Mundializacion de la ciencia y cultura nacional,* eds. Antonio Lafuente, Alberto Elena, and María Luisa Ortega. Madrid: Doce Calles, 1993, pp. 505–10.

Harding, Sandra. *Is Science Multicultural? Postcolonialisms, Feminisms, and Epistemologies.* Bloomington and Indianapolis: Indiana Univ. Press, 1998.

Harding, Sandra, ed. *The 'Racial' Economy of Science: Toward a Democratic Future.* Bloomington: Indiana Univ. Press, 1993.

Harrison, Mark. *Climates and Constitutions: Health, Race and British Imperialism in India, 1600–1850.* New Delhi: Oxford Univ. Press, 1999.

———. *Public Health in British India: Anglo-Indian Preventive Medicine, 1859–1914.* Cambridge: Cambridge Univ. Press, 1994.

———. "Tropical Medicine in Nineteenth-Century British India." *Brit. J. Hist. Sci.,* 1992, *25:*299–318.

Hartz, Louis. *The Founding of New Societies: Studies in the History of the United States, Latin America, South Africa, Canada and Australia.* New York: Harcourt, Brace and World, 1964.

Headrick, Daniel R. "Botany, Chemistry, and Tropical Development." *J. World Hist.,* 1996, *7,* 1:1–20.

———. *The Tentacles of Progress: Technology Transfer in the Age of Imperialism, 1850–1940.* New York: Oxford Univ. Press, 1988.

———. *The Tools of Empire: Technology and European Imperialism in the Nineteenth Century.* New York: Oxford Univ. Press, 1981.

Hechter, Michael. *Internal Colonialism: The Celtic Fringe in British National Development, 1536–1966.* London: Routledge and Kegan Paul, 1975.

Herle, Anita, and Sandra Rouse, eds. *Cambridge and the Torres Strait: Centenary Essays on the 1898 Anthropological Expedition.* Cambridge: Cambridge Univ. Press, 1998.

Hess, David. *Science and Technology in a Multicultural World: The Cultural Politics of Facts and Artifacts.* New York: Columbia Univ. Press, 1995.

Hill, R. D. "Some Comments on Forming New Physics Communities: Australia and Japan, 1914–1950." *Annals of Science,* 1991, *48:*583–87.

Hoare, Michael. "The Intercolonial Science Movement in Australia." *Records of the Australian Academy of Science,* 1976, *3,* 2:7–28.

Hobson, J. A. *Imperialism: A Study.* London: Unwin Hyman, 1988.

Horton, Robin. "African Traditional Thought and Western Science." *Africa,* 1992, *37:*50–71, 155–87.

Howarth, David. *Sovereign of the Seas: The Story of British Sea Power.* London: Collins, 1974.

Hulme, Peter. *Colonial Encounters: Europe and the Native Caribbean, 1492–1797.* London: Methuen, 1986.

Hunt, Susan, and Paul Carter. *Terre Napoleon: Australia through French Eyes 1800–1804.* Sydney: Historic Houses Trust of NSW/Hordern House, 1999.

Hussey, Roland D. "Traces of French Enlightenment in Colonial Hispanic America." In *Latin America and the Enlightenment,* ed. Arthur P. Whitaker. Ithaca: Cornell Univ. Press, 1961, pp. 23–51.

Huxley, Elsbeth. *A New Earth: An Experiment in Colonialism.* London: Chatto & Windus, 1960.

Hynes, William G. *The Economics of Empire: Britain, Africa and the New Imperialism, 1870–95.* London: Longman, 1979.

Inkster, Ian. *Science and Technology in History.* London: Macmillan, 1991.

———. "Science, Technology, and Economic Development—Japanese Historical Experience in Context." *Ann. Sci.,* 1991, *48:*545–63.

———. "Scientific Enterprise and the Colonial 'Model': Observations on Australian Experience in Historical Context." *Soc. Stud. Sci.,* 1985, *15:*677–704.

Izquierdo, José J. *La Primera casa de las ciencias en México, El Real Seminario de Minería, 1792–1811.* Mexico City: Ediciones Ciencia, 1958.

Jacob, James R. "Por encanto órfico: La Ciencia y las dos culturas en la Inglaterra del siglo XVII." In *La Ciencia y su público,* eds. Javier Ordóñez and Alberto Elena. Madrid: CSIC, 1990, pp. 43–69.

Jardine, N., J. A. Secord, and E. C. Spary, eds. *Cultures of Natural History.* Cambridge: Cambridge Univ. Press, 1996.

Jennings, J. N., and G. J. R. Linge, eds. *Of Time and Place: Essays in Honour of O. H. K. Spate.* Canberra: Australian National Univ. Press, 1980.

Jones, E. L. *The European Miracle: Environments, Economies, and Geopolitics in the History of Europe and Asia.* Cambridge: Cambridge Univ. Press, 1987.

Jordanova, Ludmilla. "Science and National Identity." In *Sciences et Langues en Europe,* eds. Roger Chartier and Pietro Corsi. Paris: Ecole des Hautes Etudes en Sciences Sociales, 1996, pp. 221–31.

Kargon, Robert. "Colonializing the New World and the Roots of Modern Science." In *Mundialización de la ciencia y cultura nacional,* eds. A. Lafuente, A. Elena, and M. L. Ortega. Madrid: Ediciones Doce Calles, 1993, pp. 133–37.

Karp, I., and S. Lavine, eds. *Exhibiting Cultures.* Washington, D.C.: Smithsonian Institution Press, 1991.

Karras, Alan L., and J. R. McNeill, eds. *Atlantic American Societies From Columbus Through Abolition 1492–1888.* London: Routledge, 1992.

Keenan, Philip C. "The Earliest National Observatories in Latin America." *Journal of the History of Astronomy,* 1991, *22:*21–30.

Keenan, Philip C., Sonia Pinto, and Hector Alvarez. *The Chilean National Astronomical Observatory, 1852–1965.* Santiago: Univ. of Chile, 1985.

Kloppenburg, Jack. *First the Seed: The Political Economy of Plant Biotechnology.* Cambridge: Cambridge Univ. Press, 1988.

Knapp, Jeffrey. *An Empire Nowhere: England, America and Literature from Utopia to the Tempest.* Berkeley: Univ. of California Press, 1992.

Kumar, Deepak. "Economic Compulsions and the Geological Survey of India." *Indian Journal of History of Science,* 1982, *17,* 2:289–300.

————. "Patterns of Colonial Science in India." *Indian J. Hist. Sci.,* 1980, *15,* 1:105–113.

————. "Racial Discrimination and Science in Nineteenth Century India." *The Indian Economic and Social History Review,* 1983, *19,* 1:64–82.

————. *Science and the Raj, 1857–1905.* Delhi: Oxford Univ. Press, 1995.

————. "Science in Agriculture: A Study of Victorian India." In *Science and Technology in Indian Culture: A Historical Perspective,* ed. A. Rahman. New Delhi: National Institute of Science, 1984.

————. "Science in Higher Education: A Study in Victorian India." *Indian J. Hist. Sci.,* 1984, *19,* 3:253–60.

————. "The 'Culture' of Science and Colonial Culture, India 1820–1920." *Brit. J. Hist. Sci.,* 1996, *29:*195–209.

————. "The Evolution of Colonial Science in India: Natural History and the East India Company." In *Imperialism and the Natural World,* ed. John Mackenzie. Manchester: Manchester Univ. Press, 1990.

Kumar, Deepak, ed. *Science and Empire: Essays in Indian Context (1700–1947).* Delhi: Anamika Prakashan, 1991.

Kury, Lorelai. "Les Instructions de voyage dans les expéditions scientifiques françaises (1750–1830)." *Revue d'Histoire des Sciences,* 1998, *51,* 1:65–92.

Lafuente, Antonio. "Institucionalización metropolitana de la ciencia española en el siglo XVIII." In *Ciencia colonial en América,* eds. Antonio Lafuente and José Sala Catalá. Madrid: Alianza Editorial, 1992, pp. 91–120.

Lafuente, Antonio, and José Sala Catalá. "Ciencia colonial y roles profesionales en las América española del siglo XVIII." *Quipu,* 1989, *6,* 3:387–403.

Lafuente, Antonio, and L. López Ocón. "Tradiciones científicas y expediciones ilustradas en la América hispánica del siglo XVIII." In *Historia social de la ciencia en América Latina,* ed. J. J. Saldaña. Mexico City: UNAM–M A. Porrúa, 1996, pp. 247–281.

Lafuente, Antonio, and María Luisa Ortega. "Modelos de mundialización de la ciencia." *Arbor,* 1992, *142:*93–117.

Lafuente, Antonio, and José Sala Catalá, eds. *Ciencia colonial en América.* Madrid: Alianza Editorial, 1992.

Lafuente, Antonio, and María Luisa Ortega, eds. *Mundialización de la ciencia y cultura nacional.* Madrid: Doce Calles, 1993.

Laissus, Yves. "Les Voyageurs naturalistes du Jardin du Roi et du Muséum d'Histoire Naturelle." *Revue d'Histoire des Sciences,* 1981, *34,* 3–4:260–317.

Laissus, Yves, and Jean Torlais. "Le Jardin du Roi et le Collège Royal." In *Enseignement et diffusion des sciences au XVIIIe siècle,* ed. René Taton. Paris: Hermann, 1986, pp. 261–341.

Lamb, Ursula. *Nautical Scientists and their Clients in Iberia (1508–1624): Science from an Imperial Perspective.* Lisbon: Instituto de Investigação Cientifica Tropical, 1984.

Latour, Bruno. "Give Me a Laboratory and I Will Raise the World." In *Science Observed,* eds. K. Knorr-Cetina and M. Mulkay. London: Sage, 1983, pp. 141–70.

———. *Science in Action: How to Follow Scientists and Engineers through Society.* Cambridge, Mass.: Harvard Univ. Press, 1987.

———. *The Pasteurization of France.* Cambridge, Mass.: Harvard Univ. Press, 1989.

Laurens, Henry. *L'Expédition d'Egypte, 1798–1801.* Paris: Armand Colin, 1989.

Lécuyer, M. C., and C. Serrano. *La Guerre d'Afrique et ses répercussions en Espagne: Idéologies et colonialisme en Espagne, 1859–1904.* Paris: Presses Universitaires de France, 1976.

Letouzey, Yvonne. *Le Jardin des Plantes à la croisée des chemins avec André Thouin, 1747–1824.* Paris: Muséum National d'Histoire Naturelle, 1989.

Levere, T. H. "Elements in the Structure of Victorian Science or Cannon Revisited." In *The Light of Nature,* eds. J. D. North and J. J. Roche. Dordrecht: Martinus Nijhoff, 1985, pp. 433–49.

Livingstone, David N. "Human Acclimatization: Perspectives on a Contested Field of Inquiry in Science, Medicine and Geography." *History of Science,* 1987, *25:*359–94.

———. "The Moral Discourse of Climate: Historical Consideration on Race, Place, and Virtue." *Journal of Historical Geography,* 1991, *17:*413–34.

———. "Tropical Climate and Moral Hygiene: The Anatomy of a Victorian Debate." *Brit. J. Hist. Sci.,* 1999, *32:*93–100.

Lloyd, T. *The British Empire, 1558–1983.* Oxford: Oxford Univ. Press, 1984.

Lopes, Maria Margaret. "Brazilian Museums of Natural History and International Exchanges in the Transition to the 20th Century." In *Science and Empires: Historical Studies about Scientific Development and European Expansion,* eds. Patrick Petitjean, Catherine Jami, and Anne-Marie Moulin. Dordrecht: Kluwer, 1992, pp. 193–200.

———. *O Brasil descobre a pesquisa científica.* São Paulo: Hucitec, 1997.

López Piñero, José María. *Ciencia y técnica en la sociedad española de los siglos XVI y XVII.* Barcelona: Labor, 1979.

López Piñero, José María, Thomas F. Glick, Victor Navarro Brotóns, and Eugenio Portela, eds. *Diccionario histórico de la ciencia moderna en España,* 2 vols. Barcelona: Ediciones Península, 1983.

López Piñero, José María, and José Pardo Tomás. *Nuevos materiales y noticias sobre la historia de las plantas de Nueva España, de Francisco Hernández.* Valencia: CSIC, 1994.

Lowenthal, D. "Empires and Ecologies: Reflections on Environmental History." In *Ecology and Empire: Environmental History of Settler Societies,* eds. Tom Griffiths and Libby Robin. Melbourne: Melbourne Univ. Press, 1997, pp. 229–36.

Lozoya, Xavier. *Plantas y luces en México: La Real expedición científica a Nueva España 1787–1803.* Madrid: Ediciones del Serbal, 1984.

Ly-Thio-Fane, Madeleine. "A Reconnaissance of Tropical Resources during the Revolutionary Years: The Role of the Paris Museum d'Histoire Naturelle." *Archives of Natural History,* 1991, *18,* 3:333–62.

———. *Mauritius and the Spice Trade: The Odyssey of Pierre Poivre.* Port-Louis: Mauritius Archives Publication Fund, 1958.

Mackay, David. "A Presiding Genius of Exploration: Banks, Cook and Empire, 1767–1805." In *Captain James Cook and His Times,* eds. Robin Fisher and Hugh Johnston. Seattle: Univ. of Washington Press, 1979, pp. 21–39.

———. *In the Wake of Cook: Exploration, Science and Empire, 1780–1801.* Wellington: Victoria Univ. Press, 1985.

MacKenzie, John. *Imperialism and the Natural World.* Manchester: Manchester Univ. Press, 1990.

———. *Orientalism: History, Theory and the Arts.* Manchester: Manchester Univ. Press, 1995.

MacLeod, Roy. Introduction to *Disease, Medicine, and Empire: Perspectives on Western Medicine and the Experience of European Expansion,* eds. Roy MacLeod and Milton Lewis. London: Routledge, 1988, pp. 1–18.

———. "On Science and Colonialism." In *Science and Society in Ireland: The Social Context of Science and Technology in Ireland, 1800–1950.* Belfast: Queen's Univ., 1997, pp. 1–17.

———. "On Visiting the 'Moving Metropolis': Reflections on the Architecture of Imperial Science." *Hist. Rec. Australian Sci.,* 1982, *5,* 3:1–16. Reprinted in *Scientific Colonialism: A Cross-Cultural Comparison,* eds. Nathan Reingold and Marc Rothenberg. Washington, D.C.: Smithsonian Institution Press, 1987, pp. 217–250. Subsequently reprinted as "De Visita a la 'Moving Metropolis': Reflexiones sobre la arquitectura de la ciencia imperial." In *Historia de las ciencias: Nuevas tendencias,* eds. Antonio Lafuente and Juan-José Saldaña. Madrid: Consejo Superior de Investigaciones Científicas, 1987, pp. 217–40.

———. "Passages in British Imperial Science: From Empire to Commonwealth." *J. World Hist.,* 1993, *4,* 1:2–29.

———. "Reading the Discourse of Colonial Science." In *Les Sciences hors d'Occident au XXème siècle,* ed. Patrick Petitjean, vol. 2, *Les Sciences coloniales: Figures et institutions.* Paris: ORSTOM Editions, 1996, pp. 87–98.

———. "Scientific Advice for British India: Imperial Perceptions and Administrative Goals, 1898–1923." *Modern Asian Studies,* 1975, *9,* 3:343–84.

MacLeod, Roy, and Richard Jarrell, eds. *Dominions Apart: Reflections on the Culture of Science and Technology in Canada and Australia, 1850–1945.* Special issue of *Scientia Canadensis,* 1994.

MacLeod, Roy, and Deepak Kumar, eds. *Technology and the Raj: Western Technology and Technical Transfers to India, 1700–1947.* New Delhi: Sage, 1995.

MacLeod, Roy, and Milton Lewis, eds. *Disease, Medicine and Empire.* London: Routledge, 1988.

MacLeod, Roy, and Philip Rehbock, eds. *Darwin's Laboratory: Evolutionary Theory and Natural History in the Pacific.* Honolulu: Univ. of Hawaii Press, 1994.

MacLeod, Roy, and Philip Rehbock, eds. *"Nature in its Greatest Extent": Western Science in the Pacific.* Honolulu: Univ. of Hawaii Press, 1988.

Mangan, J. A., ed. *The Imperial Curriculum: Racial Images and Education in the British Colonial Experience.* London: Routledge, 1993.

Marshall, P. J., and Glyndwr Williams. *The Great Map of Mankind: British Perceptions of the World in the Age of Enlightenment.* London: J. M. Dent, 1982.

Martín Rodríguez, Manuel. "El Azúcar y la política colonial española (1860–1898)." In *Economía y colonias en la España del 98,* ed. Pedro Tedde. Madrid: Editorial Síntesis-Fundación Duques de Soria, 1999, pp. 161–77.

McClellan, James E. III. *Colonialism and Science: Saint Domingue in the Old Regime.* Baltimore and London: Johns Hopkins Univ. Press, 1992.

———. "The Mobility of Knowledge: Local Science in World History." Unpublished paper presented at the Princeton Workshop in the History of Science, Princeton, October, 1998.

McClellan, James E. III, and Harold Dorn. *Science and Technology in World History.* Baltimore: Johns Hopkins Univ. Press, 1999.

McCracken, Donald P. *Gardens of Empire: Botanical Institutions of the Victorian British Empire.* London: Leicester Univ. Press, 1997.

McCulloch, Jock. *Colonial Psychiatry and the "African Mind."* Cambridge: Cambridge Univ. Press, 1995.

McDorman, Kathryne Slate. *Image of Empire: British Imperial Attitudes in Fiction, 1900–1939.* Ann Arbor, Mich.: Univ. Microfilms International, 1981.

Meade, Teresa, and Mark Walker, eds. *Science, Medicine and Cultural Imperialism.* New York: St. Martin's Press, 1991.

Mesa, Roberto. *El Colonialismo en la crisis del XIX español.* Madrid: Ciencia Nueva, 1967.

———. *La Idea colonial en España.* Valencia: Fernando Torres Editor, 1976.

Miller, David Philip. "Joseph Banks, Empire and 'Centers of Calculation' in Late Hanoverian London." In *Visions of Empire: Voyages, Botany, and Representations of Nature,* eds. David Philip Miller and Peter Hanns Reill. Cambridge: Cambridge Univ. Press, 1996, pp. 21–37.

Moreno, Antonio, and José Manuel Sánchez Ron. "La Ciencia española contemporánea: Del optimismo regeneracionista a la exaltación patriótica." In *Mundialización de la ciencia y cultura nacional,* eds. Antonio Lafuente, Alberto Elena, and María Luisa Ortega. Aranjuez/ Madrid: Ediciones Doce Calles/Ediciones de la Universidad Autónoma de Madrid, 1993, pp. 391–98.

Moreno de los Arcos, Roberto. *La Primera cátedra de botánica en México.* Mexico City: UNAM, 1998.

Morris-Suzuki, Tessa. "Concepts of Nature and Technology in Pre-Industrial Japan." *East Asian History,* 1991, *1*:81–97.

Morse, Richard M. *El Espejo de Próspero. Un Estudio de la dialéctica del Nuevo Mundo.* Mexico City: Siglo XXI, 1982.

Moulin, Anne-Marie. "Expatriés français sous les tropiques: Cent ans d'histoire de la santé." *Bulletin de la Société de Pathologie Exotique,* 1997, *90,* 4:221–28.

———. "Patriarchal Science: The Network of the Overseas Pasteur Institutes." In *Science and Empires: Historical Studies about Scientific Development and European Expansion,* eds. Patrick Petitjean, Catherine Jami, and Anne-Marie Moulin. Dordrecht: Kluwer, 1991, pp. 307–22.

Mundy, Barbara. *The Mapping of New Spain: Indigenous Cartography and the Maps of the Relaciones Geográficas.* Chicago: Univ. of Chicago Press, 1996.

Munteal, Oswaldo. "Todo um mundo a reformar: Intelectuais, cultura ilustrada e estabelecimentos científicos na América portuguesa, 1779–1808." *Anais do Museu Histórico Nacional,* 1997, *29*:87–108.

Nadel, George H., and Perry Curtis. *Imperialism and Colonialism.* New York: Macmillan, 1970.

Nader, Laura, ed. *Naked Science: Anthropological Inquiry into Boundaries, Power and Knowledge.* New York: Routledge, 1996.

Nanda, Meera. "The Epistemic Charity of the Social Constructivist Critics of Science and Why the Third World Should Refuse the Offer." In *A House Built on Sand: Exposing Postmodernist Myths About Science,* ed. Noretta Koertge. New York: Oxford Univ. Press, 1998, pp. 286–311.

Obregon Torres, Diana. *Sociedades científicas en Colombia: La Invención de una tradición, 1859–1936.* Bogotá: Banco de la República, 1992.

O'Brien, Patrick. "European Economic Development: The Contribution of the Periphery." *The Economic History Review,* 1982, *35:*1–18.

———. *Joseph Banks: A Life.* Chicago: Univ. of Chicago Press, 1997.

———. "The Costs and Benefits of British Imperialism, 1846–1914." *Past and Present,* 1988, *120:*163–210.

Ophir, A., and Steven Shapin. "The Place of Knowledge: A Methodological Survey." *Science in Context,* 1991, *4:*3–21.

Ordóñez, Javier, and Alberto Elena, eds. *La Ciencia y su público.* Madrid: Consejo Superior de Investigaciones Científicas, 1990.

Ortega, María Luisa, Alberto Elena, and Javier Ordóñez, eds. *Técnica e imperialismo.* Madrid: Ediciones Turfán, 1993.

Osborne, Michael A. "A Collaborative Dimension of the European Empires: Australian and French Acclimatization Societies and Intercolonial Scientific Cooperation." In *International Science and National Scientific Identity,* eds. R. W. Home and Sally Gregory Kohlstedt. Dordrecht: Kluwer Publishers, 1991, pp. 97–120.

———. "European Visions: Science, the Tropics and the War on Nature." In *Nature et environnement,* eds. Christophe Bonneuil and Y. Chatelin. Paris: ORSTOM, 1996, pp. 21–32.

———. "Histories of Science, Exploration and Scientific Institutions Around the Pacific Rim." *Forest and Conservation History,* 1991, *35:*34–5.

———. "Introduction: The Social History of Science, Technoscience and Imperialism." *Sci., Technol. Soc.,* 1999, *4,* 2:161–70.

———. "La Renaissance d'Hippocrate: L'Hygiène et les expéditions scientifiques en Egypte, en Morée et en Algérie." In *L'Invention scientifique de la Méditerranée,* eds. Marie-Noëlle Bourguet *et al.* Paris: Editions de l'École des Hautes Études en Sciences Sociales, 1998, pp. 185–204.

———. *Nature, the Exotic, and the Science of French Colonialism.* Bloomington: Indiana Univ. Press, 1994.

Owen, Norman G. *Prosperity without Progress: Manila Hemp and Material Life in the Colonial Philippines.* Quezon City: Ateneo de Manila Univ. Press, 1984.

Owen, Roger, and Bob Sutcliffe. *Studies in the Theory of Imperialism.* London: Longman, 1972.

Pacey, Arnold. *Technology in World Civilisation: A Thousand-Year History.* Cambridge, Mass.: MIT Press, 1990.

Pagden, Anthony. *European Encounters with the New World.* New Haven: Yale Univ. Press, 1993.

Palladino, Paolo, and Michael Worboys. "Science and Imperialism." *Isis,* 1993, *84*:91–102.

Pallo, G. "Some Conceptual Problems of the Centre-Periphery Relationship in the History of Science." *Philosophy and Social Action,* 1987, *13,* 1–4:27–32.

Parry, J. H. *The Spanish Seaborne Empire.* Berkeley: Univ. of California Press, 1990.

———. *Trade and Dominion: The European Overseas Empires in the Eighteenth Century.* New York: Praeger, 1971.

Paty, Michel. "Comparative History of Modern Science and the Context of Dependency." *Sci., Technol., Soc.,* 1999, *4,* 2:171–203.

Pelis, Kim. "Prophet for Profit in French North Africa: Charles Nicolle and the Pasteur Institute of Tunis, 1903–36." *Bull. Hist. Med.,* 1997, *71*:583–622.

Pérez-Grueso, María Dolores Elizalde. *España en el Pacífico: La Colonia de las Islas Carolinas, 1885–1899.* Madrid: Consejo Superior de Investigaciones Científicas, 1992.

Peset, José L. *Ciencia y libertad: El Papel del científico ante la independencia americana.* Madrid: CSIC, 1987.

Petitjean, Patrick. "Science and Colonization in the French Empire." *Ann. Sci.,* 1995, *52*:187–92.

———. "Scientific Relations as a Crossing of Supplies and Demands of Science: Franco-Brazilian Cases, 1870–1940." In *Mundialización de la ciencia y cultura nacional,* eds. Antonio Lafuente, Alberto Elena, and María Luisa Ortega. Madrid: Ediciones Doce Calles, 1993, pp. 635–649.

Petitjean, Patrick, Catherine Jami, and Anne-Marie Moulin, eds. *Science and Empires: Historical Studies About Scientific Development and European Expansion.* Dordrecht: Kluwer Academic Publishers, 1992.

Pimentel, Juan. *La Física de la monarquía: Ciencia y política en el pensamiento colonial de Alejandro Malaspina 1754–1810.* Madrid: Ediciones Doce Calles, 1998.

Pimentel, Juan, and Manuel Lucena. *Los Axiomas políticos de Alejandro Malaspina.* Aranjuez: Ediciones Doce Calles, 1993.

Pino, Fermín del, ed. *Ciencia y contexto histórico nacional en las expediciones ilustradas a América.* Madrid: CSIC, 1988.

Podgorny, Irina. "De la santidad laica del científico: Florentino Ameghino y el espectáculo de la ciencia en la Argentina moderna." *Entrepasados: Revista de Historia,* 1997, *13*:37–61.

Polanco, Xavier. *Naissance et développement de la science-monde: Production et reproduction des communautés scientifiques en Europe et en Amérique Latine.* Paris: Editions La Découverte, 1990.

———. "Science in the Developing Countries: An Epistemological Approach on the Theory of Science in Context." *Quipu,* 1985, *2*:303–18.

———. "World-Science: How is the History of World-Science to be Written." In *Science and Empires: Historical Studies about Scientific Development and European Expansion,* eds. Patrick Petitjean, Catherine Jami, and Anne-Marie Moulin. Dordrecht: Kluwer, 1992, pp. 225–42.

Pomian, Krzysztof. *Collectionneurs, amateurs et curieux.* Paris: Gallimard, 1987.

Ponting, Clive. *A Green History of the World.* Harmondsworth: Penguin, 1991.

Porter, Bernard. *The Lion's Share: A Short History of British Imperialism, 1850–1983,* 2nd ed. London and New York: Longman, 1984.

Prakash, Gyan. *Another Reason: Science and the Imagination of Modern India.* Princeton: Princeton Univ. Press, 1999.

———. "Science Gone Native in Colonial India." *Representations,* 1992, *40:*153–78.

Prakash, Gyan, ed. *After Colonialism: Imperial Histories and Postcolonial Displacements.* Princeton: Princeton Univ. Press, 1995.

Puerto, Javier. *Ciencia de cámara: Casimiro Gómez Ortega (1741–1818), el científico cortesano.* Madrid: CSIC, 1992.

———. *La Ilusión quebrada: Botánica, sanidad y política científica en la España ilustrada.* Barcelona: El Serbal/CSIC, 1988.

Puerto, Francisco Javier, and A. Gonzalez. "Política científica y expediciones botánicas en el programa colonial español ilustrado." In *Mundialización de la ciencia y cultura nacional,* eds. Antonio Lafuente, Alberto Elena, and María Luisa Ortega. Madrid: Ediciones Doce Calles, 1993, pp. 331–340.

Pulido Rubio, J. *El Piloto mayor de la Casa de Contratación de Sevilla.* Sevilla: Escuela de Estudios Hispano-Americanos, 1950.

Pyenson, Lewis. *Civilizing Mission: Exact Sciences and French Overseas Expansion, 1830–1940.* Baltimore and London: Johns Hopkins Univ. Press, 1993.

———. "Colonial Science and the Creation of a Postcolonial Scientific Tradition in Indonesia." *Akademika,* 1990, *37:*91–105.

———. *Cultural Imperialism and Exact Sciences: German Expansion Overseas, 1900–1930.* New York: P. Lang, 1985.

———. "Cultural Imperialism and Exact Sciences Revisited." *Isis,* 1993, *84:*103–8.

———. *Empire of Reason: Exact Sciences in Indonesia, 1840–1940.* Leiden: E. J. Brill, 1989.

———. "Functionaries and Seekers in Latin America: Missionary Diffusion of the Exact Sciences, 1850–1930." *Quipu,* 1985, *2,* 3:387–420.

———. "Habits of Mind: Geophysics at Shanghai and Algiers, 1920–1940." *Historical Studies in the Physical and Biological Sciences,* 1990, *21:*161–96.

———. "Prerogatives of European Intellect: Historians of Science and the Promotion of Western Civilization." *Hist. Sci.,* 1993, *31:*289–315.

———. "Pure Learning and Political Economy: Science and European Expansion in the Age of Imperialism." In *New Trends in the History of Science: Proceedings of a Conference Held at the University of Utrecht,* eds. R. P. W. Visser *et al.* Amsterdam: Rodopi, 1989, pp. 209–78.

———. "The Ideology of Western Rationality: History of Science and the European Civilizing Mission." *Science and Education,* 1993, *2:*329–43.

———. "Typologie des stratégies d'expansion en sciences exactes." In *Science and Empires: Historical Studies About Scientific Development and European Expansion,* eds. Patrick Petitjean, Catherine Jami, and Anne-Marie Moulin. Dordrecht: Kluwer Academic Publishers, 1992, pp. 211–18.

———. "Why Science May Serve Political Ends: Cultural Imperialism and the Mission to Civilize." *Berichte zur Wissenschaftgeschichte,* 1990, *13:*69–81.

Raj, Kapil. "La Construction de l'empire de la géographie: L'Odysée des arpenteurs de Sa Très Gracieuse Majesté, la reine Victoria, en Asie centrale." *Annales HSS,* 1997, *5:*1153–80.

———. "When Humans Become Instruments: The Indo-British Exploration of Tibet and Central Asia in the Mid-19th Century." In *Instruments, Travel and Science: Itineraries of Precision in the Natural Sciences, 18th–20th Centuries,* eds. Marie-Noëlle Bourguet, Christian Licoppe, and Hans Otto Sibum. Chur, Switzerland: Harwood Academic Publishing, forthcoming.

Ramos Lara, Maria de la Paz. *La Difusión de la mecánica newtoniana en la Nueva España.* Mexico: Sociedad Mexicana de Historia de la Ciencia y la Tecnología, 1994.

Rashed, Roshdi. "Is Science a Western Phenomenon?" *Fundamenta Scientiae,* 1980, *1:*7–21.

Regourd, François. "Métriser la nature: Un Enjeu colonial. Botanique et agronomie en Guyane et aux Antilles (XVIIe–XVIIIe siècles)." *Revue Française d'Histoire d'Outre-Mer,* 1999, *86:*39–63.

Rehbock, Philip. *The Philosophical Naturalists.* Madison: Univ. of Wisconsin Press, 1983.

Reingold, Nathan, and Marc Rothenberg, eds. *Scientific Colonialism: A Cross-Cultural Comparison.* Washington, D.C.: Smithsonian Institution Press, 1987.

Restrepo Forero, Olga. "El Tránsito de la historia natural a la biología en Colombia (1784–1936)." *Ciencia, Tecnología y Desarrollo* (Bogotá), 1969, *10*:181–275.

Reynolds, Henry. *Frontier: Aborigines, Settlers and Land.* Sydney: Allen and Unwin, 1987.

Rhodes, Robert, ed. *Imperialism and Underdevelopment: A Reader.* New York: Monthly Review Press, 1970.

Rich, Paul. "Race, Science and the Legitimization of White Supremacy in South Africa, 1902–1940." *International Journal of African Historical Studies,* 1990, *23*:665–86.

Risse, Guenter B. "Medicine in New Spain." In *Medicine in the New World: New Spain, New France and New England,* ed. Ronald L. Numbers. Knoxville: Univ. of Tennessee Press, 1987, pp. 12–63.

Riviere, Peter. "From Science to Imperialism: Robert Schomburgk's Humanitarianism." *Arch. Natur. Hist.,* 1998, *25,* 1:1–8.

Robinson, Ronald, John Gallagher, and Alice Denny. *Africa and the Victorians: The Climax of Imperialism.* Garden City, N.Y.: Anchor Books, 1968.

Ruiz Gutiérrez, Rosaura. *Positivismo y evolución: Introducción del Darwinismo en Mexico.* Mexico City: Universidad Nacional Autonoma de Mexico, 1987.

Safford, Frank. *The Ideal of the Practical: Colombia's Struggle to Form a Technical Elite.* Austin: Univ. of Texas Press, 1976.

Said, Edward. *Culture and Imperialism.* New York: Knopf, 1993.

———. *Orientalism: Western Conceptions of the Orient.* London: Routledge, 1978.

Sala Catalá, José. "La Communauté scientifique espagnole au XIXe siècle, et ses relations avec la France et l'Amérique Latine." In *Naissance et développement de la science-monde: Production et reproduction des communautés scientifiques en Europe et en Amérique Latine,* ed. Xavier Polanco. Paris: Editions La Découverte/Conseil de l'Europe/UNESCO, 1990, pp. 122–47.

Saldaña, Juan-José. "Ciencia y felicidad pública en la ilustración americana." In *Historia social de la ciencia en América Latina,* ed. Juan José Saldaña. Ciudad México: UNAM-M. A. Porrúa, 1996, pp. 151–207.

———. "Cross Cultural Diffusion of Science: Latin America." *Cuadernos de Quipu,* 1987, *2*:33–57.

———. "Historia de la ciencia y de la tecnología: Aspectos teóricos y metodológicos." In *Ciencia tecnología y desarrollo: Interrelaciones teóricas y metodológicas,* ed. Eduardo Martínez. Caracas: Editorial Nueva Sociedad, 1994, pp. 91–130.

———. "La Formation des communautés scientifiques au Mexique (du XVIe au XXe siècle)." In *Naissance et développement de la science-monde: Production et reproduction des communautés scientifiques en Europe et en Amérique Latine,* ed. Xavier Polanco. Paris/Strasbourg: Editions La Découverte/UNESCO, 1989, pp. 148–76.

———. "Los Orígenes de la ciencia nacional." *Cuadernos de Quipu,* 1992, *4*:9–54.

———. "Marcos conceptuales de la historia de las ciencias en Latinoamérica: Positivismo y economicismo." In *El Perfil de la ciencia en América Latina,* ed. Juan-José Saldaña. Ciudad México, DF: Sociedad Latinoamericana de Historia de las Ciencias y la Tecnología, 1986, pp. ····–····.

———. "Nuevas tendencias en la historia de la ciencia en América Latina." *Cuadernos Americanos,* 1993, *2,* 38:69–91.

———. "Teatro científico americano: Geografía y cultura en la historiografía latinoamericana de la ciencia." In *Historia social de las ciencias en América Latina,* ed. Juan-José Saldaña. Mexico City: Porrúa Ediciones, 1996, pp. 7–41.

Sandes, E. W. C. *The Military Engineer in India,* 2 vols. Chatham: The Institution of Royal Engineers, 1935.

Sangwan, Satpal. "Reordering the Earth: The Emergence of Geology as a Scientific Discipline in Colonial India." *Earth Sci. Hist.,* 1993, *12*:224–33.

———. *Science, Technology and Colonisation: An Indian Experience 1757–1857.* New Delhi: Anamika Prakashan, 1991.

Santa Rita, J. G. "A Investigação científica portuguesa nos últimos 100 anos." In *Memórias e comunicações apresentadas ao IX Congresso Colonial, in Congresso do Mundo Português,* vol. XIV. Lisbon: Comissão Executiva dos Centenários, 1940, pp. 11–30.

Scammell, G. V. *The First Imperial Age.* London: Unwin Hyman, 1989.

———. *The World Encompassed: The First European Maritime Empires c. 800–1650.* Berkeley: Univ. of California Press, 1981.

Schott, Thomas. "World Science: Globalization of Institutions and Participation." *Science, Technology and Human Values,* 1993, *18,* 2:196–208.

Schumacher, John N. "One Hundred Years of Jesuit Scientists: The Manila Observatory, 1865–1965." *Philippine Studies,* 1965, *13:*258–86.

Schumaker, Lynette L. "Fieldwork and Culture in the History of the Rhodes-Livingstone Institute, 1937–1964." Ph.D. dissertation, Univ. of Pennsylvania, 1994.

Schwartz, Stuart B., ed. *Implicit Understandings: Observing, Reporting and Reflecting on the Encounters Between Europeans and Other Peoples in the Early Modern Era.* Cambridge: Cambridge Univ. Press, 1994.

Schwartzmann, Simon. *Formação da comunidade científica no Brasil.* São Paulo: ed. Nacional; Rio de Janeiro: FINEP, 1979.

Scully, Pamela. "Rape, Race and Colonial Culture: The Sexual Politics of Identity in the Nineteenth-Century Cape Colony, South Africa." *Amer. Hist. Rev.,* 1995, *100,* 2:335–59.

Seeley, J. R. *The Expansion of England.* Chicago and London: The Univ. of Chicago Press, 1971.

Selin, Helaine, ed. *Encyclopaedia of the History of Science, Technology and Medicine in Non-Western Cultures.* Dordrecht: Kluwer, 1997.

Sellés, Manuel, José Luis Peset, and Antonio Lafuente, eds. *Carlos III y la ciencia de la ilustración.* Madrid: Alianza Editorial, 1988.

Shapin, Steven. "Placing the View from Nowhere: Historical and Sociological Problems in the Location of Science." *Transactions of the Institute of British Geographers,* n. s., 1998, *23:*5–12.

———. "'The Mind Is Its Own Place': Science and Solitude in Seventeenth-Century England." *Sci. Context,* 1990, *4:*191–218.

Sheets-Pyenson, Susan. *Cathedrals of Science: The Development of Colonial Natural History Museums during the Late Nineteenth Century.* Kingston: McGill-Queen's Univ. Press, 1988.

Shils, Edward. *Center and Periphery.* Chicago: Univ. of Chicago Press, 1975, pp. 117–30.

———. "Centre and Periphery." In *The Logic of Personal Knowledge: Essays Presented to Michael Polanyi.* London: Routledge and Kegan Paul, 1961.

Silva, Renán. *Saber, cultura y sociedad en el Nuevo Reino de Granada: Siglos XVII–XVIII.* Bogotá: Universidad Pedagógica Nacional, 1984.

Simon, W. J. *Scientific Expeditions in the Portuguese Overseas Territories (1783–1808) and the Role of Lisbon in the Intellectual-Scientific Community of the Late Eighteenth Century.* Lisbon: IICT, 1983.

Smith, Alan K. *Creating a World Economy: Merchant Capital, Colonialism and World Trade, 1400–1825.* Boulder, Colo.: Westview Press, 1991.

Smith, Bernard. *European Vision and the South Pacific, 1768–1850.* Oxford: Oxford Univ. Press, 1960.

———. *Imaging the Pacific: In the Wake of the Cook Voyages.* New Haven/London: Yale Univ. Press, 1992.

Sörlin, Sverker. "La Laponie, terre d'exploration," and "Le Réseau scientifique." In *Le Soleil et l'Etoile du Nord: La France et la Suede au XVIIIe siècle,* ed. Pontus Grate. Paris: Association Française d'Action Artistique, and Reunion des Musées Nationaux, 1994, pp. 213–15 and 216–22.

———. "National and International Aspects of Cross-Boundary Science: Scientific Travel in the 18th Century." In *Denationalizing Science: The Contexts of International Scientific Practice,* eds. Elisabeth Crawford, Terry Shinn, and Sverker Sörlin. Dordrecht: Kluwer, 1993, pp. 43–72.

———. "Opfer für einen samler." In *Wunderkammer des Abendlandes: Museum und Sammlung im Spiegel der Zeit.* Bonn: Kunst–und Ausstellungshalle der Bundesrepublik Deutschland in Bonn, 1994, pp. 150–9.

Spary, Emma. "L'Invention de l'expédition scientifique: L'Histoire naturelle, Bonaparte et l'Égypt." In *L'Invention scientifique de la Méditerranée, Egypte, Morée, Algérie,* eds. Marie-Noëlle Bourguet, Bernard Lepetit, Daniel Nordman, and Maroula Sinarellis. Paris: Editions EHESS, 1998, pp. 119–38.

————. "Making the Natural Order: The Paris Jardin du Roi, 1750–1795." Ph.D. dissertation, Univ. of Cambridge, 1993.

Stafford, Barbara Maria. *Voyage into Substance: Art, Science, Nature and the Illustrated Travel Account, 1760–1840*. Cambridge, Mass.: MIT Press, 1984.

Stafford, R. A. "A Far Frontier: British Geological Research in Australia during the Nineteenth Century." In *International Science and National Scientific Identity*, eds. R. W. Home and Sally Gregory Kolstedt. Dordrecht: Kluwer Publishers, 1991, pp. 75–96.

————. *Scientist of Empire: Sir Roderick Murchison, Scientific Exploration and Victorian Imperialism*. Cambridge: Cambridge Univ. Press, 1989.

Stearns, Peter N., Michael Adas, and Stuart B. Schwartz. *World Civilizations: The Global Experience*, 2nd ed. New York: Harper/Collins College Publishers, 1996.

Steffal, Rebecca. *Scientific Explorers: Travels in Search of Knowledge*. New York: Oxford Univ. Press, 1992.

Stein, Stanley J., and Barbara H. Stein. *The Colonial Heritage of Latin America: Essays on Economic Dependence in Perspective*. New York: Oxford Univ. Press, 1970.

Stepan, Nancy. *Beginnings of Brazilian Science*. New York: Science History, 1981.

————. "Race, Gender and Nation in Argentina: The Influence of Italian Eugenics." *Hist. European Ideas*, 1992, *15:*749–56.

————. *The Idea of Race in Science: Great Britain 1800–1960*. Hamden: Univ. of Connecticut Press, 1982.

Stokes, Eric. *The English Utilitarians and India*. Oxford: Oxford Univ. Press, 1959.

Stoking, George W. Jr. *Race, Culture and Evolution*. Chicago: Univ. of Chicago Press, 1982.

Stoler, Ann Laura. "Sexual Affronts and Racial Frontiers: European Identities and the Cultural Politics of Exclusion in Colonial Southeast Asia." In *Tensions of Empire: Colonial Cultures in a Bourgeois World*, eds. Frederick Cooper and Ann Laura Stoler. Berkeley: Univ. of California Press, 1997, pp. 198–237.

Storey, William K., ed. *Scientific Aspects of European Expansion*, vol. 6, *An Expanding World: The European Impact on World History, 1450–1800*. Aldershot: Variorum, 1996.

Struik, Dirk J. "Early Colonial Science in North America and Mexico." *Quipu*, 1984, *1*, 1:25–54.

Subrahmanyam, Sanjay. *The Portuguese Empire in Asia, 1500–1700: A Political and Economic History*. London: Longman, 1993.

Teles, Francisco Xavier da Silva. "Colonização científica e política colonial." *Revista Portugal em Africa*, 1894, *1*, 2:49–53.

Thomas, Nicholas. *Colonialism's Culture: Anthropology, Travel and Government*. Melbourne: Melbourne Univ. Press, 1994, pp. 161–36.

————. "Licensed Curiosity: Cook's Pacific Voyages." In *The Cultures of Collecting*, eds. John Elsner and Roger Cardinal. Melbourne: Melbourne Univ. Press, 1994.

Thornton, A. P. *Doctrines in Imperialism*. London: John Wiley and Sons, 1965.

————. *Imperialism in the Twentieth Century*. London: Macmillan, 1978.

————. *The Imperial Idea and its Enemies*. London: Macmillan and Co, 1959.

Thornton, L. *La Femme dans la peinture orientaliste*. Paris: Poche Couleur, 1993.

Todd, Jan. *Colonial Technology: Science and the Transfer of Innovation to Australia*. Cambridge: Cambridge Univ. Press, 1995.

Trabulse, E. "Aproximaciones historiográficas a la ciencia mexicana." In *Memorias del Primer Congreso Mexicano de Historia de la Ciencia y de la Tecnología*, vol. 1. Mexico: Sociedad de Historia de la Ciencia y de la Tecnología, 1989, pp. 51–69.

Tracy, James D., ed. *The Political Economy of Merchant Empires, State Power and World Trade, 1350–1750*. Cambridge: Cambridge Univ. Press, 1991.

Turnbull, David. "Cartography and Science in Early Modern Europe: Mapping the Construction of Knowledge Spaces." *Imago Mundi*, 1996, *48:*5–24.

————. "Cook and Tupaia, a Tale of Cartographic Méconnaissance." In *Science and Exploration in the Pacific*, ed. Margarette Lincoln. Woodbridge, Suffolk: Boydell Press, 1998, pp. 117–32.

————. "Local Knowledge and Comparative Scientific Traditions." *Knowledge and Policy*, 1993–4, *6*, 3–4:29–54.

————. "Mapping Encounters and (En)Countering Maps: A Critical Examination of Cartographic Resistance." *Knowledge and Society*, 1998, *11:*15–44.

————. *Maps Are Territories: Science is an Atlas.* Chicago: Univ. of Chicago Press, 1993.

————. "Reframing Science and Other Local Knowledge Traditions." *Futures,* 1997, *29,* 6:551–62.

Tyabji, Nasir. *Colonialism and Chemical Technology.* Delhi: Oxford Univ. Press, 1995.

Vanderpool, Christopher. "Center and Periphery in Science: Conceptions of a Stratification of Nations and Its Consequences." In *Comparative Studies in Science and Society,* eds. Sal Restivo and Christopher Vanderpool. Columbia, Ohio: Merrill, 1974, pp. 432–42.

Vázquez, J. Z., ed. *Interpretaciones del siglo XVIII mexicano: El Impacto de las reformas borbónicas.* Mexico City: Ed. Nueva Imagen, 1992.

Vernet, Juan. *Historia de la ciencia española.* Madrid: Instituto de España, 1975.

Vessuri, Hebe, ed. *Ciencia académica en la Venezuela moderna.* Caracas: Fondo Editorial, 1984.

————. *Las Instituciones científicas en la historia de la ciencia en Venezuela.* Caracas: Fundación Fondo Editorial Acta Científica Venezolana, 1987.

Vicziani, Marika. "Imperialism, Botany and Statistics in Early Nineteenth-Century India: The Surveys of Francis Buchanan (1762–1829)." *Mod. Asian Stud.,* 1986, *20,* 4:625–60.

Villar, Juan Bautista. *El Sahara español: Historia de una aventura colonial.* Madrid: Sedmay Ediciones, 1977.

Visvanathan, Shiv. *Organising for Science.* Delhi: Oxford Univ. Press, 1985.

Wallerstein, Immanuel. *The Modern World-System: Capitalist Agriculture and the Origins of the European World-Economy in the Sixteenth Century.* New York: Academic Press, 1974.

Watson, H., and David Wade Chambers. *Singing the Land, Signing the Land.* Geelong: Deakin Univ. Press, 1989.

Widmalm, Sven. "Instituting Science in Sweden." In *The Scientific Revolution in National Context,* eds. Roy Porter and Mikulas Teich. Cambridge: Cambridge Univ. Press, 1992, pp. 240–62.

Wildenthal, Lora. "Race, Gender, and Citizenship in the German Colonial Empire." In *Tensions of Empire: Colonial Cultures in a Bourgeois World,* eds. Frederick Cooper and Ann Laura Stoler. Berkeley: Univ. of California Press, 1997, pp. 263–283.

Williams, Glyndwr. *The Expansion of Europe in the Eighteenth Century: Overseas Rivalry, Discovery and Exploitation.* London: Blandford Press, 1966.

Williamson, James A. *A Short History of British Expansion: The Modern Empire and Commonwealth,* 2nd ed. London: Macmillan, 1938.

Worboys, Michael. "British Colonial Science Policy, 1918–1939." In *20th Century Science: Beyond the Metropolis.* Paris: ORSTOM, 1996, pp. 99–111.

————. "The Spread of Western Medicine." In *Western Medicine: An Illustrated History,* ed. I. Loudon. Oxford: Oxford Univ. Press, 1997, pp. 249–63.

————. "Tropical Diseases." In *Companion Encyclopaedia of the History of Medicine,* eds. W. F. Bynum and Roy Porter. London: Routledge, 1993, pp. 521–36.

Yamada, Keiji, ed. *The Transfer of Science and Technology between Europe and Asia, 1780–1880.* Kyoto/Osaka: International Research Center for Japanese Studies, 1992.

Yearley, Stephen. "Colonial Science and Dependent Development: The Case of the Irish Experience." *The Sociological Review,* 1989, *37:*308–31.

————. *Science, Technology and Social Change.* London: Unwin Hyman, 1988.

Zamudio, Graziela. "El Jardín botánico de la Nueva España y la institucionalización de la botánica en México." In *Los Orígenes de la ciencia nacional, Cuadernos de Quipu,* series no. 4, ed. Juan José Saldaña. Mexico City: Sociedad Latinoamericana de Historia de las Ciencias y la Tecnología, 1992, pp. 55–98.

Zeller, Suzanne. *Inventing Canada: Early Victorian Science and the Idea of a Transcontinental Nation.* Toronto: Univ. of Toronto Press, 1987.

————. "'Merchants of Light': The Culture of Science in Daniel Wilson's Ontario, 1853–1892." In *Thinking With Both Hands: Sir Daniel Wilson in the Old World and the New,* ed. E. Hulse. Toronto: Univ. of Toronto Press, 1999, pp. 115–38.

Notes on Contributors

Christophe Bonneuil is a researcher at the Centre Koyré d'Histoire des Sciences et des Techniques at the CNRS in Paris. Since 1990, he has published several articles and books on science policy in the French Empire and on the topic of science, colonialism, and nature. His latest book is *Mettre en ordre et discipliner les tropiques: Les Sciences du végétal dans l'empire colonial français, 1870–1940* (Paris: Edition des Archives Contemporaine/Harwood, 2000).

Wade Chambers was educated at the University of Oklahoma and received his doctorate in the history of science from Harvard University. After a postdoctoral fellowship at the Smithsonian Institution, he was for many years chair of the Social Studies of Science Program at Deakin University, Geelong, Australia, where he edited and authored more than twenty textbooks. He has written several articles on the history of colonial science in Mexico, the Caribbean, and Australia, and his recent work has focused on indigenous education. He is currently a consultant at the Institute for American Indian Art and Culture in Santa Fe, New Mexico.

Clarete Paranhos da Silva is a teacher of history in São Paulo. She is a master's degree candidate in the Department of Geosciences in Education at the State University of Campinas (Unicamp), where she works on the history of the geosciences in colonial Brazil. She has published in the Brazilian journal *History, Sciences, Health, Manguinhos* (1998).

Harriet Deacon is a medical historian with a specific interest in medicine and colonialism at South Africa's Cape Colony during the eighteenth and nineteenth centuries. She is currently working for the Robben Island Museum in Cape Town. After completing an undergraduate degree in Cape Town, she completed a Ph.D. dissertation at Cambridge University in England on the history of the medical institutions at Robben Island (1846–1910).

Alberto Elena teaches the history of science and the history of cinema at the Universidad Autónoma, Madrid. Among his latest publications are *Despues de Newton: Ciencia y sociedad durante la Primera Revolución industrial* (Anthropos, 1998) (edited with J. Ordóñez and M. Colubi) and *Los Cines periféricos (Africa, Oriente Medio, India)* (1999). He also has made recent contributions to scholarly journals including *Isis*, the *British Journal for the History of Sci-*

ence, Nuncius, and *Public Understanding of Science.*

Silvia F. de M. Figueirôa is a geologist with a Ph.D. in the history of science. Since 1987 she has worked at the University of Campinas (Inst. of Geosciences/Dept. of Geosciences Education) in Brazil. She is a member of the International Commission on the History of Geological Sciences (INHIGEO*)*, and she organized its international symposium in Brazil in 1993. She has been president (1995–1998) of the Latin American Society on the History of Science and Technology (SLHCT). Her research has concentrated on the history of the geological sciences in Brazil and its links with educational questions.

Richard Gillespie is director of the Australian Society Program at Museum Victoria, Melbourne, Australia, and a fellow in History and History & Philosophy of Science at the University of Melbourne. He has participated in a Museum Victoria survey of the invertebrate fauna of central Australia, which includes collaboration with Aboriginal communities to record their local knowledge of invertebrates.

Deepak Kumar writes on the social history of science in colonial South Asia. He has published twenty papers on different aspects of science and society in India, along with several books, including *Science and the Raj* (Delhi: Oxford University Press, 1995). He currently teaches history of science, society, and education in India at the School of Social Sciences, Jawaharlal Nehru University, New Delhi.

Antonio Lafuente holds a Ph.D. in physics and works at the Centro de Estudios Históricos (CSIC) in Madrid. He has been a visiting scholar at the University of California, Berkeley. His most recent publications are a critical edition of Voltaire's *Elements de la philosophie de Newton* (Barcelona: Circulo de Lectores, 1998); *Guía del Madrid científico: Ciencia y corte* (Madrid: Doce Calles, 1998); a book about the exhibition *Imágenes de la ciencia en la España contemporánea* (Madrid: Fundación Arte y Tecnologia, 1998); and *Georges-Louis Leclerc, conde de Buffon (1707–1788)* (Madrid: CSIC, 1999).

Maria Margaret Lopes is a geologist and holds a Ph.D. in the history of science from the Universidade de São Paulo, Brazil. In 1997, she was a visiting professor at the University of Southwestern Louisiana, and in 1998, she held a

Rockefeller Humanities Foundation Fellowship at the Museo Etnografico "Juan Ambrosetti," Universidad de Buenos Aires. She has published several papers on the history of science in Latin America and Brazil. She is the author of *O Brasil descobre a Pesquisa Científica: Os museus e as ciências naturais no século XIX* (São Paulo: Editora Hucitec, 1995).

Roy MacLeod is professor of history at the University of Sydney. He has published extensively in the social history of science, medicine, and technology. He has a particular interest in the role of scientific practices in the expansion of Europe.

James E. McClellan III is professor of history of science at Stevens Institute of Technology in Hoboken, New Jersey. He is the author of *Colonialism and Science: Saint Domingue in the Old Regime* (1992).

John Merson is a senior lecturer in the School of Science and Technology Studies at the University of New South Wales. He has written widely on the history of science and technology in the Asia Pacific, focusing specifically on environmental and development issues. He has been a consultant to the Australian government and APEC (Asia Pacific Economic Cooperation Forum) on issues of environmental technology and resource management.

Javier Ordóñez teaches the history and philosophy of science at the Universidad Autónoma de Madrid. He has written on the history of physics in nineteenth- and twentieth-century Europe. He recently co-edited with A. Elena and M. Colubi *Newton: Ciencia y sociedad durante la Primera Revolución Industrial* (Anthropos, 1998); and he has co-authored with Ana Rioja *Teorías del universo: Desde los pitagóricos a Galileo* (Madrid: Síntesis, 1999).

Michael A. Osborne, associate professor of history and environmental studies at the University of California, Santa Barbara, is the author of *Nature, the Exotic, and the Science of French Colonialism* (1994). He is currently writing a book tentatively titled *A Medicine of Race and Place: French Naval Hygiene and the Emergence of Tropical Medicine.* He is also pursuing studies of medical natural history and working on a project comparing the response of citizens and scientific experts to recent environmental disasters.

Juan Pimentel has a Ph.D. in history and works at the Departmento de Historia de la Ciencia at the Consejo Superior de Investigaciones Cientificas (CSIC) in Madrid. Previously, he was a visiting scholar in the Department of History and Philosophy of Science at the University of Cambridge, United Kingdom. His research areas are colonial science, travel literature, and expedi-

tions. He is the author of *La Física de la monarquía: Ciencia y política en el pensamiento colonial de Alejandro Malaspina (1754–1810)* (Madrid: Doce Calles/CSIC, 1998), and of *En el Panoptico del Mar del Sur* (Madrid: CSIC, 1992).

Irina Podgorny is a researcher on the staff of CONICET (Consejo Nacional de Investigaciones Cientificas y Tecnicas), and works at the Universidad de La Plata Museum, Argentina. She is the author of *Arqueologia de la educacion: Textos, indicios, monumentos* and has published on the subject of the public presentation of natural history in Argentina in the nineteenth and early twentieth centuries.

Kapil Raj, associate professor of history and sociology of science at the University of Lille III, France, is presently on a fellowship at the Centre de Recherche en Histoire des Sciences et des Techniques, La Villette, Paris. He has published extensively on the reception and practice of modern science in India. His current research is focused on the construction of the field sciences (mainly natural history and geography) through intercultural encounter in the colonial context. He is preparing a book on the history of the Survey of India, 1760–1885.

François Regourd teaches at the Université de Paris at Nanterre and is completing a doctorate at the Université de Bordeaux III under the direction of Paul Butel; his dissertation topic is "Sciences et colonisation sous l'Ancien Régime: Le cas de la Guyane et des Antilles françaises, XVIIe–XVIIIe siècles." In addition, he has published articles on colonial scientific networks, hydrography, and colonial agronomy in the eighteenth century.

Sverker Sörlin is a professor of environmental history in the Department of Historical Studies, Umeå University, Sweden. He is also the director of the Swedish Institute for Studies in Education and Research (SISTER) in Stockholm. His publications in English include numerous articles in the history of science and environmental history and two co-edited books, *Denationalizing Science: The Contexts of International Scientific Practice* (Sociology of the Sciences Yearbook, 1993), and *Sustainability—The Challenge: People, Power, and the Environment* (Montreal & New York, 1998).

Michael Worboys is professor of history and head of research in the School of Cultural Studies at Sheffield Hallam University. He has worked on the history of tropical medicine and colonial science. His most recent work has been on the history of bacteriology. His monograph *Spreading Germs: Disease Theories and Medical Practice in Britain, 1860–1900* will be published by Cambridge University Press in 2000.

Suzanne Zeller is an associate professor of history at Wilfrid Laurier University in Waterloo, Ontario, Canada. The author of *Inventing Canada: Early Victorian Science and the Idea of a Transcontinental Nation* (Toronto, 1987), she has also published *Land of Promise, Promised Land: The Culture of Science in Victorian Canada* (Canadian Historical Association Historical Booklet #56) (Ottawa, 1996) as well as many articles on the intellectual and cultural impact of science in nineteenth-century Canada. Her current research concentrates on the history of the physical sciences.

Index